Laser Chemistry
of Organometallics

ACS SYMPOSIUM SERIES **530**

Laser Chemistry of Organometallics

J. Chaiken, EDITOR

Syracuse University

Developed from a symposium sponsored
by the Division of Physical Chemistry
and the Division of Inorganic Chemistry, Inc.,
at the 203rd National Meeting
of the American Chemical Society,
San Francisco, California,
April 5–10, 1992

American Chemical Society, Washington, DC 1993

Library of Congress Cataloging-in-Publication Data

Laser chemistry of organometallics: developed from a symposium sponsored by the Division of Physical Chemistry and the Division of Inorganic Chemistry, Inc., at the 203rd National Meeting of the American Chemical Society, San Francisco, California, April 5–10, 1992; J. Chaiken, editor.

p.   cm.—(ACS symposium series, ISSN 0097–6156; 530)

Includes bibliographical references and index.

ISBN 0–8412–2687–3

1. Laser photochemistry—CHEMISTRY—Congresses. 2. Organometallic compounds—Congresses.

I. Chaiken, J., 1955–   . II. American Chemical Society. Division of Physical Chemistry. III. American Chemical Society. Division of Inorganic Chemistry. IV. American Chemical Society. Meeting (203rd: 1992: San Francisco, Calif.) V. Series.

QD716.L37L37   1993
547'.050455—dc20   93–7877
                                                  CIP

The paper used in this publication meets the minimum requirements of American National Standard for Information Sciences—Permanence of Paper for Printed Library Materials, ANSI Z39.48–1984. ∞

Foreword

THE ACS SYMPOSIUM SERIES was first published in 1974 to provide a mechanism for publishing symposia quickly in book form. The purpose of this series is to publish comprehensive books developed from symposia, which are usually "snapshots in time" of the current research being done on a topic, plus some review material on the topic. For this reason, it is necessary that the papers be published as quickly as possible.

Before a symposium-based book is put under contract, the proposed table of contents is reviewed for appropriateness to the topic and for comprehensiveness of the collection. Some papers are excluded at this point, and others are added to round out the scope of the volume. In addition, a draft of each paper is peer-reviewed prior to final acceptance or rejection. This anonymous review process is supervised by the organizer(s) of the symposium, who become the editor(s) of the book. The authors then revise their papers according to the recommendations of both the reviewers and the editors, prepare camera-ready copy, and submit the final papers to the editors, who check that all necessary revisions have been made.

As a rule, only original research papers and original review papers are included in the volumes. Verbatim reproductions of previously published papers are not accepted.

M. Joan Comstock
Series Editor

Contents

LASER-INITIATED BIMOLECULAR CHEMISTRY

viii

INDEXES

Preface

SCIENTIFICALLY, ECONOMICALLY, AND POLITICALLY, this book and the symposium on which it is based mark a turning point that's been a long time in coming. Basic and applied research in laser chemistry is being directed to a broader range of possible advantages of using lasers as agents of chemical and physical change. The initial surge of interest in lasers as "molecular scalpels" began in the 1960s and now has evolved into a list of types of laser chemistry. One result of this evolution is the perception that metal-containing systems, and in particular organometallic systems, have characteristics that are sometimes uniquely synergistic with the basic properties of laser excitation. This book appraises the current state of the laser chemistry of organometallic systems within the conceptual framework presented in the first chapter.

Fundamental research is best approached by choosing to obtain the most detailed information possible on the simplest possible chemical and physical systems with the greatest control of all relevant parameters. This ideal may be considered a luxury in the future because applied research often requires making the best judgments possible based on the limited information at hand. Fundamental research may become more difficult to pursue in the future and certainly not because all important questions have been answered or even addressed. This situation is a simple consequence of the economic and political pressure to turn our attention to the study of chemical and physical systems with near-term commercial value. For the past few decades, the U.S. government has fostered the growing perception that for economic and political purposes, it is time to harvest our cache of fundamental knowledge.

Interest in the laser chemistry of organometallic systems is not only the result of the desire for new commercial applications; it is also a natural consequence of the desire to observe new phenomena and thereby understand the applicability of our current fundamental pictures to more complex phenomena. That is, simple curiosity drives this research also. Systems that involve large, complicated molecules in dispersed and condensed media are being studied using methods previously reserved for the study of atoms and diatomics under collisionless conditions. Conceptually, the importance of studying the transition from fundamentally simple to fundamentally complex systems, for purposes of deducing underlying scaling laws and the means to manipulate matter and energy in new ways, is important to understanding why the research in this field is beginning to escalate, why the research in this field is important, and why there is a need for the information to be presented now.

This book attempts to survey a broad range of chemical and physical research, fundamental and applied, that defines the current state of laser chemistry of organometallic systems. The book's scope is intentionally expansive to provide an introduction to laser chemistry for scientists and engineers with the widest possible range of interests. One of my intentions was to provide an entry point into the primary literature of laser chemistry that spans the catalysis, ceramics, microelectronics, photonics, and materials research fields. It should be useful to undergraduate and graduate students who want an introduction to the field of laser chemistry as well as administrators and managers who want an overview into the field as it currently stands. The level of presentation should be accessible to readers with an advanced undergraduate background in physical or inorganic chemistry.

This book should be unusual and useful to many specialists in laser chemistry in the sense that I have invited the authors to speculate more than is customary in the primary literature. I have invited them to disagree, and I have made no effort to obtain a consensus on anything because the field is still rapidly evolving. I asked only that they make an effort to identify some connection with the conceptual framework I present in the first chapter.

Some presentations in the symposium were modified because of confidentiality requirements of the scientists' industrial employers. For these chapters, my judgment was in the spirit of applied research: It's better to have some information than none at all.

Acknowledgments

Many companies and organizations made generous financial contributions. Without them, the symposium would not have been possible: Coherent, Lambda Physik, Lumonics, EG&G–PAR, the Petroleum Research Fund as administered by the American Chemical Society, the Divisions of Inorganic Chemistry, Inc., and Physical Chemistry of the American Chemical Society, and Syracuse University's Vice President for Research, Ben Ware. Between the time the symposium was planned and the time it occurred, the Russian ruble was devalued three times and the Canadian government enacted a spending freeze. The generosity of the sponsors literally saved 25% of the symposium. In strategies and tactics, Eric Weitz and J. J. Valentini made many helpful suggestions throughout. I am particularly grateful to the authors and participants in the symposium who have done an excellent job in educating me and each other. Because of a severe family emergency that hampered my editing, I appreciate the understanding of all those at ACS Books who were involved in the preparation of this book.

J. CHAIKEN
Syracuse University
Syracuse, NY 13244–4100

December 1, 1992

Chapter 1

Laser Chemistry of Organometallic Species
Conceptual Framework and Overview

J. Chaiken

Department of Chemistry, Syracuse University,
Syracuse, NY 13244-4100

"Laser Chemistry" refers to chemical processes initiated by the action of laser(s) on matter. Because of the rapid advancement of relevant fundamental and applied research over the last several years, a conference to assess the state of the field is appropriate at this time. One emerging realization is that a unique synergism exists between the fundamental properties of laser excitation and the photochemistry of organometallics so Laser Chemistry of Organometallics was chosen as the specific focus of this Symposium. In succeeding chapters, accounts of research in multiphoton processes, metal atom chemistry, the chemistry of coordinatively unsaturated organometallics, thin film deposition, and the formation and chemistry of ultrafine particles and clusters in gas and condensed phases will all be presented within the unifying concept of Laser Chemistry of Organometallics. Our goal in this chapter is to describe this conceptual framework.

The properties of lasers and laser light allow reaction mixtures to be energized selectively with respect to species, quantum state(s), and temporal and/or spatial distribution. The composition and physical properties of organometallic chemical systems can lead to unique phenomena and processes when lasers are used to energize systems. In the context of specific forms of laser chemistry, we shall discuss some particular examples in which organometallics provide the possibilities for unique chemistry. The following chapters describe research which bears on the processes and phenomena connected in the flowchart in Figure 1. Some of the fundamental and practical questions which must be answered to discover the extent of possible applications are evident.

Introduction to Laser Chemistry

The term "Laser Chemistry" is nearly as old as the laser itself with reviews(1,2,3,4) and patents(5,6) first appearing in the literature in the sixties(7) and continuing(8) today. Of the good recent reviews available, we choose to mostly follow the

0097-6156/93/0530-0001$07.00/0
© 1993 American Chemical Society

categories and terminology used by Woodruff(1). To survey the unique possibilities
of laser chemistry as applied to organometallics, we first present a very short
introduction to laser chemistry in general, then a short digression into the established
single(9) and multiphoton(10)chemistry of organometallics, and finally a brief
summary of some very recent results of our attempts to synthesize clusters and
ultrafine particles using laser chemistry of organometallics. For clarity in
presentation, we mostly choose specific examples from our own research to illustrate
our main points. No attempt is made to be exhaustive of the literature.

The flowchart on the next page is meant to suggest a certain temporal
separation in a sequence of events, which taken together, could be called "laser
chemistry of organometallics." The first stage is the laser excitation process. The
second stage is the chemistry of the nascent distribution of species and states. A final
stage describes still longer time scale chemistry and is convenient if one is concerned
about producing materials which are subsequently removed from a reaction vessel
and subjected to some other processing or aging in the course of characterization.
This could also correspond to sintering in the case of ceramic particle chemistry or
possibly to post-deposition annealing in air in the case of transparent metal film
deposition. Further discussion and examples of each stage follow.

The chart is labeled to suggest that the process can occur either
homogeneously in gas or condensed phases, or heterogeneously involving a number
of phases. These questions of phase set the context of the mass and energy transfer
characteristics for the laser chemistry. Thus terms like "film-phase" arise in the
application of the flowchart to processes like laser chemical vapor deposition.
Although not necessarily involving an organometallic, the same scenario can be
applied to the use of laser ablation to rapidly produce gas phase metal atoms from a
metal rod(see Hackett's chapter). Although the laser provides a unique form of
excitation, much of the chemistry which occurs after laser induced species begin to
exist is dictated by kinetics and thermodynamics consistent with those mass(11) and
energy transfer(12) conditions. Bauerle's chapter give an introduction and analysis of
some of these issues with regard to modelling such systems. Singmaster's chapter
presents the current state of the well studied process of laser chemical vapor
deposition of metal films using metal carbonyl precursors.

The net result of the laser excitation is to convert an organometallic into either
metal atoms, metal ions, or coordinatively unsaturated species, or some mixture of
all of the above and ligands, which can then participate in materials chemistry or
catalysis. Questions associated with mechanisms for unimolecular
chemistry(13,14,15,16,17) are therefore of paramount importance during the first
stage because they determine the internal state distributions of the atoms which then
may participate in bimolecular chemistry. If ionic species are to be used as reactants,
they could be generated directly via a uv-visible multiphoton process. Laser induced
discharges(18,19) or other methods(20), are also available in which the ions are
mostly formed by electron impact phenomena. The internal state and species
distribution of these nascent ionic species would often be interpreted in terms of
quasi-equilibrium theory(QET). In any case, some recently explored spectroscopy of
organometallic ions is described in Kimura's chapter.

In the single to few photon range, IR excitation can only lead to the kinds of
laser chemistry Woodruff calls either selective heating or vibrationally enhanced.
Thermodynamics dictates that IR laser excitation can only lead to dissociation via
nearly simultaneous absorption of at least a few photons and such multiphoton
dissociation *tends* to produce neutral fragments. Visible and shorter wavelength
excitation *tends* to produce ions in the multiphoton limit. In the single photon limit,

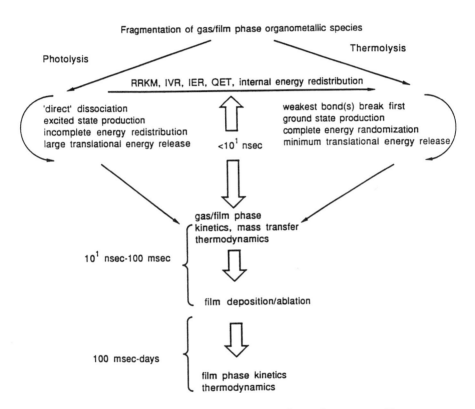

Figure 1.   Flowchart describing laser chemistry of organometallics as a crudely separable sequence of events.

uv-visible excitation can lead to selective ligand labilization. In general(16,17), processes are termed "thermal" when their products are formed by rupture of the weakest bonds of the precursor and produce "reactants" for subsequent chemistry which are relatively cool translationally. Alternatively, "direct" dissociation processes are conceivable in which non-RRKM like behavior competes with the more commonly observed thermalized processes.The distinction between thermalized and direct processes is summarized in Figure 1.

One type of study which bears on the intrinsic propensity of organometallics to internally redistribute energy is the multiphoton ionization experiment. The chapters of Gerrity, Wight and Belbruno review this field and present some new results. Hossenlopp's chapter gives new data on some of the largest organometallics ever studied in this way and Garvey's chapter describes multiphoton processes involving clusters containing organometallics. Given the net effects of size and the presence of loosely bound "solvent" molecules, these chapters would seem to give some idea of the variation in multiphoton processes as the medium which is being excited becomes more "condensed".

Of the general types of laser chemistry schemes, a few may be particularly relevant to organometallics. The idea of "bond selective chemistry", in which lasers are used to cleave one or more chemical bonds, while leaving the others in the same molecule unaffected, has been the basis for a variety of fundamental(21,22) and applied studies. Early research in the bond selective chemistry field focussed on the study of isomerization reactions of small organic molecules and infrared, i.e. vibrational state selective, excitation. These studies ultimately showed that intramolecular vibrational relaxation(IVR) will almost always cause unimolecular chemistry consistent with RRKM theory. Regardless of these intramolecular processes, collisions with many species are also very effective in causing energy redistribution with a variety of manifestations. Jackson's hapter presents a review and a revealing case study of the effects of such collisions on the multiphoton dissociation of the organometallics.

The photochemistry of organometallics, in particular those having carbonyl ligands, is dominated by ligand labilization. Laser chemistry based on classical organometallic photochemistry(21,22) is rationalized on the basis of ligand labilization being a strong function of the electronic state to which the organometallic is excited. Some cases do suggest the possibility of non-RRKM type behavior but the general question of whether electronic state selective ligand labilization occurs in organometallics remains to be completely settled. In Woodruff's terminology, bond/mode specific chemistry is unimolecular and involves only a single reactant. Koplitz's and Bartz's chapters give an indication of the type of measurements which are at the current state of the art. Unimolecular photodissociation is a case where the intrinsic properties of organometallics would *seem* to be adequately synergistic with laser chemistry to lead to commercializable processes.

Purely unimolecular photochemistry such as isomerization or fragmentation to form coordinatively saturated species may have some practical significance, but the likelihood is small because the net quantum yield of the first step cannot exceed unity. In the context of laser chemical synthesis of catalysts, calculations(23) which include all costs, including the high relative cost of laser photons, suggest that only "net" yields far in excess of unity will be commercially viable processes. The net yield includes the product of the unimolecular yield and the turnover number of the catalyst. Equivalently, laser synthesis of materials must produce a particularly high value added product to achieve economic viability for the whole enterprise.The work Rice, Bauerle and Comita described in their chapters in this book are possible examples of such catalysts or value added materials. Comita's work is particularly novel because the organometallic precursor is a type of organometallic polymer.

The practical significance to observing bond/mode selective chemistry seems much more likely to reside in the chemistry which occurs subsequently involving the reaction products. Woodruff refers to the preparation of unique reactants for subsequent chemistry which are unique by virtue of their identity as "nonspecific laser chemistry." Photogenerated catalysts are a prime example of a material which potentially has a quantum yield for the net chemistry initiated by the laser far in excess of unity. Here the econmoic viability must reflect the costs of the starting materials and energy going into catalyst preparation and the turnover of reactants into products promoted by the catalyst. On the other hand, if a catalyst can be prepared which has unique properties(see Rice's chapter), e.g. selectivity, then the price differential can be recalculated and may be much more favorable.

A different example in the general genre of laser chemistry involves the idea of "selective heating" of one component of a mixture and is the basis for laser isotope separation$(24,3)$. In cases of state or species selective excitation, macroscopically significant amounts of matter are often necessary for large scale synthetic applications. Although fundamental research can often be conducted with relatively small amounts of materials, the question of scale-up is but one which must be addressed to determine the commercial viability of a potential application. In a different type of example, ultrafine particles can be synthesized using organometallics in nozzles and a laser supplying energy to the reaction mixture by addition of a sensitizer to the flow. Selectively introducing energy into a complex mixture is a means for enhancing the chemistry of one component of the mixture.

So-called "classical photochemistry" with lasers offers the potential for preparation of unique quantities of reactants, with unique spatial and temporal properties. Because reasonably high vapor pressures can be obtained with organometallics, laser pyrolysis can be used to produce large gas phase quantities of highly refractory metals. As in the production of ceramics with the selective heating, metal clusters can be formed directly from the organometallics by thermalized laser induced processes. It is also possible that electronically excited metal atoms have an enhanced tendency to nucleate$(25,26)$. The effect of the initial steps of the cluster growth process must not be underestimated. For example, Xe_n^- clusters(27) are unstable to destruction by evaporation for n<6. However, *once formed,* Xe_6^-, allows formation of a broad distribution of stable sizes by a continuation of the coalescence process. As a means for utilizing metal atoms as reactants for small scale synthetic chemistry or fundamental studies, there is a synergism involving the photochemical properties of organometallics and laser excitation. The chapters by Hackett, Rayner, Mitchell, and Weisshaar give an idea of the type of research being conducted presently concerning metal atoms and clusters. The bimolecular chemistry of coordinatively unsaturated organometallics with free ligands and other coordinatively unsaturated organometallics is described in the chapters by Weitz and Weiller.

In addition to the distinction between multi- and single photon processes being particularly important in the context of organometallic laser chemistry, the distinction between "single and multiple" kinetic pathays is also particularly important. Weitz's pioneering use of kinetics properties for the assignment of transient IR spectra produced after laser excitation of a gas metal carbonyl, relys on supressing reactions between coordinatively unsaturated organometallics, by addition of an excess of free carbon monoxide. The reactions between these species, which leads to clustering and is an example of what might be called multiple kinetic pathways. When one has reaction conditions which only leads to one set of reactions between the nascent photoproducts and some other added component, one is in the single kinetic pathway limit. Clusters and ultrafine particles are produced by multiple kinetic pathway processes, i.e. agglomeration, and molecular species are produced by single kinetic pathway processes. Since lasers can be used with organometallics to produce high vapor pressures of very refractory metals, multiple pathwyas quickly become important.

Other potential types of laser chemistry have been proposed but we have restricted our attention to those which we currently suspect might possess some special potential or synergism for involvement of organometallics. The remainder of this chapter will survey characteristics of some of the types of research being done to investigate the first two stages of laser chemistry of organometallics. One goal is to focus attention on the meaning and value of some commonly used assumptions used to rationalize laser chemical experiments. Opportunities for advancement of fundamental science and potential applications are evident from this survey. As a new example from our own recent research, we describe how the chemistry of metal cluster formation provides a context for comparing the properties of laser chemistry of organometallics to those of other currently used synthetic methods. To this end, we present a short description of a model for systematizing and interpreting cluster and particle size distributions.

Multiphoton vs. Single Photon

The following chapters present a few histories of the earliest attempts(10) at measuring the effects of propagating a laser through an organometallic. In this overview, we only consider the most general picture and we begin by asking how we can anticipate the major effects of laser irradiating a chemical system containing an organometallic molecule. Thermodynamics, i.e. standard energetics from a variety of sources, provides a set of expectations relating the amount of energy transferred from the laser pulse to the reacting system, to the product(s) which such a system can subsequently form. Given adequate thermodynamic information of sufficient precision, comparison of products detected with the energy content of a single photon allows a calculation of the number of photons which had to have somehow been imparted to the chemical system. An independent approach, based on rate equation treatments of the multiphoton excitation/dissociation process(28), involves interpretation of a measurable quantity known as the power index for the process.

A variety of criteria are employed to determine the power index of laser induced excitation. By definition, a measure of excitation must be identified, e.g. a reaction product, and the scaling of its efficiency for production with increasing applied laser fluence determined. When a power law dependance is evident, the slope of a log-log plot of that type of data is what is commonly referred to as the "power index." In the limit of rate equations being a valid treatment(29), the power index will be equal to the number of photons absorbed during the rate limiting absorption step needed to produce the observed net excitation. Knowing the number of photons required to obtain a specific level of excitation is an important first step in describing the precise mix of species produced by the laser pulse. An index of 3/2 is appropriate for the simple expansion of the focal volume(30) with increasing energy content. This interpretation of the power index is only valid above the threshold power needed to induce the multiphoton process in question.

For practical purposes, general rules for predicting the threshold for a particular effect would be very useful. The threshold for single photon excitation is easily obtained from simple absorption spectroscopy. We describe a simple experiment, which has been described in detail elsewhere*(31)*, which shows that for organometallics, the threshold for inducing multiphoton processes can often be surprisingly small. Consider the one color multiphoton excitation experiment depicted in Figure 2. A *very loosely* focussed laser beam intersects a supersonic expansion containing benzene chromium tricarbonyl(BCT) thereby initiating the multiphoton process depicted in Figure 3. To obtain this data, a 1 meter lens focused an initially 4 mm diameter laser pulse from a Lambda Physik Fl2002 dye laser. The interaction of laser beam with the symmetry axis of the molecular beam occurs ≈ 10 cm from the center of the laser focal volume. We detect whatever excitation was imparted by monitoring emission from a well defined observation zone downstream roughly ≈ 5 cm downstream. A set of scans of the $^7S_3 <--- ^7D_J$ transition(≈ 354 nm) using different laser powers is shown in Figure 4.

It is clear that substantial amounts of a net three photon (minimum) process can be driven by as little as a few tens of microjoules of uv-visible light delivered in a loosely focussed twenty nanosecond duration pulse. The atomic transition is resonant with the laser wavelength but is nominally forbidden. The wavelength of the incident photons in this one-color experiment was chosen such that the first photon absorbed is resonant with at least some allowed rovibronic transitions of the parent molecule. Since there only broad, spectrally congested absorption features*(32)*, it was assumed and later indirectly confirmed*(33)*, that over moderate ranges of laser wavelength, a few nanometers and away from any resonances, the power index is constant. The observed power index for this transition in the low power limit was only ≈ 1. In addition to showing that the power index cannot be invoked with impunity, these data provide empirical proof that at least in some cases, large volumes of static gases/reactants can be processed by either one or multiphoton processes. Given commercially available tunable lasers capable of producing multi-millijoule, nanosecond duration pulses, volumes into the 10^1 cm^3 range can be used for reactions. Excimer lasers produce 10^1-10^3 millijoules per pulse and it is clear that in some cases, the processes which are induced are not very sensitive to the exact choice of wavelength.

The above excitation process was observed under collisionless conditions(excitation occurred>50 nozzle diameters from the source) and still posed interpretative problems. Single color multiphoton dissociation(MPD), multiphoton ionization(MPI) experiments*(34,35)* are the simplest types of multiphoton excitation spectra one can produce. In the MPD case, the excitation is detected by observing the presence of neutral fragments and, in the MPI case, the excitation is manifested by the production of ions from the irradiated molecules and atoms. The emission type of experiment like that described in the preceding paragraphs is an example of an MPD experiment. Effectively, ground state metal atoms, produced by a MPD process, are detected by laser induced fluorescence. Such experiments, either as static bulk gas or flows of various kinds, also correspond to conditions which could easily be inexpensively commercialized.

Figure 5 shows the power index for the static gas MPD/MPI of benzene chromium tricarbonyl as a function of the total added argon pressure. The behavior of the ionization signal itself as a function of added buffer gas has been described*(18,19)* and is less sensitive to the effect of the buffer gas on the multiphoton process than on the laser induced discharge following the MPI events. On the other hand, the power index is an important measurement because it contains dynamical information in spite of the complicated sets of processes which can occur

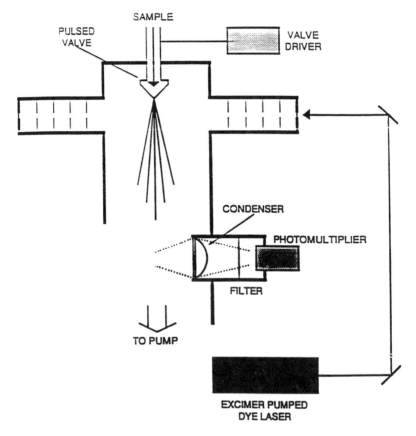

Figure 2.   Apparatus for observing long time emission from metal atoms formed by MPD of organometallic in a supersonic expansion.

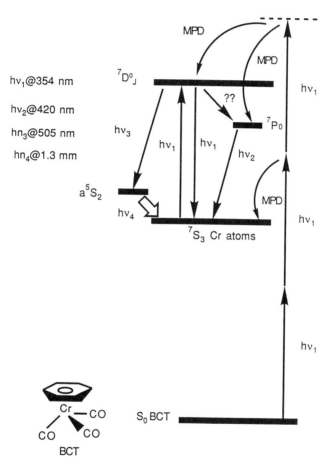

Figure 3.   Energy level representation of excitation sequence
corresponding to MPD formation of ground state chromium atoms from
benzene chromium tricarbonyl near 354 nm.

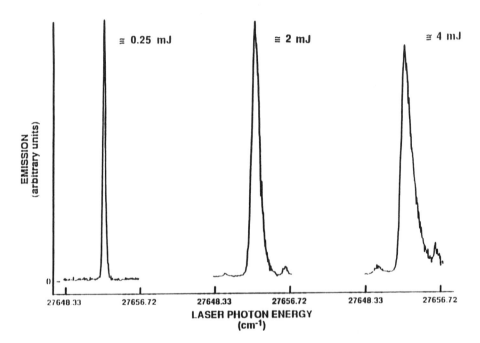

Figure 4.   Scans of ^5D_J<---$^7S_3$ transition of Cr atoms formed by MPD of benzene chromium tricarbonyl. The average laser pulse energy is as indicated.

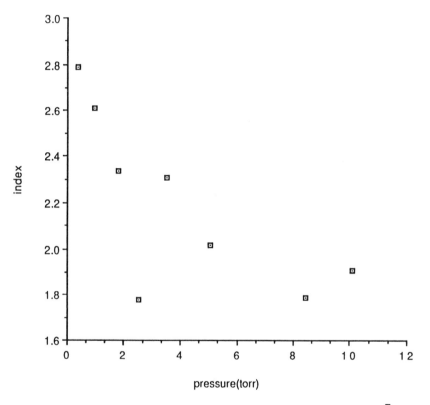

Dependence of power index for MPI of benzene chromum tricarbonyl
on pressure of added buffer gas(Ar) at 27936 cm-1

Figure 5.   Graph of power index corresponding to formation of $^7P_3$ Cr
atoms from MPD of benzene chromium tricarbonyl as a function of pressure
of added Ar buffer gas.

subsequent to the passage of the laser pulse through the system. The chance exists
that it can be interpreted physically. The problem here is that there are a variety of
parameters which are necessary to construct a model. Again we see that finding the
number of photons needed to produce an excitation is not as simple as simply
measuring a power index. The power index is clearly a function of buffer gas
pressure. To illustrate the kinds of complexities that can arise, consider that in cluster
formation using laser chemistry of organometallics, buffer gas pressure is an
important process parameter which is defined to insure appropriate post-MPD mass
transport conditions. However, probing the effect of collisions on the multiphoton
process itself, is an important first step towards anticipating the interaction between
process control parameters. Such probing would also be an initial step towards an
analytic approach to describing multiphoton processes in condensed phases.

Even if one knows the power index, it is an empirical fact that it is nearly
impossible to attain a perfectly uniform spatial illumination pattern. Thus the power
index is a function of the spatial position within the laser focal volume. In an attempt
to explore this effect, consider the experiment depicted in Figure 6. The laser pulse is
probed after it initiates an ensemble of MPI events. The pulse is magnified by a
factor of roughly 50 and projected onto a screen marked with a grid. A single
detector is placed at different positions on the grid and a power index is measured.
Thus the correlation between the total amount of ions produced by the entire laser
pulse and the amount of light present in that quadrant/grid space can be measured.

Using thousands of laser pulses for each determination, the power index is
largest in the center of the laser pulse, as high as ≈ 3.5, and diminishes to 0.0 as the
detector probes the light near the edges of the focal volume. The correlation
coefficient behaves similarly and that is what is shown in Figure7. While it is
possible to obtain extremely even illumination, it is still true that during the duration
of a standard nanosecond laser pulse many products can be formed early in the laser
pulse which are transported either into or out of the focal volume. Species which
move into the focal volume after an initial multiphoton event have a greater chance of
undergoing another event and thereby increasing the net power index. So the
variation in power index need not be solely due to a greater power density near the
center of a laser focus.

Although the examples given pertain mostly to use of nanosecond laser
pulses, it is clear that some of the fundamental considerations employed bear on the
use of other lasers with other types of samples. Very different behaviors can be
observed and some of these will be discussed in the chapters which follow. Without
any attempt to be exhaustive, it is clear that multiphoton as well as single photon
processes are probably possible on commercially meaningful scales. On the other
hand, it is also reasonably clear that a variety of processes can and usually will occur
simultaneously in laser chemical systems and unravelling these various processes
can be challenging. For example, the selectivity of multiphoton processes in
processing organometallics will be discussed extensively in several of the chapters
which follow. As a species selective method of promoting gas phase reactions
involving refractory metals, laser chemistry of organometallics presents some
unrivaled possibilities. The possibility of adding state selectivity, achieved by either
judicious choice of organometallic precursor(36) or excitation conditions, is still a
controversial proposition.

Stage Two: Laser Chemical Cluster Synthesis As An Irreversible Fractal Aggregation Phenomenon

Rapidly transforming a gas phase organometallic containing mixture to atoms and
other fragments sets the stage for unique chemistry to occur. Some of the chapters

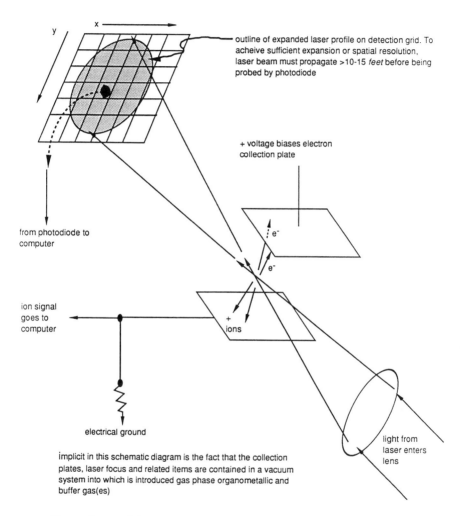

Figure 6.   Schematic diagram showing method for correlating spatial
region within laser focus with observed multiphoton processes.

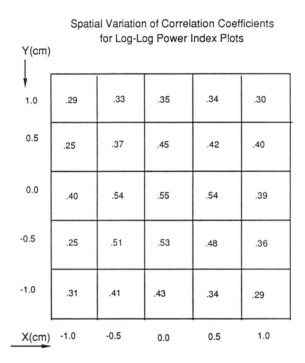

Figure 7   Typical grid of correlation coefficients for linear fits to log log plots of ion signal vs. laser pulse energy.

which follow are concerned with the type of "single kinetic pathway" chemistry involving metal atoms, ions and coordinatively unsaturated organometallics which occurs in that complex laser produced mixture. In this, the final section of this chapter, we describe one type of process, cluster and ultrafine particle production, which was recognized very early on as a potential use of laser chemistry involving organometallics. This type of chemistry, which we referred to above as being comprised of "multiple kinetic pathways" and for which the phenomenology and experimental aspects are well described by Puretzky, starts off exactly where the chemistry described by Weitz, Weiller and others stops. Because these authors do such a good job with describing the "single kinetic pathway" limit, no further mention will be made except to note that Figure 1 in Hackett's chapter is a good basis for depicting the distinction between the two limits.

The earliest work on what Friedlander(37) calls "gas to particle conversion" began appearing in the mid-1970s. Tam(25), Bauer(42), Haggarty(38), Yardley(39), Rice(40), Yabazuki(26), Scheive(41), and others all contributed to a picture which is still evolving. At the time, attempts to measure, model, and to fundamentally understand the process, employed concepts of supersaturation(42), nucleation, and bimolecular interactions. The effect of the electronic state of the monomers on the nucleation process was immediately recognized as was the importance of internal cooling of the nascent products of aggregation. The internal cooling is required to keep the clusters from evaporating off smaller fragments. Without such energy transfer from the clusters there would be no cluster formation. Whereas simple statistics supports the proposition that bimolecular interactions are the most important, the relative role of monomers, dimers, trimers etc. in the aggregation process has not been elucidated. Perhaps the most important measurement missing from these early studies of laser produced particles and clusters was size distribution histograms. Given the extreme fundamental(43,44) and applied interest(45,46,47,48,49,50)in clusters and ultrafine particles, and the expected tendency of electrical, optical and chemical properties to be size dependant, such histograms would seem to be very valuable when obtainable.

Recently, production of metal and other clusters by entrainment of metal atoms into nozzle beam expansions has produced many(51,52) such distributions. Whether the metal atoms are produced using ovens, laser evaporation of metal rods or some other methods, mass spectrometry has allowed a large increase in the number of distributions which can be catalogued and studied. Earlier attempts(53) at obtaining size distributions using electron microscopy were the most successful when the clusters were made using the evaporating wire method. Of course, mass spectrometrically determined distributions present the possibility of artifacts marring the measured histograms due to fragmentation competing with ionization. The electron microscopic approach offers the possibility of artifacts(54,55) arising from the coalescence of smaller clusters after they have been collected from their nascent location, depth of focus effects or even other sources.

We became interested in such distributions after obtaining a few size histograms of platinum clusters produced using laser chemistry of organometallics(56). We found that it is possible to use the lognormal distribution to compare distributions from all the methods used to produce gas phase clusters. The lognormal distribution has been in common use for small particle characterization for nearly 150 years(57). It's value was essentially empirical at best since there were no analytic mathematical connections between it and the gas to particle conversion process that did not employ random numbers. Curiously, it became apparent that except for some histograms produced using the bell-jar evaporating wire approach, all the other distributions had a common variation from the standard lognormal distribution. They all produce a distribution deficient in larger clusters compared to that consistent with a pure lognormal distribution. In fact, with appropriate scaling, despite the fact that the distributions ranged from ensembles of small clusters containing at most some dozens of atoms/monomers, to ensembles in which the smallest clusters contained hundreds and thousands of atoms, all the distributions were essentially identical in overall shape.

For all the distributions which have been measured experimentally, and the painstaking on-going search for "magic number" cluster sizes within those distributions, there has been almost no work whatsoever in analyzing what the size distribution function should be assuming the *absence* of magic number effects. The remainder of this section is devoted to describing our recent work toward achieving a method for analyzing nascent cluster size distributions. In our view, this is a purely kinetics problem which we believe must be solved satisfactorily before much of the data which already exists concerning gas phase clusters can be interpreted in such a way as to separate the measurement of the clusters' properties from the intrinsic properties of the cluster formation process.

The simplest set of assumptions concerning the aggregation process are contained in Smoluchowski's[58] model which is now almost 80 years old. We shall embrace this approach in which only binary collisional processes are considered relevant. We will assume that larger clusters are formed irreversibly by the coalescence of smaller clusters. We assume that the only method for cluster destruction is when clusters coalesce with one another to make larger clusters. Thus we have **1**, the Smoluchowski equation for the time rate of change of a k-mer population in terms of the current populations of all cluster sizes.

$$\frac{dn_k}{dt} = \frac{1}{2} \sum_{j=1}^{k-1} K_{j,k-j} n_j n_{k-j} \; - \; n_k \sum_{j=1}^{\infty} K_{j,k} n_j \qquad\qquad \mathbf{1}$$

The $K_{i,j}$ are called kernels and contain the crossections and relative velocities to combine with the explicitly indicated number densities, n_i and n_j, to obtain the rate of collision between the i-mer and j-mers. There cannot be any assumptions of steady state because the distribution simply gets broader and shifted to higher average cluster sizes with increasing time. Assumptions regarding the lack of boundary interactions, changes in temperature and depletion of reactants are ignored in this first approximation. The problem is sufficiently complicated even without these potentially important issues being raised.

Ignoring any coalescence/destruction channels is probably wrong if we are to form a quantitative model of the cluster formation process. Earlier work coming from the literature of homogeneous nucleation[59] nearly always focussed on cluster growth by successive addition of monomers and not by cluster-cluster coalescence. To obtain the essence of the problem, including formation and destruction by all binary collision channels seems to impart sufficient detail to form a basic quantitatively useful model. It is now clear, however, that to properly analyze magic number relevant measurements, evaporation/destruction pathways for at least a few cluster sizes just larger than certain magic number sizes must be included.

What would we hope to obtain from measurements of cluster size distributions? Structures, thermodynamic properties, and measurements of other internal properties which would mediate the interactions between clusters and other species in various chemical and physical environments are the goals of this research.

The Smoluchowski model only introduces one type of parameter and that is the crossection for collisions between i-mers and k-mers. Since every collision is assumed to lead to coalescence, the physical size of the crossection may have no obvious connection with the size of the cluster. The only parameters which enter the model are the crossections, temperatures and number densities which determine the mass and energy transfer regime in which the aggregation phenomenon is assumed to occur. The latter two parameters are not really adjustable since they are fixed by experimental conditions whereas determining the crossections is one goal of the model. The crossections are the connection between dynamical interactions, e.g. collisions, and the static energetic, i.e. thermodynamic, properties of the endpoints of the trajectories.

If the aggregation phenomenon is assumed to occur in a low total pressure regime, then the mass transfer is said to occur by "ballistic" interactions in which the usual terms from gas kinetic theory are relevant, i.e. mean free path, collision frequency, to describe the means by which the coalescing partners find each other. In this limit, the mean free path is long compared to the average cluster dimensions. On the other hand, the term "diffusive regime" is used to signify the fact that under higher total pressure conditions, the coalescence takes place by partners finding each other by diffusion. Analytical expressions in either regime can be constructed to implement the Smoluchowski treatment, i.e. calculate the kernels, but at some point it becomes necessary to make at least some assumption regarding the behavior of the crossections.

As Friedlander showed nearly twenty years ago, even the simplest assumption made regarding the variation of coalescence crossections with varying sizes of the coalescing species has important consequences. One method to obtain the size of a cluster is obtained by summing the volumes of all the monomers and calculating the radius of a sphere with an equivalent volume. In this case, there is an implicit assumption that the cluster is globular and other possibilities also exist for simple minded assumptions. The most important consequence of these considerations is that the kernels which drive the Smoluchowski equation are a homogeneous function of cluster size. This homogeneity is best summarized as in equation **2**.

$$K(\lambda i, \lambda j) = \lambda^{2\omega} K(i,j) \qquad\qquad \textbf{2}$$

Given the homogeneity condition specified above, it is possible to obtain general solutions to the Smoluchowski equation in the long-time, large cluster size limit. Jullien*(60)* has shown that the size distribution function is of the form given in equation **3**.

$$n_k \rightarrow A\, k^a\, e^{-bk} \qquad\qquad \textbf{3}$$

The parameter ω, defined in terms of a and b in **3**, is called the homogeneity exponent and can be decomposed into a few other parameters such as the fractal dimension of the clusters, the fractal dimension of the trajectories the clusters execute between coalescence events, the scaling of the clusters translational energies with increasing cluster sizes, and the dimensionality of the space in which the coalescence phenomenon is assumed to occur. We have recently described these parameters in detail elsewhere*(61)* but it is worthwhile to point out that regardless of the ethereal quality of their monickers, they are real parameters which can be rigorously defined from first principles.

By some rather simple algebra, it is possible to show that Jullien's solution very closely mimics the lognormal distribution.

$$n_k \propto C\, e^{-\left(\frac{\ln k - \ln c}{\sigma}\right)^2}$$

The most significant deviation occurs at larger cluster sizes, in agreement with our observation that all the observed experimental distributions were deficient in large cluster sizes relative to the pure lognormal distribution. We suspect that the historical significance of the lognormal distribution derives from its resemblance to the asymptotic solution to the Smoluchowski equation. It should be pointed out that although Jullien was able to guess the solution based on a mathematical homomorphism between classical statistical mechanics and the the highly interconnected kinetic pathways that permeate the Smoluchowski picture, such a homomorphism is not at all needed to show that the solution is correct. Thus, the homomorphism therefore need not have any physical significance.

To immediately appreciate the utility of our analysis, we have plotted data taken from the literature concerning aluminum clusters using logarithmic scales. A lognormal distribution plotted in this way is rigorously an inverted parabola. This is clearly evident in the plot shown in Figure 8. The point of this plot is to show how to obtain the most probable cluster size from an experimental histogram. Jullien showed that using equation **4**, it is possible to use such an experimentally obtained histogram to obtain the homogeneity parameter ω. The average cluster size is obtained directly from the measured distribution.

$$\frac{\langle k \rangle}{k_m} = \frac{a+1}{b} = \frac{2\omega - 1}{2\omega} \qquad\qquad \mathbf{4}$$

Using these equations it is possible to catalogue values of ω which is supposed to only be a function of the nature of the monomer and the reaction mass and energy transport conditions.

The mere fact that the model we have described is applicable to experimental data suggests that the basic premise of homogeneity is at least loosely obeyed. This means that the smooth variation in clusters size and shape and mass with increasing number of monomers is more important in determining the behavior of the aggregation process than are the presence of any magic numbers or special shapes. For a given k-mer, the larger k is, the more structures exist and, although possibly a number of structures and conformations exist, only the average behavior is important. The details of interaction potential surfaces are not very important unless a cluster size is being considered which could correspond to a magic number. Much more discussion of these concepts(60,61), as well as a description of the fractal dimensions listed above, can be found elsewhere.

We choose as a final demonstration of the utility of this model to consider the formation of a magic number cluster(62). Whether a number is "magic" because of a shell closing or because a bonding mode within the cluster can be found which fully saturates the coordination capacity of all the monomers, the crossection for addition of any specie to the fully saturated magic number cluster is probably much smaller than would otherwise be expected based on the homogeneity exponent. As an extreme example, we assigned all kernels involving the coalescence, and thereby destruction, of 60-mers and 70-mers with any other cluster size to be zero. We assigned all other kernels consistent with $\omega = -0.7$. As can be seen in Figure 9, while the overall distribution looks much like it did without the arbitrary breaking of the

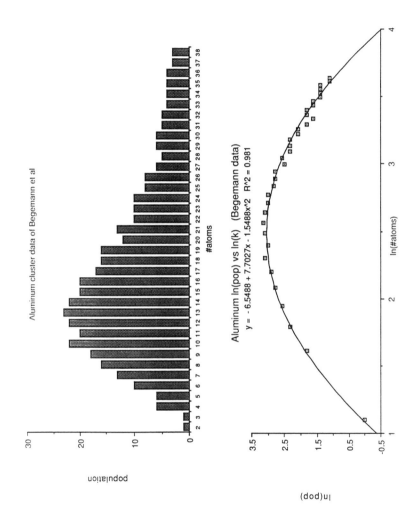

Figure 8.   Cluster population data of Begemann is shown above log-log plot of same data. Parabolic fit is shown. Log-log plot of standard lognormal distribution is parabolic.

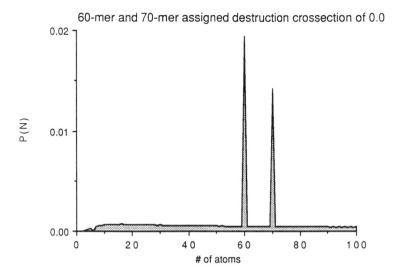

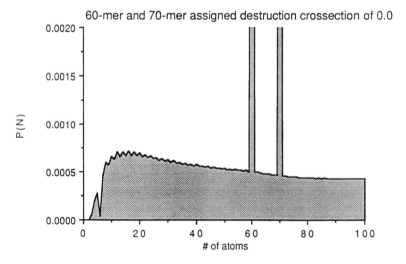

Figure 9.   Results of integrating Smoluchowski equation with all kernels for 60-mer and 70-mer assigned to zero. Upper plot shows population of 60-mers and 70-mers compared to rest of distribution. Bottom plot is same simulation but with a expanded scale so that shape of overall distribution can be seen. The oscillations at small cluster sizes are an artifact of the particular integration method used for solving the Smoluchowski equation. They disappear when more appropriate methods are used.

homogeneity, the population of 60-mers and 70-mers are clearly much larger than it would otherwise be. Note that if only monomer addition was important for cluster formation, then all coalescence would stop at the formation of a 60-mer. Since experimental data showing formation of clusters on both sides of magic numbers is quite common, it is clear that only cluster-cluster coalescence allows direct formation of clusters larger than magic numbers.

This model allows extraction of a single parameter from experimental data, ω, which relates the size distribution function to the nature of the monomer and the reaction conditions. It rationalizes the effects of kinetics and transport phenomena on the formation process. We are very optimistic that this model will allow analysis of evaporation effects, systematic categorization of various monomers in terms of their bonding/valence modes and their capacity to transfer substantial amounts of internal energy to various types of collision partners. The ability to describe reactions in nozzle beams, involving oven-generated metal atoms or atoms from laser heated rods, and similar reactions involving laser chemistry of organometallics, using a single formalism, only one parameter, and employing the stages described by the flowchart in Figure 1, provides a conceptual context for laser chemistry of organometallics.

Summary and Concluding Remarks

Laser chemistry of organometallics can be described using a flowchart in which the overall process is considered as three crudely separable stages. We proposed there are potential synergies between the photo- and physical chemical properties of organometallics and the types of processes most advantageously initiated/promoted using lasers. We presented data suggesting that organometallic laser chemistry can involve either single or multiphoton processes and that at least moderate scale-up is probably possible in either limit. Once multiphoton processes are employed, then the excitation process is necessarily not spatially uniform. This may or may not be of consequence depending on the type of laser chemistry being attempted. The same parameters which affect the mass and energy transport characteristics of the reaction mixture can also have an effect on the order of the excitation process. Again, this may or may not be of consequence depending on the type of laser chemistry being attempted but it is an example of a type of chemical nonlinearity which can arise and ruin our attempt to separate, if only conceptually, the overall process of laser chemistry into various stages. Finally, we sketched the major components of a model for laser chemical gas to particle conversion. This model allows for the formation of magic number clusters, predicts the relationship of the observed distribution function to the lognormal distribution and begins to suggest meaningful new ways to think of the structure of clusters in terms of the properties of the monomers and the aggregation conditions.

Acknowledgements

Our research is supported by Rome Laboratory and the Donors to the Petroleum Research Fund as Administered by the American Chemical Society. I gratefully appreciate the efforts of Dan Rooney, B. Samoriski, J. M. Hossenlopp, M. J. Casey, M. Villarica and Matt Cote.

Literature Cited

1   Woodruff, W. H. *Inorganic Chemistry: Toward the 21st Century*; ACS
    Symposium Series, Vol. 211, pp. 473-508, (American Chemical Society,
    Washington, D.C. 1983)
2   Yablonovitch, E. Springer-Series Solid State Sci., **18**(Relax. Elem.
    Excitation)pp. 197-205(Cambridge, Mass USA, 1980)
3   Outhouse, A. ,Lawrence, P., Gauthier, M. and Hackett, P. A., *Appl. Phys. B.*
    1985,36, pp.63-75
4   Letokhov, V. S. *Nonlinear Laser Chemistry: Multiple-Photon Excitation;*
    Springer Series in Chemical Physics, Springer-Verlag, Berlin, 1983; Vol. 22
5   Garbuny, M., US Patent 4,176,024(**1979**)
6   Ronn, A. M., US Patents 4,328,303 and 4, 343,687(**1982**)
7   Rousseau, D. L., *J. Chem. Ed.* **1966**, 43, pp.566-570, Wilson, M., *Ann. N. Y.*
    Acad. Sci. **1970**, 168, Art.3; pp.615-620, Moore, C. B., *Annu. Rev. Phys.*
    Chem., **1971**, 22, pp. 387-428
8   Ronn, A. M. *Sci. Am.* **1979**, 240, pp.114-128, Duxbury, G. *Annu. Rep.*
    Prog. Chem., Sect. C, 78(C), pp.31-61, Hall, R. B.; Kaldor, A.; Cox, D. M.;
    Horseley, J. A.; Rabinowitz, P.; Kramer, G. M.; Bray, R. G.; Maas, E. T.
    Adv. Chem. Phys. **1981**,47, pp.639-659, Katayama, A. *CEER, Chem. Econ.*
    Eng. Rev. **1986**, 18(1-2), pp.14-19
9   Wrighton, M.; Geoffroy, G. *Organometallic Photochemistry,*; Wiley-
    Interscience, New York, 1979
10 Gedanken, A.; Robin, M. B.; Kuebler, N. A. *J. Phys. Chem.* **1982**, 86,
    pp.4096-4107
11 Kodas, T. T.; Comita, P. B. *Acc. Chem. Res.* **1990**, 23, pp.188-194
12 Flygare, W. H. *Acc. Chem. Res.* **1968**, 121, pp.1
13 Laidler, K. J. *Chemical Kinetics*,Third Edition, Harper and Row, New York,
    1987
14 Bernstein, R. B.; Levine, R. D.; *Molecular Reaction Dynamics*, Clarendon
    Press, Oxford, 1974
15 Bernstein, R. B.; *Chemical Dynamics Via Molecular Beam and Laser*
    Techniques, Clarendon Press, Oxford, 1982
16 MacDonald, J. D. *Ann. Rev. Phys. Chem.* **1979**, 30, p.29
17 Crim, F. F. *Ann. Rev. Phys. Chem.* **1984**, 35, pp.657-691
18 Hossenlopp, J. M.; Chaiken, J. *J. Phys. Chem.* **1987**, 91, p.2825
19 Hossenlopp, J. M.; Chaiken, J. *J. Phys. Chem.* **1992**, 96, pp.2994-3000
20 Jervis, T. R.; Newkirk, L. R. J. Mat. Res. **1986**, 1, pp.420-424
21 Turner, J. J.; Poliakoff, M. *Inorganic Chemistry: Toward the 21st Century*;
    ACS Symposium Series, Vol. 211, pp. 35-37, (American Chemical Society,
    Washington, D.C. 1983)
22 Boxhoorn, G.; Shoemaker, G. C.; Stufkens, D. J.; Oskam, A. *Inorg. Chim.*
    Acta, **1979**, 33, p.215, Boxhoorn, G.; Shoemaker, G. C.; Stufkens, D. J.;
    Oskam, A.; D. J. Darensbourg *Inorg. Chem.* **1980**, 19, p.3455
23 Bauer, S. H. *Proc. SPIE* **1984**,458, p.1
24 Horsley, J. A.; Rabinowitz, P.; Stein, A.; Cox, D.; Brickman, R.; Kaldor, A.
    IEEE J. Quant. Elec. **1980**, , pp.412-419, Jensen, R. J.; *Report, LA-UR-76-*
    499,9 pp. avaial NTIS from *ERDA Energy Res. Abstr.* **1977**, 2(2) Abstr. No.
    2452
25 Tam, A. C.; Moe, G.; Happer, W. *Phys. Rev. Lett.* **1975**, 35, pp.1630
26 Yabuzaki, T.; Sato, T.; Ogawa, T. *J. Chem. Phys.* **1980**, 73, pp.2780-2783
27 Haberland, H.; Kolar, T.; Reiners, T. *Phys. Rev. Lett.* **1989**, 63, pp.1219-22

28   Zakheim, D. S.; Johnson, P. M. *Chem. Phys.***1980**, 46, pp.263-272
29   Ackerhalt, J. R.; Shore, B. W. *Phys. Rev. A.* **1977**, 16, 277-282, Ackerhalt, J.
     R.; Eberly, J. H. *Phys. Rev. A.* **1976**, 14, pp.1705-1710, Schek, I.; Jortner,
     J. *Chem. Phys.* **1985**, 97, pp.1-11
30   Gandhi, S. R.; Bernstein, R. B. *Chem. Phys.* **1986**, 105, pp.423-430
31   Villarica, R. M.;Wiedeger, S.; Samoriski, B. *Proc.SPIE* **1991**, 1412, pp.12-
     18
32   Rooney, D.; Driscoll, C.; Chaiken, J. *Inorg. Chem.* **1987**, 26, pp.3939-3945
33   Hossenlopp, J. M.; Rooney, D.; Samoriski, B.; Chaiken, J. *J. Chem. Phys*
     1986, 86, p.3331
34   Hossenlopp, J. M.; Rooney, D.; Samoriski, B.; Chaiken, J. *Chem. Phys. Lett.*
     1985, 116, p.350
35   Samoriski, B.; Chaiken, J. *J. Chem. Phys.* **1989**, 90, pp.4079-4090
36   Hossenlopp, J. M.; Rooney, D.; Samoriski, B.; Chaiken, J. *J. Chem. Phys.*
     1986, 86, pp. 3326-3330
37   Friedlander, S. K. *Smoke, Dust, and Haze,* Wiley Interscience, New York,
     1972
38   Cannon, W. R.; Danforth, S. C.; Haggarty, J. S.; Marra, R. A. *J. Am. Cer.
     Soc.* **1982**, 65, pp.331-335
39   Gupta, A.; Yardley, J. T. *Proc. SPIE* **1984**, 458, pp.131-139
40   Rice, G.; Woodin, R. L. *J. Am. Cer. Soc.* **1988**, 71, pp.C181-C183
41   Zurek, W. H.; Schieve, W. C. *J. Chem. Phys.* **1978**, 68, pp.840-846, *J. Phys.
     Chem.* **1980**, 84, pp.1479-1482
42   Freund, H. J.; Bauer, S. H. *J. Phys. Chem.* **1977**, 81, pp.994-1000
43   Kubo, R. *J. Phys. Soc. Japan,* **1962**, 17, p.975
44   Gor'kov, L. P.; Eliashberg, G. M. *Zh. Exsp. Teor. Fiz.* **1965**, 49(english
     translation *Sov. Phys.-JETP* **1965**, 21, p.940)
45   Das, P. *J. Phys. Chem.* **1985**, 89, pp.4680-4687
46   Hanamura, E. *Phys. Rev. B.* **1988-II**, 37, pp.1273-1279
47   Neeves, A. E., Birnboim, H. *Opt. Lett.* **1988**, 13, pp.1087-1089
48   Drube, W.; Himpsel, F. J. *Phys. Rev. Lett.* **1988**, 60, pp.140-142
49   Pool, R. *Science,* **1990**, 248, pp.1186-1188
50   Carver, G. E.; Divrechy, A., Karbal, S., Robin, J.; Donnadieu, A. *Thin Solid
     Films,* **1982**, 94, pp.269-278
51   Haberland, H.; Kornmeier, H.; Langosch, H.; Oschwald, M.; Tanner, G. J. *J.
     Chem. Soc. Faraday Trans.* **1990**, 86, pp.2473-2481
52   Kappes, M.; Leutwyler, S.in *Atomic and Molecular Beam Methods* , Editor
     G.Scoles, Oxford University Press, 1988, Vol.1, Chapter 15
53   Granqvist, C. G.; Buhrman, R. A. *Solid State Commun.* **1976**, 18, pp.123-
     126, *J. Catal* **1976**, 42, pp.477-479
54   Tence, M.; Chevalier, J. P.; Jullien, R. *J. Physique* **1986**, 47, pp.1989-1998
55   Koide, K.; Yatsuya, S.; Yoshimura, I. *Jap. J. Appl. Phys.* **1980**, 19, pp.367-
     369
56   Chaiken, J.; Casey, M. J.; Villarica, M. *J. Phys. Chem.* **1992**, 96, pp.3183-
     3186
57   Siano, D. B. *J. Chem. Ed.* **1972**, 49, pp.755-757
58   Smoluchowski, M.V. *Phys. Z.* **1916**, 17, p.557
59   Zettlemoyer, A. C. *Nucleation,* Marcel Dekker, New York, 1969 or Andres,
     R. P. *Nucleation Phenomena,* ACS Symposium Series, Amercian Chemical
     Society, Washington, D. C. 1966

60   Jullien, R. *New J. Chem.* **1990**, 14, pp.239-253 and Botet, R.; Jullien, R. *J. Phys. A.* **1984**, 17, pp.2517-2530

61   Villarica, M., Casey, M. J., Goodisman, J.; Chaiken, J. *J. Chem. Phys.* **1993**, 98, pp.1-16

62   see for example Martin, T. P.; Bergmann, T.; Golich, H.; Lange, T. *J. Phys. Chem.* **1991**, 95, pp.6421-6429, or Knight, W. D.; Clemenger, K.; deHeer, W. A.; Saunders, W.; Chou, M. Y.; Cohen, M. L. *Phys. Rev. Lett.* **1984**, 52, pp.2141-2143

RECEIVED March 29, 1993

DISSOCIATION AND OTHER PHOTOPROCESSES

Chapter 2

Parity State Selectivity in the Multiphoton Dissociation of Arene Chromium Tricarbonyls

Daniel P. Gerrity and Steve Funk

Department of Chemistry, Reed College, Portland, OR 97202

This chapter discusses results of laser multiphoton dissociation (MPD) studies of various gas phase chromium carbonyl complexes, and presents evidence that the MPD mechanism involves parity selectivity in the production of Cr(I). Multiphoton ionization (MPI) and fluorescence excitation spectroscopy have been employed in previous studies to probe the electronic state population distribution of product metal atoms in order to gain understanding of the MPD mechanism. This paper presents results of a new study of the 355 nm MPD of toluene-, ethylbenzene-, and propylbenzene-Cr(CO)$_3$. The product population behavior of several emitting as well as non-emitting atomic states have been measured, and evidence that the MPD mechanism involves parity selectivity in the production of Cr(I) is presented and discussed.

The multiphoton dissociation (MPD) of vapor-phase organometallics has been actively studied for several years (1-14). Early investigations determined that photodissociation of chromium and iron carbonyls, and other organometallics, efficiently produces neutral metal atoms in ground and excited electronic states, while formation of ionic metal atoms is not significant (1-4). The electronics industry is already exploring potential applications for producing atomic metal vapors via these reactions, and these metal vapors could also prove very useful as reactants for an early stage of what could be termed "Laser Chemistry of Organometallics". The mechanism involved in the dissociation process remains uncertain, however.

The multiphoton dissociation behavior of these metal complexes was quite surprising when first observed, because it had been well established that the result of focusing intense, pulsed laser light on the organic compounds that had been studied up to that point in time led to ionization of the parent compound. One explanation for the difference in behavior for these organometallics is that the rate of dissociation out of the excited electronic states competes very effectively with the rate of further up-pumping to the ion. This is not too surprising, since single photon studies have shown that these species are incredibly photolabile, having very rapid dissociations, with high dissociation quantum efficiencies (*15-16*). One-photon dissociation of $Cr(CO)_6$ in solution results in the loss of only one carbonyl (*17-19*). In contrast, photodissociation of gas phase hexacarbonyls can result in removal of more than one ligand, depending on the photolysis wavelength. Eric Weitz's time-resolved infrared work has shown that at 355 nm, gas phase photodissociation leads to loss of only one CO, but as the photolysis wavelength gets shorter and shorter, more and more ligands are removed, i.e., as many ligands can be removed as there is energy from the photon (*20*). The simultaneous absorption of two or more near-UV photons would provide enough energy to remove all the ligands; can we think of the multiphoton dissociation mechanism of these species as being equivalent to the photodissociation brought about by a single, vacuum UV photon of equivalent energy? It is hoped that these studies will help answer this question.

One may also ask if the dissociation proceeds directly, with essentially simultaneous loss of all ligands, or does it involve the sequential loss of ligands, proceeding through a sequence of electronically excited fragments. Whereas the MPD of chromium hexacarbonyl using visible lasers undoubtedly involves a true MP absorption, at least in the initial step, MPD using UV radiation could proceed as a series of one-photon absorption/dissociation steps since the parent compound has significant one-photon absorption intensity in the UV. That is, absorption of one photon could lead to partial dissociation, followed by further absorption of a photon by the intermediate, followed by loss of more ligands, until finally all ligands are removed. Recent work (*8, 9*) at 248 nm suggests that one pathway may include a combination of these two descriptions, with the parent complex simultaneously absorbing two photons, partially dissociating into some intermediate, which itself undergoes absorption of one or more photons and then dissociates into the final products. This work will be discussed shortly.

An important theme of these studies was built around the exploration of the possibility and conditions for non-statistical unimolecular reaction dynamics in these molecules. By monitoring the nascent product state distribution of the chromium atom, these studies hope to elucidate the influence of intramolecular vibrational energy redistribution, and the consequences of the internal structure of the molecule, on the photodissociation dynamics. One of the questions we hope to answer is, can one control the distribution with laser wavelength, laser flux, buffer gas pressure, or by choosing the appropriate ligand? Being able to control the metal atom product energies could be very beneficial to their use as reactants in subsequent chemistry. Ultimately, one hopes knowledge of the mechanism(s) involved in the dissociation will enable us to control the photochemical reactions and nonstatistical unimolecular reaction dynamics. This chapter begins with a brief history of what has been done to determine these distributions for chromium hexacarbonyl, the first organometallic complex for which multiphoton dissociation into neutral fragments was demonstrated to be the predominant MPD pathway. Then, results of recent studies on a series of arene-chromium tricarbonyl complexes are presented.

Historical Background

$Cr(CO)_6$. A little over a dozen years ago, we, as well as a number of other groups, showed that pulsed lasers with very short pulse durations, on the order of nanoseconds, could be used to remove all of the ligands from various gas-phase transition metal complexes (1-4). The total bond energy that had to be overcome far exceeded the energy of one of these photons, making it clear that the photodissociation process required at least two, and in some cases the absorption of at least five, photons, to account for the product energies involved. The high photon fluxes resulting from focusing these short-pulsed lasers makes the simultaneous absorption of more than one photon quite probable.

Resonance-enhanced multiphoton ionization (REMPI) and fluorescence excitation spectroscopy have both been employed to test the electronic state population distribution of product metal atoms in order to gain understanding of the MPD mechanism. The original MPD studies of $Cr(CO)_6$ utilized a tunable dye laser with 10 ns pulse duration and 0.3 mJ/pulse to cause multiphoton dissociation and to simultaneously ionize the resultant Cr(I) atoms via resonance-enhanced multiphoton ionization (2). As can be seen from the scan shown in Figure 1, the non-resonant background ion signal is much less than the ion signal produced from the resonant ionization of a particular electronic state of the neutral atom. I.e., the ion signal is small unless the laser is in resonance (1, 2, or 3 photons) with an electronic transition of the neutral Cr(I) atom. Therefore, there are a lot more neutral atoms being formed than ions from the MPI of the parent or its photofragments; the predominant multiphoton dissociation products are neutral atoms and molecules, not ions as had been observed for purely organic molecules. Many other groups have shown that this is a general property of organometallic complexes; they tend to dissociate rather than ionize, in contrast to the multiphoton behavior of purely organic species.

One of the drawbacks of the REMPI detection method is that it is difficult to get quantitative information about the nascent relative population distributions of the product. However, only signal originating from even parity states of the Cr atom product was observed over the region probed (363-585 nm), which might suggest some sort of parity selectivity occurs in the MPD process. Since the atomic states of Cr(I) up to 23,000 cm^{-1} are all even parity states (i.e., have electronic configurations with an odd number, or no, electrons in p orbitals), this was not necessarily the case (Figure 2 shows the lower electronic levels of atomic chromium). However, transitions originating from even parity states up to 30,000 cm^{-1} were observed, and there are a number of odd parity states below this energy. We decided to look for fluorescence from odd parity states, since fluorescence from many of the lower-energy odd parity states down to the ground state ($a\,^7S_3$) or low-lying $a\,^5S_2$ and $a\,^5D_J$ states (~8,000 cm^{-1}) should be dipole-allowed. In addition, fluorescence intensities can be used to determine the nascent population distributions of the electronically excited Cr(I) product.

Figure 3 shows a portion of the emission spectrum produced from the MPD of Cr(CO)$_6$ (5). As the data in Table I shows, emission from many odd parity states with multiplicities of 5 and 7 were observed, as well as one even parity state ($e\,^7D_5$). Emission was observed from states with energies up to 45,700 cm^{-1}; this excess energy exceeds the energy of one dissociation photon! The sum of all the emission intensities originating from a particular level can be used as the relative population of that level, after correcting for detector response and degeneracy. This is because all of the emission signal is collected from each state, using a boxcar window which is much longer than the lifetimes of the states observed. MPD emission data was collected for laser wavelengths across the visible and near-UV spectrum. Most of these measurements utilized a tunable dye laser with 10 ns pulse duration and about 1 mJ/pulse. Relative populations of all levels observed through emission were calculated at 359 and 500 nm. At the wavelengths studied, these relative populations fit a Boltzmann distribution (i.e., plots of $ln(N_i/g_i)$ vs. E_i were linear), implying that perhaps some kind of statistical dissociation process was occuring. No selectivity in spin (S), orbital (L), or total (J) angular momentum was observed. MPD at 355 nm using a 25-ps laser pulse was also investigated, and the resulting Boltzmann plot of the data is show in Figure 4. The observation of anomalously high population in the $y\,^7P^o$ levels due to laser-induced up-pumping from the ground state shows that a significant amount of multiphoton dissociation to the ground state occurs within 25 ps, since up-pumping can only occur while the laser pulse is still on. Thus, the total dissociation process can occur within 25 ps, at least under these conditions, and any intermediates requiring more photons to reach complete dissociation would have to do so within this time.

A study of the laser flux dependence was done at 359 nm using 20-ns MPD pulses; the results are shown in Figure 5. The plots show distributions obtained at three different laser fluxes, where the highest flux differs from the lowest by about a factor of 20. Each of the plots has a different slope, and therefore a different corresponding temperature, and the temperature of the distribution increases as the laser flux increases. This was unexpected, since the amount of excess energy available to the

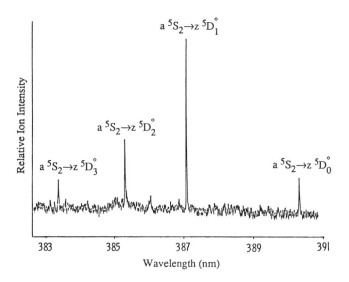

Figure 1. MPD/MPI spectrum of $Cr(CO)_6$, showing one-photon resonance enhanced MPI transitions originating from the $a\,^5S_2$ state of neutral atomic chromium.

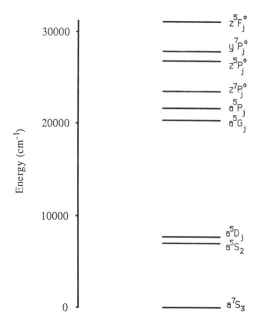

Figure 2. A selection of atomic states for Cr(I).

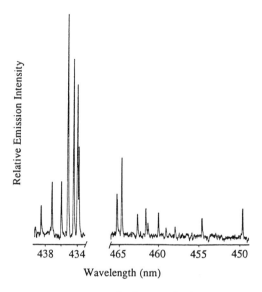

Figure 3. Emission spectra of $z\,^5F^{\,o}$ and $y\,^5P^{\,o}$ levels of Cr(I) produced by photodissociation of $Cr(CO)_6$, showing multiplet structure. (Adapted from ref. 5.)

Table I. Levels Observed in Emission Following MPD of $Cr(CO)_6$

desig- nation	J	energy level, cm^{-1}	desig- nation	J	energy level, cm^{-1}
$z\,^7P^{\,o}$	2	23,305	$z\,^5F^{\,o}$	1	30,787
	3	23,386		2	30,859
	4	23,499		3	30,965
				4	31,106
$z\,^5P^{\,o}$	3	26,788		5	31,280
	2	26,796			
	1	26,802	$z\,^5D^{\,o}$	0	33,338
				1	33,424
$y\,^7P^{\,o}$	2	27,729		2	33,542
	3	27,820		3	33,672
	4	27,935		4	33,816
$y\,^5P^{\,o}$	1	29,421	$e\,^7D$	5	42,261
	2	29,585			
	3	29,825	$y\,^5H^{\,o}$	3	45,566
				4	45,615
				5	45,663
				6	45,707
				7	45,741

Adapted from ref. 5.

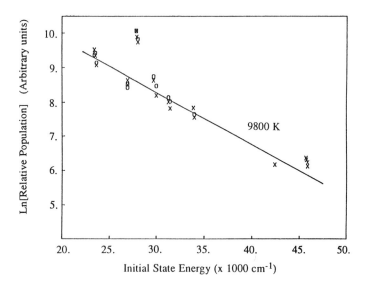

Figure 4. Boltzmann plot of the relative populations of Cr(I) excited states resulting from the MPD of $Cr(CO)_6$. A focussed, 25-ps laser at 355 nm was used for the photolysis. Two sets of data are plotted, both obtained under similar conditions. The population in the $y\ ^7P^o$ level (~27,800 cm^{-1}) is anomalously high because the laser wavelength is close to the transition from the ground state to this level. (Adapted from ref. 5.)

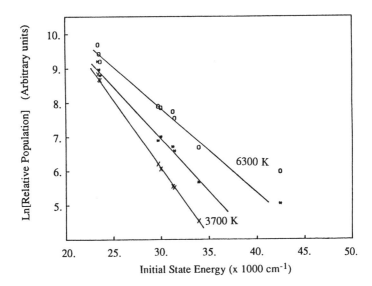

Figure 5. As in Figure 4, but with a 20-ns laser at 358.6 nm for photolysis. The three plots represent different laser fluxes, as explained in the text (X's are the lowest and 0's are the highest flux). The population in the $z\,^5P^{\circ}$ and $y\,^7P^{\circ}$ levels (~26,800 and 27,800 cm^{-1}) is anomalous because the laser is tuned close to transitions involving these atomic levels. These points were omitted from the plot and the analysis. (Adapted from ref. 5.)

products should be dictated by the energy, and therefore the wavelength, of the photon, not by the number of photons in the beam. This result can be explained if the atoms are formed via more than one MPD pathway, with at least some paths requiring different total numbers of photons. Thus, as the laser flux is increased, the channels requiring more photons will be favored because of their higher power dependences, and so the average product energy will increase because these higher-photon pathways should yield more excess energy than the channels utilizing fewer photons. Even if each channel gave rise to a statistical product distribution, with higher temperatures associated with channels involving more photons, one might expect that the overall distribution would not give a linear Boltzmann plot. However, if one calculates the net distribution resulting from the addition of three thermal distributions, each differing by one photon worth of excess energy, the net population distribution can approximate that of a single, strongly flux-dependent "temperature" over the energy range of states we monitored (23,000 to 43,000 cm^{-1}). For example, 2000, 5000, and 8000 K with weighting factors of 1, 5.5 x 10^{-5}, and 3 x 10^{-7}, respectively, gives a linear Boltzmann plot over the region monitored with a "temperature" of 4600 K, while weighting factors of 1, 1 x 10^{-3}, and 5 x 10^{-4} gives a "temperature" of 7,600 K. These calculations illustrate that it does not take much of a higher-photon process to significantly alter the distribution over this range of energies. Alternatively, the existence of a large number of *non-statistical* channels can also give rise to a net population distribution that appears statistical. Finally, in a stepwise photodissociation process, a flux-dependent temperature could also arise if intramolecular energy redistribution (IER) can compete with further photodissociation of intermediate fragments at lower laser fluxes, resulting in less excess electronic energy in the Cr(I) product.

The characteristic temperature for the 500 nm MPD (~1500 K) was much lower than for the 359 nm distribution (6300 K) obtained at about the same laser flux, indicating that there is a wavelength dependence to the photodissociation process.

These MPI and fluorescence studies of the MPD of $Cr(CO)_6$ indicate that the dissociation does not involve spin-differentiated Cr(I) states, at least for the production of the spontaneously emitting odd parity excited states, and that both even and odd parity states have significant populations. There is no evidence for parity selectivity in the photodissociation, although the fluorescence studies only include population data from one even parity state. To get a more complete picture of what is happening in the dissociation process requires investigating a large number of both emitting and non-emitting states at the same time, under the same conditions.

A few years ago, George Tyndall and Robert Jackson used this same atomic fluorescence detection method to extend these MPD studies of $Cr(CO)_6$ into the ultraviolet, at 351, 248 and 193 nm (8). As in our studies, they observed statistical distributions of the high-lying, spontaneously emitting odd parity atomic states at all three dissociation wavelengths, with a different characteristic temperature for each wavelength. However, their 248 nm photolysis laser was in resonance with two allowed transitions originating from low-lying atomic states ($a\ ^5S_2$ and $a\ ^5D_J$), both around 8,000 cm^{-1} in energy, to very high energy states which readily fluoresced. By measuring the intensities of these laser-induced emissions, they were able to monitor the nascent populations of these low-lying states. They used buffer gas studies to investigate the effects of collisions on the dissociation process. They observed that, whereas the upper state populations (those that showed significant spontaneous emissions) were unaffected by the addition of up to 100 Torr of buffer gas into the cell, the laser-induced emissions from these lower-lying states decreased sharply as the pressure increased. This difference in behavior led them to propose that there were at least two different dissociation channels; one in which a fragmentation intermediate is susceptible to collisional relaxation (the sequential channel), and one in which fragmentation is complete within about 1 ns, before many collisions can occur at 100 Torr pressure (the direct channel). The direct channel would involve dissociation via three photons and lead to production of both high- and low-lying atomic states, whereas the sequential dissociation process, which is susceptible to collisional relaxation, would involve only two photons and therefore lead to production of only ground and low-lying energy states. With only two photons involved in the sequential dissociation, there would only be enough excess energy to produce Cr(I) atoms in the ground and low-lying (~8,000 cm^{-1}) states, and thus only these state populations would decrease with added buffer gas, as observed. Further support for this proposed two-channel mechanism was provided by results of studies of the MPD behavior of a series of arene chromium tricarbonyl complexes, which will be discussed in the next section.

Arene-Cr(CO)$_3$ Complexes.   As Joe Chaiken described in his introductory chapter, a series of ligands with different densities of vibrational states can be used to investigate the influence of intramolecular energy redistribution (IER) on the dissociation process. They used REMPI to explore the MPD behavior of a series of n-alkyl substituted arene chromium tricarbonyl (ACT) complexes in the wavelength range of 369-348 nm (7). Features associated with transitions originating from the $a\ ^7S_3$ ground state and a broad feature assigned to transitions originating from $b\ ^3F_4$ (33,113 cm^{-1}), $b\ ^3H_4$ (35,871 cm^{-1}) and $b\ ^3H_6$ (35,934 cm^{-1}) were compared for several ACTs. They observed a decrease in the excited state peak intensity relative to those of the ground state as ligand complexity increased, which indicates that IER does play a role in the dissociation process of these compounds. Although the metal-ligand bond energies are the same for the methyl-, ethyl- and propylbenzene chromium tricarbonyl complexes, the branching ratio between formation of ground state atoms and excited state atoms depends on the n-alkyl chain length; the longer the alkyl chain, the higher the ratio of ground state population to excited state population.

Tyndall *et al.* then investigated the MPD behavior of this series of arene chromium tricarbonyl complexes at 248 nm (*9*), and showed that the results were consistent with the two-channel mechanism they had proposed for $Cr(CO)_6$ multiphoton dissociation. Again, they used atomic fluorescence signal to obtain relative population information for the spontaneously emitting states. As was the case for $Cr(CO)_6$, all three ACTs (benzene, toluene, and propylbenzene derivatives) showed a Boltzmann population distribution of product energy states for the odd parity emitting states. Also, the resultant temperature was roughly the same for all three compounds, indicating that IER did not affect the dissociation process for formation of these higher-energy product states. This result fit very nicely with the fast, direct dissociation pathway proposed for $Cr(CO)_6$ for the formation of these high-lying atomic states. Just as for $Cr(CO)_6$, the excited state emission was unaffected by added buffer gas pressures up to about 25 Torr.

In contrast, the laser induced emission from the low-lying $a\,^5S_2$ and $a\,^5D_2$ atomic product states decreased in intensity as the buffer gas pressure was increased, just as was the case for $Cr(CO)_6$. The pressure studies indicated that less than 5% of the $a\,^5S_2$ state population was due to the direct process. Finally, they also observed that the amount of induced emission from the low-lying states decreased relative to the uninduced emission from the high-lying states as the ligand complexity increased; i.e., the $a\,^5D_2/y\,^7P^{\circ}$ intensity ratio, normalized to 1.00 for the toluene ACT complex, decreased from 1.61 for benzene chromium tricarbonyl down to 0.60 for propylbenzene chromium tricarbonyl. Thus, the sequential mechanism they proposed for the MPD of $Cr(CO)_6$ leading to low-lying atomic product states was also consistent with the observed pressure-dependence data for the ACTs, and the dependence on ligand complexity could be explained by proposing that the fragmentation intermediate, which was susceptible to collisional relaxation, was also susceptible to IER. As the ligand complexity increases, the rate of IER should increase, resulting in energy initially in electronic states localized predominantly on the metal atom being redistributed throughout the molecule. The result would be less excess energy in the metal atom, and therefore a decrease in the population of the states probed by laser-induced fluorescence at around 8,000 cm^{-1}. This proposed mechanism could also explain the REMPI results on ACTs at ~360 nm, which showed the reverse trend, i.e., an increase in ground state MPI signal relative to signal initiating from high-lying states as the ligand complexity increased. As the IER rate increased, the excess energy, and therefore the temperature, of the product via the sequential pathway would decrease, resulting in more ground state population and thus a higher ground/excited state population ratio (the populations of the high-lying excited states, including the reference state, are assumed to be unaffected by ligand complexity because the direct pathway is too rapid to be influenced by IER or collisions). The proposed two-channel 248 nm MPD mechanism is illustrated in Figure 6.

Recent Results on the MPD of ACTs

We decided to test the MPD mechanism proposed by Tyndall and Jackson by observing a wider range of non-emitting states to see if the results conform to their mechanism. In particular, we wanted to see if the ground state population would increase relative to the populations of the low-energy excited states (~8,000 cm^{-1}) as ligand complexity increased. It also occured to us that, since all the low-lying atomic states were even parity states, and all the high-lying states they monitored were odd parity states, perhaps it was the difference in parity, not energy, that was important here. To do this, we simply needed to add a second, tunable laser beam to the experiment, so that we could use laser-induced flourescence to study whatever state we wanted. Finally, we wanted to see if the mechanism Tyndall *et al.* used to explain their $Cr(CO)_6$ and ACT MPD results at 248 nm would hold for other UV photolysis wavelengths as well. The studies we have completed so far have all been done at 355 nm; studies at 266 nm will be underway shortly.

Experimental. Our current experimental set-up (see Figure 7) is similar to that used by Steve Mitchell *et al.* in the visible MPD studies of $Fe(CO)_5$ (11) and $Cr(CO)_6$ (12), and allows us to observe the relative population behavior of the ground state and a number of non-emitting excited states, which complements the population data obtained for several spontaneously emitting states. The third harmonic (355 nm) of a pulsed Nd:YAG laser is used to bring about photolysis as well as to pump a dye laser. The tunable dye laser (470-530 nm) generates the probe beam used to excite product metal atoms from a particular even parity electronic state into an emitting state.   The two beams are recombined, and their focii are spatially overlapped using an adjustable beam expander; the probe beam arrives 1 to 2 ns after the photolysis beam. The beams are expanded before focusing into the cell with a 100 mm focal length lens in order to minimize burning at the entrance window. The UV MPD beam was either 1 or 2 mJ/pulse for the measurements described below, with a 7-ns pulse duration. The dye laser probe beam energy is about 0.5 mJ/pulse, which is enough to saturate the LIF transitions.

The atomic emission (both induced and spontaneous) is collected and then dispersed via a 0.6 meter monochromator. Since the actual atomic emission linewidth is considerably narrower than the monochromator bandwidth of 1.3 Å, the peak heights could be used for the relative emission intensity. The photomultiplier tube signal is sent to a boxcar integrator for averaging. The boxcar window is set to a width of 200 ns, which is much longer than the fluorescence lifetimes of the atomic states we studied.

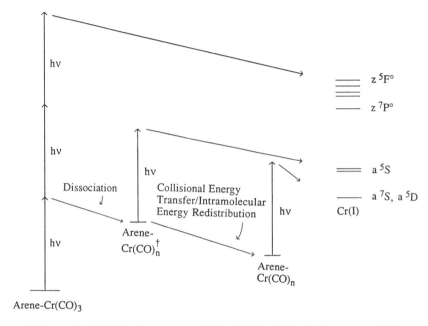

Figure 6. Mechanism proposed by G. Tyndall, C. Larson, and R. Jackson for the 248 nm MPD of arene chromium tricarbonyl complexes. (Adapted from ref. 9.)

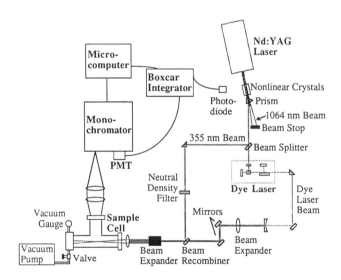

Figure 7. Schematic of the experimental apparatus currently used by the authors to investigate the MPD of metal carbonyls.

Thus, for the uninduced emissions, the total emission intensity of all transitions originating from a particular state can be used as the relative nascent excited state population of that state. A photodiode is used to trigger the boxcar window. The averaged signal from the boxcar is then sent to a microcomputer for storage.

Because of the small monochromator slit width and magnification due to the collection optics, only light from a very short length of the photolysis beam is collected. We observe significant differences in relative intensities from different electronic states when signal is collected from the focal region versus a few millimeters away from the laser focus; in fact, at the laser energies we use, the fluorescence signal from the reference state is a maximum about 6 mm on either side of the focus. In the experiments to be described below, all measurements were made with the focusing lens moved back so that the reference state signal was maximized; i.e., the emission signal was collected from a small region about 6 mm in front of the focal point of the photolysis beam. The laser fluence is estimated to be only about 5 mJ/cm^2 for this loosely focused 1 mJ/pulse 355 nm photolysis beam.

The toluene chromium tricarbonyl was purchased from Aldrich, and the ethyl and propyl chromium tricarbonyls were synthesized according to procedures described in the literature (*21, 22*). Each ACT sample is placed in a separate vacuum cell because of the difficulty in completely cleaning out the cell between compounds. During data collection, the cell pressure is maintained at 200 milliTorr.

In order to make meaningful comparisons between relative intensities of peaks associated with different ACTs, the laser powers and dye laser wavelength are untouched until data is collected from all three cells; in fact, a second set of data is collected as each cell is again rotated into position, and the results of both measurements are averaged. Non-induced emission from $z\ ^7P_3\,^\circ$, which served as the reference state, is collected every time a cell is rotated into position. Background spectra for induced emissions are measured by blocking the dye laser; in all cases, the uninduced signal from these very high-lying states is negligible compared to the intensity with the dye beam present. That is, the nascent population of these high-energy states is much smaller than the population produced via laser-induced excitation from the lower states.

Table II gives transitions used in the LIF studies of the non-emitting states. Atomic energy level data was obtained from Reference 23. Note that a two photon excitation is used to monitor the ground state population. Also, the $z\ ^7P_3\,^\circ$ reference state is probed via laser-induced fluorescence to compare with results from spontaneous emission measurements of the same state.

Results and Discussion. In the ion collection schemes used for REMPI, ion signal from a fairly large region of space is collected, not just from the focal point of the laser. Thus, population information from REMPI studies

Table II. Transitions Used for LIF Probing of Cr(I) Electronic State Population Distributions

State Probed	Probed State Energy cm^{-1}	Induced State	Induced State Energy cm^{-1}	Wavelength of Induced Transition Å	Ending State for Fluorescence	Fluorescence Wavelength Å
$a\,^7S_3$	0	$e\,^7D_4$	42258	4732.8	$z\,^7P^o_3$	5298.0
$a\,^5S_2$	7593	$z\,^5P^o_3$	26788	5209.9	$a\,^5D_4$	5411.4
$a\,^5G_5$	20524	$y\,^5F^o_4$	41225	4832.8	$a\,^5D_4$	3038.8
$a\,^5G_6$	20520	$y\,^5F^o_5$	41393	4789.9	$a\,^5D_4$	3019.8
$a\,^5P_3$	21841	$x\,^5D^o_4$	42909	4745.3	$a\,^5D_4$	2889.2
$z\,^7P^o_4$	23499	$e\,^7D_4$	42258	5331.6	$z\,^7P^o_3$	5299.0

may pertain to a wide range of laser fluxes. In contrast, the fluorescence collection optics used in our studies limit collection of light to a very small region; therefore, signal from an area of fairly uniform flux can be collected. The fluorescence signal from every state we monitored reached its maximum level not at the focal point of the photolysis laser, but rather at least several millimeters on either side of the laser focus. This is presumably because, up to this distance, the decrease in signal due to decreased laser flux is more than offset by the increase in the number of molecules irradiated. Furthermore, we observe that the position of maximum fluorescence signal depends on the energy of the state being probed; for the spontaneously emitting odd parity states, fluorescence from higher energy excited states is maximized when the light is collected closer to the focal point of the photolysis laser. This is very consistent with the flux-dependent temperature behavior we observed in the MPD of Cr(CO)$_6$; the laser flux is much higher near the focal point, favoring pathways requiring a larger number of photons and resulting in a higher average product energy distribution.

A rough estimate of the electronic temperature of the Cr(I) product distribution was calculated for each compound using non-induced emissions from $z\,^7P_J{}^o$ and $z\,^5P_J{}^o$, where we assumed all the high-lying, spontaneously emitting states fit a Boltzmann population distribution as observed in previous studies on Cr(CO)$_6$ at 359 nm (5) and on arene chromium tricarbonyls at 248 nm (9). The resultant temperature, 1500 K at 1mJ/pulse photolysis, was about the same for each of the three ACTs just as Tyndall et al. observed in their 248 nm study.

While absolute state populations were not obtained, relative populations could be compared from one compound to another by using the $z\,{}^7P_3{}^\circ$ state emission intensity for a reference. We looked for changes in other emission intensities (both induced and non-induced) relative to this one; i.e., we scaled the relative fluorescence intensity vs. wavelength plot of each compound so that this reference peak was the same size, and we looked for differences in the heights of the other peaks from one compound to the next (see, for example, Figure 8). Tables III-V summarize the dependences of the relative populations of various atomic states on the ligand complexity; for Tables IV and V, all population ratios are normalized with respect to the toluene complex, which is given a normalized relative population of 1. The experiments were done at two different UV intensities, 1 and 2 mJ/pulse.

Table III. Relative Population Data[a] from the Spontaneous Emissions of Cr($z\,{}^7P^\circ_{2,4}$) Produced in the MPD of a Series of ACTs

State, Energy	MPD Pulse Energy	Toluene-Cr(CO)$_3$ Rel. Pop.	Ethylbenzene-Cr(CO)$_3$ Rel. Pop.	Propylbenzene-Cr(CO)$_3$ Rel. Pop.
$z\,{}^7P_2{}^\circ$ 23,305 cm^{-1}	1 mJ	1.01±.03	1.07±.04	0.99±.12
	2 mJ	1.03±.03	1.04±.04	1.06±.22
$z\,{}^7P_4{}^\circ$ 23,498 cm^{-1}	1 mJ	1.08±.05	1.05±.03	1.11±.15
	2 mJ	1.08±.05	1.03±.04	1.11±.11

[a] Populations are relative to the reference state, Cr($z\,{}^7P_3{}^\circ$).

Table IV. Averaged Relative Population Data[a] for a Series of ACTs

State, Energy	MPD Pulse Energy	Toluene-Cr(CO)$_3$ Norm. Rel. Population	Ethylbenzene-Cr(CO)$_3$ Norm. Rel. Population	Propylbenzene-Cr(CO)$_3$ Norm. Rel. Population
$z\,{}^5P^\circ_{1\text{-}3}$ ~26,800 cm^{-1}	1 mJ	1.0	1.21±.10	1.41±.20
	2 mJ	1.0	1.28±.25	1.17±.27

[a] The relative populations are arbitrarily normalized to 1.0 for the ratio of emission intensities observed in the MPD of Toluene-Cr(CO)$_3$.

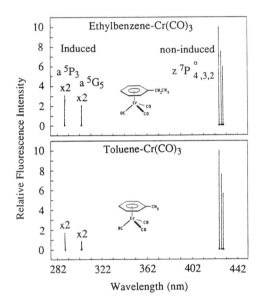

Figure 8. Sample LIF and spontaneous emission spectra from the chromium atoms produced in the MPD of two ACTs.

Table V. Normalized Relative Population Data for Cr States
Probed by LIF

State, Energy	MPD Pulse Energy	Toluene-$Cr(CO)_3$ Norm. Rel. Population	Ethylbenzene-$Cr(CO)_3$ Norm. Rel. Population	Propylbenzene-$Cr(CO)_3$ Norm. Rel. Population
$a\,^7S_3$ ground state	1 mJ	1.0	1.02±.25	1.6±.5
	2 mJ	1.0	0.8±.3	0.6±.3
$a\,^5S_2$ 7,593 cm^{-1}	1 mJ	1.0	1.51±.20	2.3±.4
	2 mJ	1.0	1.52±.20	2.2±.4
$a\,^5G_6$ 20,520 cm^{-1}	1 mJ	1.0	2.19±.25	1.8±.4
	2 mJ	1.0	2.16±.04	3.2±.8
$a\,^5G_5$ 20,524 cm^{-1}	1 mJ	1.0	2.1±.5	2.5±.6
	2 mJ	1.0	2.5±.7	2.7±1.2
$a\,^5P_3$ 21,841 cm^{-1}	1 mJ	1.0	1.69±.07	1.9±.4
	2 mJ	1.0	1.6±.3	1.9±.3
$z\,^7P_4{}^\circ$ 21,841 cm^{-1}	1 mJ	1.0	1.1±.4	0.98±.05
	2 mJ	1.0	0.99±.10	1.00±.10

Relative population data for the spontaneous emissions of $z\,^7P_2{}^o$ and $z\,^7P_4{}^o$ are listed in Table III. It can be seen that the relative populations of the individual J states behaved similarly as the ligand complexity increased; this was also true for the J states of the $z\,^5P^o$ level, so the relative population data for these states are presented as averages of the three J states (see Table IV). Comparison of the ratio of high-lying odd parity states $z\,^5P_J{}^o$ to $z\,^7P_J{}^o$ show only a small change for the different compounds, which is consistent with the results observed by Tyndall $et\ al.$ using 248 nm photolysis. That is, the distribution of high-lying emitting states does not depend strongly on the complexity of the arene ligand.

We checked our LIF technique by comparing the spontaneous signal from the $z\,^7P_4{}^o$ reference state to an LIF signal originating out of this same state; the induced to non-induced intensity ratios should remain constant regardless of ligand, and this is indeed observed, within the precision of the data (see Table V).

Contrary to what Tyndall $et\ al.$ observed at 248 nm photolysis, our 355 nm photolysis study shows that the population of the $a\,^5S_2$ state $increases$ relative to the reference state as the ligand complexity increases (see Table V). The $a\,^7S_3$ ground state population behavior is less clear; the changes relative to the reference state are in different directions for the two different laser powers, and in both cases is only comparable to the standard deviation. In their REMPI study of this series of ACTs in the wavelength range 348-369 nm, Hossenlopp et al. observed an increase in the ground state population relative to a cluster of excited state populations as the ligand complexity increased.

The most striking result of this study is in the population behavior of high-lying even parity states, which were probed for the first time. Two even parity states, the $a\,^5G$ and the $a\,^5P$, have energies very near the energy of the reference state (all three levels are within 2,800 cm^{-1}), yet they both show large population increases relative to this reference state when the ligand is changed from toluene to ethylbenzene or propylbenzene (see Figure 8 and Table V). The observation of such different dependences on ligand complexity from states so close in energy is evidence for a non-statistical mechanism for the production of these atoms; i.e., there is a very different IER dependence from states of very similar energy. The two odd parity states we monitored conformed roughly to a single electronic temperature regardless of ligand, while the populations of most of the even parity states increased significantly relative to the odd reference state as the ligand complexity increased. This dependence on parity is especially apparent when the relative population behaviors of the $a\,^5P$ and the $z\,^5P^o$ states are compared (Tables IV and V), since parity is the only difference between them. The even parity states appear to be formed via a different mechanistic pathway than the odd parity states, and one of the two pathways is much more sensitive to the complexity of the arene ligand. At least at 355 nm, IER influences the odd parity/even parity branching ratio, and so the changes in relative populations are not due to a "cooling" effect on one channel as Tyndall *et al.* proposed to explain their 248 nm MPD data. This is supported by the data at two different dissociation energies; although no significant difference is observed in the normalized relative populations when the laser intensity is doubled, the ratio of the induced even parity intensities to the odd parity reference state emission decreased between 20-30% as the laser power was doubled.

Although absolute comparisons of even and odd parity populations have not been made yet, the intensities of the LIF emission from low-lying even parity states was less than or only comparable to the intensity from the spontaneous emission from the odd parity states, even though the LIF signal was saturated with respect to the dye laser intensity. This would not be expected if the mechanisms Tyndall *et al.* proposed to explain their 248 nm results were applicable at 355 nm photolysis as well, since then the low-lying populations should be several orders of magnitude larger than the populations of the high-lying states; this would have to be the case in order for the direct channel to account for less than 5% of the lower state population. Our proposal that the direct channel for 248 nm photolysis only leads to odd parity states seems much more consistent with the observed low-lying versus high-lying state (induced versus spontaneous) emission intensities. The source of this parity-selectivity is still uncertain and more information is necessary to propose a mechanism to explain all of the data. Perhaps a configuration selectivity is actually involved, since the odd parity states all have either $3d^5\ 4p$ or $3d^4\ 4s\ 4p$ electronic configurations, whereas the even parity states have either $3d^5\ 4s$ or $3d^4\ 4s^2$ configurations. This configuration selectivity could occur in the initial multiphoton absorption of the parent molecule; perhaps alternate initial steps involving simultaneous absorption of two, three and four photons all occur, each leaving the predominantly metal electrons in different configurations. The metal electronic configuration may then be conserved during the dissociation process for each pathway.

In the studies done by Tyndall *et al.* on the the MPD of ACTs, statistical distributions for high-lying odd parity states were reported, while high-lying even states were not studied. It is conceivable that high-lying even state populations would not have conformed to the same thermal distribution as the odd parity states. We note that the experimental conditions used in this study differ substantially from those used by Tyndall *et al.* Whereas our dissociation wavelength was 355 nm, they used one of 248 nm. Also, we believe the laser intensity is much lower for our experiments, since the electronic temperature calculated for the odd states we tested was less than 2000 K. Tyndall *et al.* calculated an electronic temperature of 8500±1000 K at 248 nm, which is probably due mostly to the shorter photolysis wavelength, but their 350 nm MPD temperature for $Cr(CO)_6$ was 4000 K, over twice as high as our 355 nm temperature. It can be seen that these differences have important effects on the results. While Tyndall *et al.* noticed that as the arene ligand complexity increased, the relative populations of $a\ ^5S_2$ and $a\ ^5D_2$ decreased, we noticed an increase in the relative population of $a\,^5S_2$. Despite the apparent dependence of results on the experimental conditions, it seems likely that the parity-selectivity observed by us is an inherent aspect of the MPD process, and occurs under the conditions used by other groups, but has gone unnoticed. If parity selectivity does occur at 248 nm, one would have to explain why the odd parity channel is less favored when IER increases at 248 nm, but more favored at 355 nm. Perhaps the two pathways Tyndall *et al.* proposed to explain their 248 nm results are reversed at 355 nm; a sequential pathway, susceptible to IER, leads to formation of odd parity atomic states, and a direct channel leads to only even parity states. It is worth noting that three photons at 355 nm have an energy fairly close to two photons at 248 nm, i.e., 118 nm vs. 124 nm, but the dipole selection rules are different. Also, four 355 nm photons are not very different in energy from three 248 nm photons; 89 nm vs. 83 nm. Again, the dipole selection rules would be different for an odd number of photons vs. an even number of photons.

More information is needed before we can formulate a mechanism to explain the results of these experiments. We are currently investigating the dependence of the relative populations on buffer gas pressure to determine the sensitivity of intermediates to collisional quenching. Also, since the LIF absorption steps were saturated by the dye laser, it is possible to determine the relative populations of even parity states for comparison with the odd parity populations. We hope to determine whether or not the odd parity distributions fit a Boltzmann distribution, and if so, to compare the temperature with that of the odd parity states. This should tell us whether IER effects the temperature of the even parity distribution, or just the odd/even branching ratio, for 355 nm MPD. Finally, we plan to look at the behavior of the relative populations, including the high-energy even parity states, at 266 nm MPD, which is much closer to the wavelength used in the ACT studies of Tyndall *et al.* .

Conclusions

Clearly, MPD of $Cr(CO)_6$ and of the arene chromium tricarbonyl complexes occur via multiple dissociation channels. For $Cr(CO)_6$, it has been shown that at least some of these channels have different laser power dependences, presumably because they involve different numbers of photons. Even with the rather loose focusing conditions used in our ACT studies, there is still evidence of a flux dependence to the final population distribution, indicating that there are multiple pathways with different flux dependences at work here as well. For the ACTs, at least one channel shows evidence of parity (or configuration) selectivity, at least at 355 nm photolysis, and the branching ratio between this channel and the channel producing the odd parity product states is influenced by IER. As the *n*-alkyl length increases, populations of even parity states ($3d^5\,4s$ or $3d^4\,4s^2$) increase relative to odd parity states ($3d^5\,4p$ or $3d^4\,4s\,4p$). The terms from the different electronic configurations may result from adiabatic dissociation of different excited states of the same fragment or states of different intermediate fragments. Comparison between our results at 355 nm and the results at 248 nm by Tyndall *et al.* indicate that there is a wavelength dependence to the photodissociation process. MPD of ACTs at 355 nm results in an increase in the population of the $a\,{}^5S_2$ state (~7,600 cm^{-1}) relative to the reference state as chain length increases, in contrast to the decrease observed with 248 nm MPD. Also, behavior predicted from the mechanisms proposed by Tyndall *et al.* to explain 248 nm MPD results does not occur at 355 nm; lower-lying state distributions do not show a "cooling" as ligand complexity increases, and high-lying states of similar energy show very large changes in relative populations as the arene ligand is changed from toluene to ethylbenzene.

These recent results illustrate the difficulties in comparing results from different studies. One must be very cautious in making comparisons between studies done using different photolysis wavelengths and/or focusing conditions, and in particular when comparing results probed by REMPI, which may arise from a range of laser fluxes, to results using fluorescence probing of the atomic product state population. These results also show that it is possible to control the Cr(I) product distribution by choosing the appropriate ligand, laser wavelength and laser flux. This may prove very useful as a source of electronically excited metal atoms for subsequent chemistry.

Acknowledgments

The early MPI and fluorescence MPD work on Cr(CO)$_6$ was done with Lewis Rothberg and Veronica Vaida at Harvard University. The recent work on the MPD of ACTs done at Reed College involved undergraduate senior thesis students Sakae Suzuki, Matthew Blackwell, and Reginald Tanaban. Acknowledgment is also made to the Donors of the Petroleum Research Fund, administered by the American Chemical Society, and to Research Corporation, for the partial support of the research at Reed College.

Literature Cited

1. Karny, Z.; Naaman, R.; Zare, R. N. *Chem. Phys. Lett.* **1978**, *59*, 33.
2. Gerrity, D. P.; Rothberg, L. J.; Vaida, V. *Chem. Phys. Lett.* **1980**, *74*, 1.
3. Engelking, P. C. *Chem. Phys. Lett.* **1980**, *74*, 207.
4. Fisanick, G. J.; Gedanken, A.; Eichelberger IV, T. S.; Kuebler, N. A.; Robin, M. B. *J. Chem. Phys.* **1981**, *11*, 5215.
5. Gerrity, D. P.; Rothberg, L. J.; Vaida, V. *J. Phys. Chem.* **1983**, *87*, 2222.

48   LASER CHEMISTRY OF ORGANOMETALLICS

6. Hossenlopp, J. M.; Rooney, D.; Samoriski, B.; Bowen, G.; Chaiken, J. *Chem. Phys. Lett.* **1985**, *116*, 380.
7. Hossenlopp, J. M.; Samoriski, B.; Rooney, D.; Chaiken, J. *J. Chem. Phys.* **1986**, *85*, 3331.
8. Tyndall, G. W.; Jackson, R. L. *J. Chem. Phys.* **1988**, *89*, 1364.
9. Tyndall, G. W.; Larson, C. E.; Jackson, R. L. *J.Phys.Chem.* **1989**, *93*, 5508.
10. Rieger, R.; Rager, B.; Bachmann, F. *Mat. Res. Soc. Symp. Proc.* **1989**, *129*, 91.
11. Mitchell, S. A.; Hackett, P. A. *J. Chem. Phys.* **1990**, *93*, 7813.
12. Parnis, F. M.; Mitchell, S. A.; Hackett, P. A. *J. Phys. Chem.* **1990**, *94*, 8152.
13. Buntin, S. A.; Cavanagh, R. R.; Richter, L. J.; King, D. S. *J. Chem. Phys.* **1991**, *94*, 7937.
14. Tyndall, G. W.; Jackson, R. L. *J. Phys. Chem.* **1991**, *95*, 687.
15. Wrighton, M. *Chem. Rev.* **1974**, *74*, 401, and references therein.
16. Perutz, R. N.; Turner, J. J. *J. Am. Chem. Soc.* **1975**, *97*, 4791.
17. Welch, J. A.; Peters, K. S.; Vaida, V. *J. Phys. Chem.* **1982**, *86*, 1941.
18. Kelly, J. M.; Long, C.; Bonneau, R. *J. Phys. Chem.* **1983**, *87*, 3344.
19. Church, S. P.; Grevels, F.; Hermann, H.; Schaffner, K. *Inorg. Chem.* **1985**, *24*, 418.
20. Seder, T. A.; Church, S. P.; Weitz, E. *J. Am. Chem. Soc.* **1986**, *108*, 4721.
21. Gilbert, J. R.; Leach, W. P.; Miller, J. R. *J. Organomet. Chem.* **1973**, *49*, 219.
22. Mahaffy, C. A. L.; Pauson, T. L. *Inorg. Synth.* **1978**, *19*, 154.
23. Sugar, J.; Corliss, C. *J. Phys. Chem. Ref. Data* **1977**, *6*, 317.

RECEIVED January 19, 1993

Chapter 3

Multiphoton Dissociation Dynamics of Organoiron and Organoselenium Molecules

Joseph J. BelBruno

Department of Chemistry, Dartmouth College, Hanover, NH 03755

The non-linear photochemistry of selenium and iron containing organometallic molecules, both experimental and computational, is reviewed. It is shown that the mechanisms for the two different metals are not related. Organoiron complexes are typical of organometallics while the organselenium molecules exhibit a dissociation mechanism related to that typical of organic molecules.

An important step in gas phase organometallic chemistry is the production of free metal atoms or ions. Nonlinear photochemistry or multiphoton dissociation (MPD) provides a convenient method of production using visible or near uv radiation. If coupled to a sensitive detection technique such as multiphoton ionization or fluorescence, the products may be analyzed *in the same laser pulse*. In these **Laser Chemistry of Organometallics** experiments, the following questions are posed:

(1) What are the reactive species?
(2) What is the mechanism that results in the observed products?
(3) How do laser properties control the reaction?
(4) Can any novel and/or useful chemistry be derived from MPD?

In this review, we primarily discuss the first three questions. However, a significant body of research describing applications of this chemistry is available. The subject matter is restricted to the title molecules, but again the reader is advised that the literature contains numerous references to other organometallic molecules.

At the focus of an intense laser field, an organometallic molecule typically dissociates prior to ionization (whereas most organic molecules preferentially ionize prior to dissociation). Such photophysical/photochemical behavior implies that organometallic complexes are convenient sources of gas phase metal atoms or ions. This inviting prospect has led to a broad range of fundamental studies into the photodecomposition of this class of molecules. Prominent among the many organometallic molecules used in this research have been the carbonyls and, in particular, $Fe(CO)_5$. Alkylmetals of the semiconductor groups have played a much smaller role as have substituted carbonyls such as $Fe(CO)_3L$. The effort in our laboratory has been concentrated on substituted carbonyls and the semiconductor groups.

0097–6156/93/0530–0049$06.00/0

Historical Review

Early reports involving MPD of iron carbonyls were from Zare and co-workers.*(1)* A catalog of observed emission lines from iron carbonyl was reported. A sequential absorption scheme was postulated for excitation at both 193nm and 249nm. The efficiency of metal atom production from $Fe(CO)_5$ was confirmed by Smalley and co-workers*(2)*, who reported Fe^+ at laser intensities as low as $10^5 W$ cm^{-2}. Engelking subsequently reported that the MPI spectrum of ferrocene consisted entirely of atomic iron lines.*(3)* All three of these studies used multiphoton excitation into the charge-transfer (CT) band of the iron complex. Grant and co-workers excited $Fe(CO)_5$ into a mixed ligand field(LF)/charge transfer state.*(4)* They proposed a dissociation model involving excited states with lifetimes between 0.6 and 1 ps. A follow -up study by Zare and co-workers*(5)* cataloged numerous atomic iron transitions via MPI through the CT states. These authors also presented the first experimental evidence of the presence of unsaturated fragments such as $Fe(CO)^+$ and $Fe(CO)_2^+$, but only at levels of ~1% that of Fe^+. Engelking and co-workers later proposed a the following mechanism for the MPD dynamics of ferrocene.*(6,7)* Excitation of the molecular target occurred through sequential single-photon absorption steps resulting in production of the lowest electronic state of iron after dissociation, via absorption of a final photon, to a repulsive potential energy surface. The distribution of product states was attributed to surface crossings with this surface. This was clearly in contrast to the statistical mechanism routinely invoked for $Fe(CO)_5$ dissociation. Kimura*(8)* employed MPD/MPI with photoelectron spectroscopy (PES) to characterize the photodissociation of a series of iron containing molecules; however, most of these molecules were salts of Fe^{2+} rather than organometallics of neutral iron. Recently, Wight, Armentrout and co-workers*(9)* used MPD-MPI in tandem with TOF-PES to characterize the dissociation process of $Fe(CO)_5$. Multiphoton excitation into the CT band resulted in production of Fe in a distribution of states up to 32,000 cm^{-1} in energy, but preferentially populating the lowest two states. Mitchell and Hackett*(10)* used two-photon excitation into the CT band and drew similar conclusions regarding the population of Fe quantum states. Collisional quenching of the two lowest excited states was also reported for a variety of gases. The curve crossing mechanism was invoked to explain the Fe state distribution and intermultiplet transitions were determined to be the main source of collisional loss of the a^5F and a^3F states.

MPD has been less widely applied to dialkylmetal molecules in general and to dialkylselenium compounds, in particular. The only kinetic studies related to dimethylselenium dissociation employed thermolysis not photochemistry.*(11,12)* However, MPD has been applied to the related Te molecules.*(13-15)* For $(CH_3)_2Te$, only Te was detected by LIF, but the parent molecular ion was also detected via MPI. The ability to readily detect the parent ion as well as the unsaturated organic fragments, distinguishes the dialkylmetal dissociation from that of the carbonyls, where with the exception of small inorganic ligands such as CO or NO, ligands are not readily observed spectroscopically.

Our studies have focused on the production of Fe atoms as a function of the organic ligand(L) in molecules of the general class $Fe(CO)_3 L$ and on the conditions needed to generate Se or Se_2 from dialkyl selenium molecules.*(16-19)* Although our initial interest was purely fundamental, the iron dissociation studies are currently perceived as precursors to a study of iron atom reactivity (as a function of electronic state) and the selenium experiments are expected to improve semiconductor film and powder production techniques.

Experimental

Nonlinear photochemical experiments in the vapor phase may be performed in either a mass selective or bulk regime. In our laboratory, the former involve a linear time-of-flight mass spectrometer of the Wiley-McLaren design.*(20)* The sample is admitted to the source region via a molecular leak or a supersonic jet. The pressure in the source region when sample is admitted is always less than 10^{-5} torr, resulting in an estimated minimum mean free path of better than 0.25 m. The drift tube pressure is maintained in the 10^{-6} torr range at all times. Ions are detected by means of an electron multiplier. An ion arrival time spectrum is sent directly to a transient recorder for signal averaging and manipulation. Ion arrival time spectra are converted to mass spectra by calibration using a mixture of chlorofluoro- and bromochlorofluorocarbons.

REMPI spectra may be recorded under either static or free jet conditions; however, given the specificity of these spectroscopic experiments, bulk results may be nearly identical to the mass spectrometric studies. The REMPI spectra reveal the neutral fragments produced by the multiphoton dissociation process. Current vs. wavelength signals for samples with pressures of 10-100mtorr are amplified, sent to a boxcar averager and subsequently stored in a microcomputer, which also controls the scanning of the laser wavelength.

Finally, bulk nonlinear photochemistry may be observed by use of the laser analog of the pump and analyze technique. In this procedure which is simple bulk photochemistry, the sample is irradiated for a measured length of time and the mixture analyzed for product by whatever technique, but typically GC, GC-MS or FTIR spectroscopy, is convenient. All three methods are used in our studies.

Results

Alkylselenium Compounds. Metal alkyls have been the targets of research interest as precursors for vapor phase deposition of metal atoms and alloys. It is necessary to characterize the mechanism of the homogeneous gas phase photochemistry of these materials in order to improve the systematic application of chemical vapor phase deposition (CVD). Gas phase by-products, especially organic materials, may deposit or become trapped as impurities in the CVD film and change the characteristics of the desired material. The dynamics of the photodissociation of dimethylselenide (DMS), as an example of one of these types of molecules, have been examined and the mechanism (wavelength dependence, laser intensity dependence, pressure dependence, etc.) of the production of Se ascertained.*(18)* The energies levels of this system are schematically presented in Figure 1. A related study is exploring the MPD dynamics of dimethyl diselenide(DMDS), especially the feasibility of Se_2 production.

The photodissociation of DMS was observed via resonant excitation through the linear $^1A_1$ excited state and by means of non-resonant excitation at selected wavelengths both within this absorption band and in other regions of the visible spectrum. The TOF spectra, see Figure 2, show signals both atomic and molecular reflecting the multiple isotopes of selenium. The selenium containing ions include $(CH_3)_2Se^+$, CH_3Se^+ and Se^+. In addition, there are strong hydrocarbon signals assigned to $C_2H_x^+$ and to CH_x^+. In general, the extent of fragmentation increases with increasing laser frequency, that is, the relative ratio, Se^+: CH_3Se^+: $(CH_3)_2Se^+$, favored Se^+ at shorter wavelengths. The CH_3^+ signal was intense at all pump wavelengths and the signal assigned to the C_2H_x ions was observed at all frequencies. Contrary to results for the Group IIB dialkylmetals, the parent ion was observed at all excitation wavelengths.

Irradiation of the sample with visible light may only lead to photodissociation via a two-photon process. For the one-photon wavelengths used in our study, 355nm < λ < 435nm, coherent two-photon absorption into the "linear" state of $^1A_1$ symmetry was the initial step in the photofragmentation process. Analysis of the uv-vis absorption spectrum indicated that absorption in this spectral region (or in the region of the $^1B_1$, bent, state) resulted in a $(CH_3)_2Se$ molecule with significantly weakened bonds since the excited electron was being promoted from a non-bonding to an antibonding MO. REMPI spectra provide an indication of the neutral products arising from the MPD of the $(CH_3)_2Se$ parent molecule. In these spectra, the only peaks

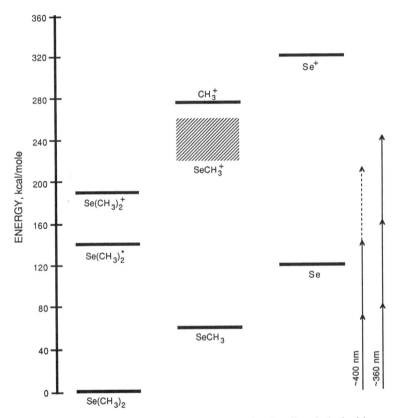

Figure 1. Schematic energy level representation for dimethylselenide. Reprinted with permission from Reference 18.

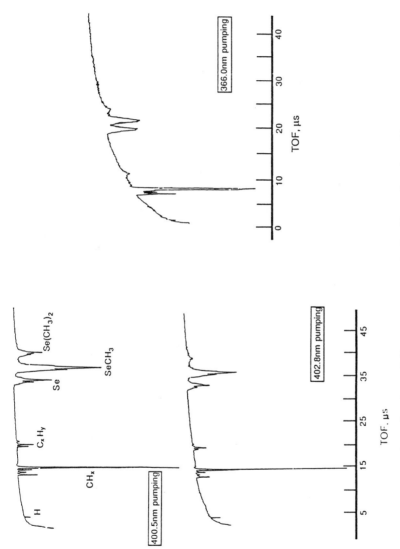

Figure 2. Time-of-flight mass spectra for DMS as a function of laser wavelength. Reprinted with permission from Reference 18.

assigned to atomic Se transitions originated in the M_J states of the ground electronic state. Se is a neutral dissociation product and only ground state atoms were produced in the dissociation. REMPI spectra at shorter wavelengths, ~390nm, contain resonances attributed to $(CH_3)_2Se$. However, the TOF spectra indicated that the fraction of parent ion was quite small. The transitions assigned to Se^+ and CH_3^+ depended on the sixth power of the laser intensity, I^6. This is the minimum number of photons necessary to produce ground state Se and CH_3 by MPD and detect their formation by MPI. The spectra in the region around 50,000cm^{-1} exhibited an I^4 photon dependence. Only three photons are necessary to ionize the parent molecule. However, production and ionization of the CH_3Se fragment is energetically possible with four photons for wavelengths at or below 400nm. The TOF fragment laser energy dependence studies indicated that MPD and MPI processes were readily saturated. For example, the linear laser intensity dependence exhibited by Se^+ appears to reflect saturation of all but one of the steps in the chain of photochemical processes beginning with absorption by CH_3Se and proceeding through MPI of ground state selenium.

Over the 50nm range of the study, a significant change in the yield of the atomic Se product was observed. For ~400nm irradiation, whether or not resonant with a parent molecule transition, the major photoproduct was CH_3Se and the Se mass peak was only 1/4 to 1/3 of its signal. However, for radiation in the 350-370nm range, Se is a major product. At the lower end of the wavelength range, it became the predominant fragment. At all pump wavelengths, the relative magnitude of the parent TOF peak was small; however, there appeared to be a significant difference between resonant (400.5 and 405nm) and nonresonant excitation (402.8nm). The parent mass signal was vanishingly small at the latter wavelength, presumably due to photodecomposition of the parent molecular ion produced by three photon nonresonant ionization. To summarize: (1) there were indications for fragmentation proceeding through both ionic and neutral mechanisms; (2) the yield of atomic product increased with increasing pump laser frequency and pump laser pulse intensity and (3) the resonant mechanism appeared to be initiated by excitation of a non-bonding electron into a σ^* orbital.

The DMDS uv-vis absorption is similar to that of DMS except that the separation between the structured and unstructured transitions, the "A" and "B" electronic states, is quite small. Effectively, the molecule absorbs continuously from 240 nm to the instrumental cutoff at 185nm. Our recent studies attempted MPI detection of both the molecular fragments and atomic selenium; however, the experiments have detected only molecular products. TOF spectra indicate that DMDS is unstable under 355nm irradiation and fragments across the Se-Se bond to produce the monomethylselenide. An example of the TOF-MS of DMDS is shown in Figure 3. As indicated, the major fragment is due to CH_3Se^+. Overall the TOF spectrum bears a striking similarity to that of DMS (compare the assigned fragments in Figures 2 and 3) and one may safely speculate that the CH_3Se formed by molecular cleavage of DMDS decomposes via the same process as the fragment originating with DMS.

Tricarbonyl Iron Complexes. Our MPD experiments using this photosystem have been intended to characterize the dissociation process as a function of the initial excitation and to explore the feasibility of employing ligand variations to control the distribution of final product states. The iron carbonyl complexes included in this study(*16*) were $Fe(CO)_3(\eta\text{-}C_4H_6)$, $Fe(CO)_3(\eta^4\text{-}C_6H_8)$, $Fe(CO)_3(\eta^4\text{-}C_8H_8)$, two isomers of $Fe(CO)_3C_8F_8$ and $Fe(CO)_4(\eta^2\text{-}C_8F_8)$. The vis-uv spectra indicate that the two lowest energy excitations occur in the near uv. We have chosen to use the third harmonic of the Nd:YAG laser to excite S_1 since this wavelength is in the tail of the absorption band. This is the ligand field (LF) transition of the complex. The S_2 state was obtained using coherent two-photon excitation. In this case the photon energy was ~22,000cm^{-1} (450nm). The initial absorption resulted in the promotion an

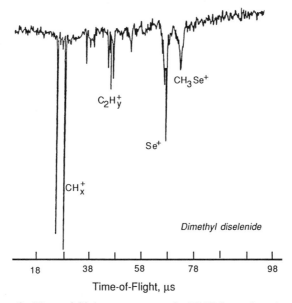

Figure 3. Time-of-flight mass spectra for DMDS as a function of laser wavelength.

electron to a π^* orbital of the molecule via a Fe $\rightarrow$ π^* CO transition, the CT transition. The two excitation cases are discussed separately below and the schematic energy levels are shown in Figure 4.

For the CT band experiments at least six photons are required to supply sufficient energy to both dissociate Fe from the complex and ionize the atom. Measured power indices indicated that one or more steps in the process were saturated. Relative populations of the photoproduced Fe electronic states were then compared as a function of organic ligand to discern the dynamical constraints on the photosystem.

The most intense REMPI spectral lines were observed between 447.2 and 448.4 nm and quantitative efforts were concentrated in this region of the spectrum. The values of the integrated areas of the lines belonging to the $e^5D_J \leftarrow a^5D_J$ multiplet, relative to the $e^5D_4 \leftarrow a^5D_4$ were independent of parent molecule. A plot of $\ln(N_J/g_J)$ vs. E_J indicated that, within experimental error, the ground state multiplet population was statistical and characteristic of a temperature of ~900K, regardless of the molecular precursor. The population of the excited levels, a^5F and a^3F, is molecule specific. No other Fe electronic states were observed in these experiments. The available energy would not be sufficient to populate higher quantum states.

The relative population of the ^3F state in Fe(CO)$_3$(η^4-L) decreased with increasing complexity of L. Structurally, these are equivalent molecules with only a different organic ligand. One may, for the first three, use the number of ligand modes as a crude measure of the density of states. In addition, the energy of C-F vibrational motions is considerably less than that for C-H and the density of states in the C_8F_8 complex may be roughly assigned a value several times greater than that of C_8H_8. Therefore, the initial conclusion is that the relative populations of the accessible electronic levels of Fe are indeed controlled by the availability of "bath" modes for redistribution of the initial excitation energy. This view is reinforced if one examines the total dissociation energies for the molecules in question. The bond dissociation energy is expected to influence the final energy disposal, perhaps as significantly as geometry or the nature of the ligand substituent. The bond enthalpies for Fe(CO)$_3$(η^4-L) where L = C_4H_6, C_6H_8 and C_8H_8 are 5.8, 5.7 and 5.6eV, respectively.[21] A crude argument based solely on the metal-organic ligand bond energy would predict a small effect, but in the opposite direction of that which was experimentally observed.

The argument presented above appears to support the theory that the density of vibrational states is the controlling factor in the disposition of initial excitation energy and that the redistribution of that energy throughout the organic ligand leads to selectivity in the final atomic energy levels. However, the data allowed for a further test of the importance of bond dissociation energy. Three molecules contained only a C_8F_8 ligand. However, the detailed molecular bonding in these complexes was very different. The first was an η^4-carbonyl with an estimated bond dissociation energy of 5.7eV. The second had the identical molecular formula, but may be regarded as the oxidative product of the first. The two bonds were actually an η^3-allyl interaction and a metal-ligand σ-bond. The total bond dissociation energy for this molecule is very difficult to estimate, but a value of 7.5eV is conservative. Finally, the last molecule was an η^2-olefin complex and the Fe-C bond energy is approximately one-half of that for an η^4 molecule. The total bond enthalpy was estimated to be 5.9eV. In all three of these molecules the density of vibrational states is approximately the same and, to the first approximation, any differences in product state distributions should be attributed to the changes in the energy required to completely dissociate the complex. The results bear out this prediction. The populations of both the ^5F and ^3F levels depend upon the bond energy, increasing with a concomitant decrease in total bond enthalpy.

The following mechanism is consistent with our CT data and the earlier reports[4,22] on organometallic MPD, both of which also involved excitation of the CT bands. Initial coherent two-photon excitation populates S_2 in the parent molecule.

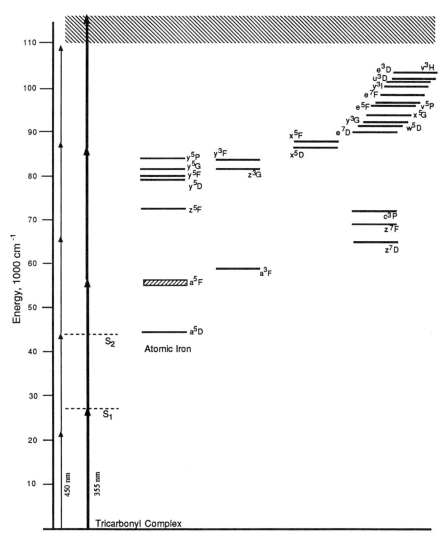

Figure 4. Schematic energy level representation for Fe(CO)$_3$C$_4$H$_6$.

The most likely scenario is that the CO ligands are boiled off leaving a Fe-L fragment behind. Absorption of an additional photon by the Fe-L fragment results in formation of iron atoms which then experience REMPI. Given similar bond energies for the Fe-L fragment, the transfer of energy to the ligand becomes the determining factor in the availability of energy for production of excited state Fe. The mechanism is consistent with the absence of any significant laser energy dependence (at modest energies). The total intensity of any transition depends on the laser energy, but the population ratios are independent of this variable. Since the areas of the various lines examined in this study vary with precursor and all exhibit power indices of ~2, the most likely controlling step in the dissociation (the unsaturated step) is the initial two-photon absorption. The fate of the ligands in the MPD/REMPI scheme is unknown.

In order to fully characterize the possible MPD pathways, excitation into the LF bands of the $Fe(CO)_3C_4H_6$ complex was also investigated. The third harmonic of the Nd:YAG laser lies in the long wavelength tail of this band. Initial excitation was clearly a one-photon process and the input of the energy necessary for bond breakage was attributed to sequential single photon absorption events. For the low pressures employed in these studies (300mtorr) a large number of atomic iron emission lines were observed. Nearly all of the lines could be assigned to known transitions and emission from levels nearly 100,000 cm^{-1} above the initial energy were observed. The emission intensity was found to be dependent upon the pressure of the $Fe(CO)_3C_4H_6$ complex, but also sensitive to the nature and pressure of background gas added to the cell. Pressures of up to 1 atm of He were found to increase the intensities of all of the emission lines and to cause additional lines to grow into the spectrum. The addition of Ar caused mixed effects in which some lines increased, some decreased and new lines appeared. Nitrogen caused a significant decrease in the intensity of all lines. Emission was only observed from atomic iron. However, analysis of the cell contents by infrared spectroscopy following the emission studies indicated the presence of free butadiene. Photolysis of $Fe(CO)_3C_4H_6$ was also found to produce a polymer not observed with a hexadiene complex is substituted in its place. The exact sequence of events leading to these results is unclear, but is indicative of the intriguing chemistry observed with photoexcitation of organometallic molecules. Additional experimental studies are underway.

To test the proposed model for CT excitation, a series of RRKM calculations were performed for the entire series of complexes.(17) In these calculations, we attempted to ascertain whether the mechanism of iron production was consistent with a statistical mechanism or a series of statistical processes. An RRKM calculation in which three photons are assumed to be absorbed, followed by "explosive" decomposition of the complex to produce ground state Fe yields rate constants which are much too small to allow for the observation of any product during the time scale of the experiment (the laser pulse width of ~15 ns). A calculation based on a series of three sequential statistical decays after the absorption of two quanta of energy, with subsequent absorption of a third photon by the remaining FeL fragment and statistical decomposition of that fragment as well was also performed. The results yielded rate constants for FeL production within the range which could ultimately produce a quantity of iron atoms easily detected in the MPI detection scheme employed. However, the statistical decay of the FeL fragment upon absorption of a third photon was only realistic within the time frame of the experiment in the case of the smallest ligands and the mechanism for all of the complexes is identical.

The RRKM calculation has ruled out a purely statistical production of all fragments, leaving direct photodissociation as the only alternative of the "pure" mechanisms. The production of triplet atomic products is inconsistent with a direct transition to a dissociative potential surface leading to the observed photofragments. The latter would be symmetry forbidden. Earlier research has shown that the initial excitation of the parent pentacarbonyl at these energies is a singlet-singlet transition, but that rapid intersystem crossing occurs. For the complexes currently under study, extrapolation of this analysis implies that the RRKM decomposition to the FeL fragment occurs along a triplet potential surface*(23)* which correlates with the ground states of all of the unsaturated fragments. An FeL triplet fragment so formed absorbs an additional photon to populate an excited, dissociative state which directly correlates with the ^3F iron - singlet organic ligand products. In this model, the ground state is populated by curve crossings of the potential surface which leads to ^5D iron atoms with the directly prepared triplet surface of FeL. Further support of the proposed mechanism arises from the relative populations of the ^3F atoms reported in our experimental work. The excited state was over populated in comparison to the temperatures extracted from the ground state populations. The most reasonable rationale for such a distribution is the statistical decomposition-direct excitation scheme proposed above. In retrospect, the proposed mechanism is a synthesis of those proposed for iron pentacarbonyl*(24)* and ferrocene*(6,7)*. Such behavior is entirely reasonable since the complexes employed in our CT excitation study were intermediate between these two extremes of carbonyl substitution.

Model calculations indicate that the molecules excited into the LF band follow a similar reaction pathway. A model based upon the density of available states for the product channel, which is similar to the maximum entropy formulation, does accurately predict the observed emission rates. A series of RRKM calculations, involving channels such as explosive decomposition, sequential release of the ligands and combinations of these two, indicate that the only feasible fully statistical mechanism is one in which a single photon absorption is followed by loss of two CO ligands, a second absorption leads to loss of the third CO and a final single photon event promotes loss of the organic ligand. It is as yet impossible to distinguish this mechanism from one in which the final step is a direct dissociation as discussed above.

Literature Cited
1.   Karny, Z.; Naaman, R.; Zare, R.N. *Chem. Phys. Lett.*, **1978,***59*, 33.
2.   Duncan, M.A.; Dietz, T.G.; Smalley, R.E. *J.Phys. Chem.*, **1979**, *44*, 415.
3.   Engelking, P.C. *Chem. Phys. Lett.*, **1980**, *74*, 207.
4.   Whetten, R.L.; Fu, K.J.; Grant, E.R. *J. Chem. Phys.*, **1983**, *79*, 4899.
5.   Harrison, W.W.; Rider, D.M.; Zare, R.N. *Int. J. Mass Spectrometry Ion Processes*, **1985**, *65*, 59.
6.   Liou, H.T.; Ono, Y.; Engelking, P.C.; Moseley, J.T. *J.Phys. Chem.,* **1986**, *90*, 2888.
7.   Liou, H.T.; Engelking, P.C.; Ono, Y.; Moseley, J.T. *J.Phys. Chem.*, **1986**, *90*, 2090.
8.   Nagano, Y.; Achiba, Y.; Kimura, K. *J.Phys. Chem.*, **1986**, *90*, 1288.
9.   Niles, S.; Prinslow, D.A.; Wight, C.A.; Armentrout, P.B. *J. Chem. Phys.*, **1990**, *93*, 6186.
10. Mitchell, S.A.; Hackett, P.A. *J. Phys. Chem.,* **1990**, *93*, 7813.
11. Didenkulova, I.I.; Dyagileva, L.M.; Tsyganova, E.I.; Aleksandrov, Y.A. *Zh. Obsh. Khim.* , **1984, 54**, 2288.
12. Yablokov, V.A.; Dozorov, A.V.; Zoron, A.D.; Feshchenko, I.A.; Ronina, O.V.; Karataev, E.N. *Zh. Obsh. Khim.,* **1986**, *56*, 1571.
13. Connov, J.; Greig, G.; Strausz, O.P. *J. Amer. Chem. Soc.* , **1969**, *91*, 5695.
14. Stuke, M. *Appl. Phys. Lett.,***1984**, *45*, 1175.
15. Brewer, P.D. *Chem. Phys. Lett.,* **1987**, *141*, 301.
16. BelBruno, J.J.; Kobsa, P.H.; Carl, R.T.; Hughes, R.P. *J.Phys. Chem.*, **1987**, *91*, 6168.
17. BelBruno, J.J. *Chem. Phys. Lett.*, **1989**, *160*, 267.
18. BelBruno, J.J.; Spacek, J.; Christophy, E. *J. Phys.Chem.*, **1991**, *95*, 6928.
19. BelBruno, J.J. in *Research Trends in Physical Chemistry*, **1992**, *2*, 185.

20. Wiley, W.C.; McLaren, J.H. *Rev. Sci. Instr.*, **1955**, *26*, 1150.
21. Connor, J.A.; Demain, C.P.; Skinner, H.A.; Zafarani-Moattar, M.T. *Organomet. Chem.*, **1979**, *170*, 117.
22. Samoriski, B.; Hossenlopp, J.; Rooney, D.; Chaiken, J. *J. Chem. Phys.*, **1986**, *85*, 3326; 3331.
23. Seder, T.A.; Ouderkirk, A.J.; Weitz, E. *J. Chem. Phys.*, **1986**, *85*, 1977.
24. Waller, I.M.; Davis, H.F.; Hepburn, J.W. *J. Phys. Chem.*, **1987**, *91*, 506.

RECEIVED January 26, 1993

Chapter 4

Laser Photoionization Probes of Ligand Binding Effects in Multiphoton Dissociation of Gas-Phase Transition-Metal Complexes

Charles A. Wight and P. B. Armentrout

Department of Chemistry, University of Utah, Salt Lake City, UT 84112

In this review, we summarize some of the mechanistic consequences of metal-ligand binding on multiphoton dissociation/multiphoton ionization (MPD/MPI) pathways of organometallic complexes. A statistical model of MPD is strongly supported by rotational state distributions of free nitric oxide ligands derived from $Co(CO)_3NO$ and a series of closely related complexes in which one CO is replaced by various trialkylphosphine ligands. The same series of compounds has been used to understand branching between doublet and quartet spin states of the central cobalt atom following scission of all the metal-ligand bonds. A model is developed to explain the surprising result that lower laser fluences and/or substitution of trialkylphosphine ligands enhances the selectivity of forming the excited doublet states at the expense of the lower energy quartets. Lastly, a recent report of pulsed laser MPD/MPI of $VOCl_3$ reveals an unusual case in which the metal ion, V^+, is formed by photodissociation of VO^+ rather than by ionization of the neutral metal atom. This observation confirms that the strength of metal-ligand bonds has a major influence on the branching between photoionization and photodissociation pathways in the MPD/MPI mechanism.

During the past dozen years or so, several new laser-based techniques have been developed for investigating the photochemistry and photophysics of gas phase organometallic complexes, often metal carbonyls. This research is motivated partly by a desire to understand fundamental interactions between radiation and molecules that have a high density of vibrational and electronic states. In addition, gas phase photodissociation provides a means of producing coordinatively unsaturated metal

0097–6156/93/0530–0061$06.00/0

62   LASER CHEMISTRY OF ORGANOMETALLICS

atoms and clusters that are catalytically active, and many studies are aimed at
controlling and optimizing the reactive properties of these materials.
Early on, researchers discovered that focusing a tunable pulsed laser into a chamber
containing a few millitorr of organometallic vapor generates intense metal ion signals
that can be detected merely by collecting the charges on a pair of electrodes in the cell.
Much of the work of this type on metal carbonyls was reviewed by Hollingsworth and
Vaida in 1986 (*1*). Analyses of metal atom state distributions (particularly in laser-
excited atomic fluorescence experiments) showed that there is no obvious selectivity
in the processes leading to formation of bare metal atoms (*2*). In fact, the observations
are completely consistent with a statistical partitioning of available energy among
product degrees of freedom.

Chemical trapping experiments of Yardley and co-workers (*3*) on $Fe(CO)_5$
showed that UV excitation results in extensive fragmentation by stepwise scission of
the Fe-CO bonds. This is in stark contrast to solution studies in which loss of only
one CO ligand is the major photodissociation channel. Although many other studies
of $Fe(CO)_5$ fragmentation have been reported since then, perhaps the most sophisticated
is the recent report of Ryther and Weitz, in which time-resolved infrared diode laser
spectroscopy was used to identify the coordinatively unsaturated fragments and to
monitor the kinetics of their subsequent gas phase reactions in the photolysis cell (*4*).

In 1986, Chaiken and co-workers reported a study of Cr atom state distributions
from multiphoton dissociation/multiphoton ionization (MPD/MPI) of a homologous
series of n-alkyl arene chromium tricarbonyl complexes (*5*). They found that the metal
atom distributions are directly correlated with the size and molecular structure of the
organic ligand, and that the main effect of increasing the size of alkyl substituents on
the arene ligand was to promote intramolecular vibrational relaxation in the molecule.
This study represents one of the first and clearest examples of the effects of ligand
substitution on the mechanism (and therefore, the product state distributions) of
MPD/MPI of an organometallic complex. The conclusions reinforced ideas emerging
at that time that the nature of photodissociation processes in organometallics is
governed largely by statistical considerations.

Although many early studies of pulsed laser MPD of organometallics examined
states of the metal atom fragments, it wasn't until the mid-1980's that the first reports
of ligand fragment state distributions began to appear (*6,7*). Using a one-color
multiphoton ionization scheme, Georgiou and Wight measured the rotational and
vibrational distributions of NO molecules arising from pulsed laser MPD of CpNiNO
($Cp \equiv \eta^5\text{-}C_5H_5$).(*8*) The results strongly implicated a mechanism of metal-ligand bond
dissociation in which the energy of the UV photon initially absorbed by the molecule
is statistically distributed among all degrees of freedom of the dissociated fragments.
No evidence was found for a direct or impulsive dissociation mechanism.

Hepburn and co-workers investigated the CO vibrational and rotational state
distributions from $Fe(CO)_5$ by VUV laser-induced fluorescence of the CO fragments
(*9*). They also found that the observed state distributions were consistent with
statistical partitioning of the available energy among all possible degrees of freedom
in the dissociated fragments. One of the difficulties encountered in this study was that
UV photolysis of organometallic complexes often results in loss of more than one CO
ligand. The results therefore consisted of convolutions of state distributions for up to

four different CO loss channels. Vernon and co-workers (*10,11*) reported time-of-flight CO photofragment translational energy distributions from $Fe(CO)_5$ that nicely complement the internal state distributions reported by Hepburn.   While these translational distributions were subject to the same difficulty as above (i.e., they contain contributions from multiple reaction channels), they were nonetheless consistent with statistical energy partitioning among the photofragments.

In this paper, we review some of the work from our laboratories during the past several years on MPD/MPI spectroscopy of organometallic complexes. The focus of this review is on studies that address the effects of ligand substitution on the internal state distributions of photolysis products.   We first examine nitric oxide (NO) rotational state distributions that arise from pulsed laser photodissociation of transition metal nitrosyl complexes and the effects of "spectator" ligand substitution on those distributions.   Then we look at the effects of ligand substitution on spin state distributions of the central metal atom, as probed by resonance-enhanced multiphoton ionization (REMPI) spectroscopy. Finally, we examine a system for which branching between photodissociation and photoionization takes an unusual turn (for organometallics) due to the presence of a very strong metal-ligand bond in the complex.

Nitric Oxide Ligand State Distributions

Since most metal-ligand bond dissociation energies in organometallic complexes are small compared with the energies of UV photons typically used to carry out pulsed laser MPD, it is unlikely that convolutions of state distributions due to multiple reaction pathways can be completely avoided. We therefore adopted a somewhat different strategy for investigating ligand state distributions. This new approach is typified by our study of NO state distributions from a series of closely related cobalt nitrosyl complexes: $Co(CO)_3NO$ and $Co(CO)_2(PR_3)NO$ (where R = methyl, ethyl, n-propyl and n-butyl) (*12*).   The basic philosophy is similar to Chaiken's earlier experiment (*5*), but applied to ligand state distributions rather than that of the central metal atom.   Specifically, we set out to determine how the NO state distributions (determined by REMPI and detection of NO^+) are affected by substituting progressively larger trialkylphosphines for one of the CO ligands. We postulated that if statistical partitioning of energy throughout the parent molecule was an important factor in determining the product state distributions, then this type of "spectator ligand" substitution should have a dramatic affect on the experimental results. On the other hand, if the state distributions were influenced mainly by features of the potential energy surface that are local to the metal-nitrosyl bond, then the results for all the different compounds should be similar.

We found that there are indeed major differences in the experimental results for these five closely related parent compounds. MPD of $Co(CO)_3NO$ resulted in a NO rotational distribution in the v=0 level that is bimodal (i.e., was consistent with two different populations of NO molecules, each of which is characterized by a Boltzmann distribution of rotational states at different temperatures), as shown in Figure 1. The trimethylphosphine- and triethylphosphine-substituted cobalt complexes each yielded an NO rotational state distribution that was characterized by a single temperature (900

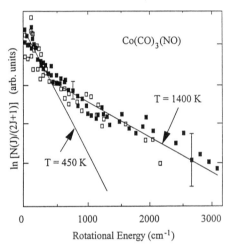

Figure 1. A semi-log (Boltzmann) plot of rotational state populations (divided by the degeneracy) as a function of rotational energy (including spin-orbit contributions) for NO from Co(CO)$_3$NO. The open squares are data obtained using a 1 m focal length lens, whereas the filled squares represent data obtained at somewhat higher laser fluence using a 0.5 m focal length lens. The error bars represent ±1 standard deviation uncertainty in the relative population for each state. The solid lines represent rotational temperatures which are characteristic of the high-J and low-J regions. (Adapted from ref. 12).

and 600 K, respectively). The distributions became progressively colder with increasing size of the parent molecule (relative to the "high temperature" part of the distribution observed from the tricarbonyl derivative), as illustrated in Figure 2. Also, the intensity of the NO^+ ion signal decreased dramatically with increasing size of the parent molecule. For the tripropylphosphine-substituted complex, the NO distribution was qualitatively colder than any of the others, but the signal was too small to extract a quantitative state distribution. The tributylphosphine complex did not yield an observable NO^+ ion signal.

Although we presented evidence that the NO molecules observed in this experiment arise from one-photon dissociation of the complexes, we had no reason to expect that the metal-NO bond would be the *first* one cleaved in the photodissociation mechanism. In fact, all of the available evidence based on thermochemical bond strengths (*13,14,15*), matrix studies of $Co(CO)_3NO$ (*16*), solution phase studies of various carbonyl complexes (*17,18*), and our own subsequent study of visible laser MPD of $Co(CO)_3NO$ (*19*) shows that loss of CO is the first step in the reaction, followed by subsequent loss of other ligands.

The rotational state distribution results also indirectly support this conclusion. We calculated rotational state distributions that are expected for NO molecules produced as the second or third ligand lost from the various complexes based on the assumption that all of the available energy is partitioned statistically among all possible degrees of freedom of the products. These results are presented in Table I. The expected rotational temperature for NO produced as the second ligand lost from $Co(CO)_3NO$ is 1900 K, which is not too different from the "high temperature" portion of the observed rotational distribution (1400 K). The "low temperature" portion of the observed distribution (450 K) is also qualitatively consistent with the calculated distribution (200 K) expected for NO loss from $Co(CO)NO$ (tertiary fragmentation). The observed temperatures for the trimethylphosphine and triethylphosphine derivatives agree with the predicted values for secondary loss of NO.

Table I. RRKM Lifetimes and Rotational Temperatures for NO Fragments.[a]

Dissociating Complex	Avg. E_{int} (kJ/mol)	RRKM Rate Constant (s^{-1})	Rotational Temp.(K)
$Co(CO)_2$ - NO	331	2×10^{12}	1900
$Co(CO)(PMe_3)$ - NO	378	1×10^9	950
$Co(CO)(PEt_3)$ - NO	386	2×10^7	750
$Co(CO)(PPr_3)$ - NO	394	3×10^5	-
$Co(CO)(PBu_3)$ - NO	398	2×10^4	-
$Co(CO)$ - NO	166	$\approx 10^9$	200
$Co(PMe)_3$ - NO	231	$\approx 10^7$	-

[a] Details of the calculations are presented in reference (*12*).

We also calculated unimolecular lifetimes for metal-ligand bond dissociation for the various channels. The reason for this calculation is that in order to observe free NO ligands in our experiment, the rate of bond scission must exceed the reciprocal of the laser pulse duration (about 10^8 s^{-1}). For large parent molecules, the calculated dissociation rates are too slow to produce a strong NO ligand fragment signal, and we believe that this is the principal reason for the disappearance of the NO^+ ion signal for the larger phosphine-substituted complexes in this study. The dissociation rates for $Co(CO)_3NO$ are fast enough that both secondary and tertiary loss of NO should be observable (thereby producing the high and low-temperature portions of the state distribution. However, dissociation of NO as the third ligand lost from the phosphine-substituted complexes is slow enough that no "low temperature" portion of distribution is observed.

In summary, all of the available results for this family of closely related cobalt nitrosyl complexes are consistent with a mechanism that involves initial loss of CO from the parent followed by unimolecular dissociation of other ligands. In the examples considered, the state distribution of the NO fragment is qualitatively consistent with equipartitioning of the available energy among all vibrational and rotational degrees of freedom of the product fragments. This analysis considers dissociation only along a single potential energy surface. Although subsequent work in our laboratory (discussed in the following section) more clearly reveals the importance of excited electronic states in the overall dissociation process, there is not enough detailed information about the ground and excited state bond energies of the various species to justify a more detailed analysis of the final NO state distributions at this time.

Spin State Distributions of the Metal Atom

The same series of cobalt complexes was subsequently used to characterize the effects of "spectator" ligand substitution on state distributions of the central cobalt atom (20). REMPI spectra of $Co(CO)_3NO$ show that under unfocused laser conditions, most of the observed resonances can be associated with Co atoms which are formed in the three lowest energy terms. An energy level diagram for these terms and their spin-orbit components is shown in Figure 3. Interestingly, the most intense atomic REMPI lines are associated with the second excited term ($a^2F_{7/2}$ and $a^2F_{5/2}$ states) rather than the lower-lying quartet states. We reasoned that the selectivity for this term was likely due to a propensity for electron spin conservation. After all, the parent complex has a singlet ground state, each of the CO ligands must be lost as singlet molecules, and the free NO ligand fragment is formed as a doublet, so overall spin conservation demands that the neutral cobalt atom also be formed in a doublet state *regardless of the order of ligand loss.*

Substitution of one CO ligand by one of the trialkylphosphine groups (also singlet molecules) has the effect of increasing the selectivity for producing Co atoms in the second excited ^2F term, as summarized in Table II. In fact, under relatively low-intensity pulsed laser conditions it was not possible to observe *any* REMPI transitions associated with the quartet states for the tributylphosphine-substituted compound. This is a somewhat surprising result because increasing the number of degrees of freedom

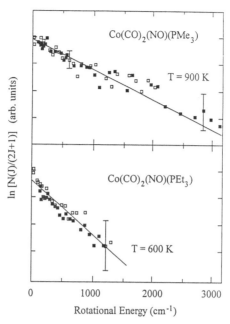

Figure 2. Boltzmann plot similar to figure 1, except for the trimethylphosphine-substituted derivative (upper panel) and the triethylphosphine-substituted compound (lower panel). In both graphs, the open symbols refer to bands associated with the $^2\Pi_{1/2}$ spin-orbit state, whereas the filled symbols refer to $^2\Pi_{3/2}$ states. The straight line fits correspond to rotational temperatures of 900 K (upper panel) and 600 K (lower panel). (Adapted from ref. 12).

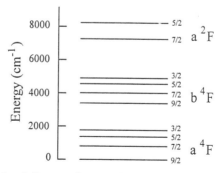

Figure 3. Energy level diagram for atomic Co states. The most intense atomic REMPI lines from MPD of $Co(CO)_3NO$ are associated with Co atoms formed in J levels of the second excited ^2F term.

in the molecule causes the metal atom state distribution to become "hotter" in purely energetic terms.

Changes in the selectivity for forming the excited doublet states of the Co atom can be understood by considering the scheme presented in Figure 4. There are almost no thermodynamic data for these cobalt complexes, so the details of this diagram (including the energies of the intermediates and position of the surface crossing) are not quantitatively correct, but the qualitative picture is useful for understanding the unusual variations in spin-state distributions. We start by recognizing that the low-spin reaction pathway (starting with the singlet state of the parent molecule and ending with the excited doublet states of the Co atom) must cross the high-spin reaction pathway (starting with the excited triplet state of the parent and ending with the ground quartet state of the Co atom). Removal of one or two ligands from the parent compound decreases the splittings between ground and excited states by weakening the ligand field, and allows the high-spin pathway to gain population by internal conversion or intersystem crossing events in the early stages of reaction. As additional ligands are lost, the overall density of states becomes small enough that the distribution between low and high-spin pathways becomes "locked in", and the probability of spin-changing transitions at the crossing of the two pathways becomes relatively small.

Substitution of complex ligands such as tributylphosphine affect the spin distribution early in the dissociation process in two ways. First, the high density of vibrational states associated with the bulky ligands reduces the amount of internal energy per degree of freedom. Second, when these ligands are lost from the complex, they remove large amounts of energy in the form of internal vibrations and rotations. Both effects lower the energy on a per mode basis, which tends to keep molecules on the lower energy surface (i.e., on the low-spin pathway) based on its greater overall state density. Because this spin population becomes locked in prior to the crossing, the larger phosphine ligands have a propensity for selective formation of the spin-allowed (excited) doublet states of the Co atom.

Increasing the laser intensity (e.g., by reducing the beam diameter with a long focal length lens) has just the opposite effect. Under these conditions the populations of quartet states of the Co atoms are increased relative to the higher lying doublet states. This is presumably because the increased rate of photon absorption (relative to the rate of bond scission) increases the average internal energy of the intermediate complexes on the pathway to dissociation. Increased population of high-spin states early in the dissociation process leads directly to an enhancement in the populations of Co atom quartet states.

When Co atoms are formed by MPD of cobaltocene, $Co(\eta^5-C_5H_5)_2$, the selectivity is completely destroyed, even at low laser fluence. Using this precursor, the most intense REMPI transitions are associated with the ground quartet term (21). This is because the cyclopentadienyl ligands in cobaltocene have doublet ground states, so formation of the ground ^4F state *becomes a spin-allowed pathway.*

A somewhat different situation arises when Fe atoms are formed by MPD of $Fe(CO)_5$. The most intense REMPI transitions observed are associated with the ground a^5D and first excited a^5F terms (22). This process clearly does not conserve spin, because the ground state of $Fe(CO)_5$ is a singlet, as are all the ligands. However, it is now well-established that $Fe(CO)_4$ has a triplet ground state (23), and there are

Table II. Relative intensities of atomic REMPI lines arising from low-lying doublet and quartet states of Co for several different organometallic precursors.

λ_{obs} (nm)	Assignment	$Co(CO)_3NO$	$Co(CO)_2(PR_3)NO$ R = Me	R = n-Pr	R = n-Bu	Cp_2Co
445.074	$a^4F_{3/2}\to e^2F_{5/2}$	0	0	0	0	2
445.610	$a^2F_{5/2}\to e^2D_{3/2}$	73	18	0	29	3
446.213	$a^2F_{7/2}\to e^4D_{3/2}$	38	11	0	8	4
446.611	$a^4F_{9/2}\to e^4F_{9/2}$	15	6	0	0	100
447.232	$a^2F_{7/2}\to e^4G_{7/2}$	13	0	0	0	4
447.288	$a^2F_{7/2}\to e^2G_{9/2}$	34	18	10	23	40
447.642	$a^2F_{7/2}\to e^4H_{9/2}$	76	26	16	8	20
447.716	$a^2F_{7/2}\to e^2H_{11/2}$	96	50	100	100	39
447.906	$a^2F_{7/2}\to f^2F_{7/2}$	97	76	90	59	30
448.157	$a^2F_{7/2}\to g^4F_{5/2}$	47	30	13	41	14
448.525	$a^2F_{7/2}\to e^4P_{3/2}$	85	100	68	38	20
448.667	$b^4F_{7/2}\to f^4F_{5/2}$	0	0	3	0	3
448.777	$a^4F_{3/2}\to e^4F_{3/2}$	2	0	0	0	11
449.250	$a^4F_{5/2}\to e^2F_{7/2}$	0	0	0	0	8
449.342	$a^2F_{5/2}\to f^2F_{5/2}$	90	37	59	29	12
449.738	$a^4F_{5/2}\to e^4F_{5/2}$	2	0	0	0	13
450.416	$a^2F_{5/2}\to h^4F_{9/2}$	5	10	0	9	5
450.497	$a^2F_{5/2}\to e^2G_{7/2}$	92	82	77	61	17
450.565	$b^4F_{5/2}\to f^4F_{3/2}$	1	8	0	0	3
451.313	$a^2F_{5/2}\to e^2H_{9/2}$	100	82	34	29	22
451.432	$a^2F_{5/2}\to g^2F_{7/2}$	8	15	0	21	2
451.569	$a^4F_{7/2}\to e^4F_{7/2}$	1	2	0	0	25
451.917	$a^2F_{5/2}\to e^4H_{7/2}$	20	10	23	18	7
452.753	$a^2F_{5/2}\to e^4D_{1/2}$	1	0	0	0	1
Avg. Int.	^4F terms	3	2	0.4	0	21
Avg. Int.	^2F terms	55	35	31	30	15

strong indications that $Fe(CO)_3$ and $Fe(CO)_2$ have triplet ground states as well (*4*). This means that crossing to the triplet dissociation pathway can occur early in the overall photodissociation process. Transition to the quintet dissociation pathway may also occur prior to scission of the last Fe-CO bond, because Fe(CO) is calculated to have a low-lying $^5\Sigma^-$ state which may in fact be the ground state (*24*). An additional factor that strongly favors formation of the a^5D and a^5F terms of Fe is that the lowest energy triplet states lie 1.48 eV above the ground state.

Many of the results for spin-state distributions of bare metal atoms formed by MPD of organometallic complexes can therefore be interpreted in terms of an overall scheme in which relaxation to spin states of higher multiplicity can occur during intermediate steps in the MPD process. For carbonyl complexes of cobalt, it appears that crossings of the doublet and quartet surfaces occurs sufficiently late that the low spin states cannot efficiently cross to the quartet surface. The result is a surprisingly high selectivity for production of excited metal atoms in states that conserve overall spin in the MPD process.

Branching Between Photodissociation and Photoionization in VOCl₃

Virtually all MPD/MPI spectra of organometallic complexes have been consistent with a mechanism in which all of the metal-ligand bonds are cleaved before ionization of the central metal atom takes place. This so-called Type B behavior (25) is in stark contrast to the MPI mechanism in many organic molecules, which generally ionize first and subsequently fragment into various daughter ions. The Type A behavior of organic molecules is generally attributed to the greater bond dissociation energies of typical organic molecules in comparison with the weak metal-ligand bond energies of organometallic complexes. Thus, the strength of the chemical bonds appears to play a crucial role in determining the branching between photodissociation and photoionization in the intermediate steps of MPD/MPI processes for all molecules.

What happens when metal ligand bond energies are strong? In a recent study, we re-examined the pulsed laser REMPI spectrum of VOCl₃ in order to clarify the answer to this question (26). The REMPI spectrum of this complex is unusual in that both VO⁺ and V⁺ ions are formed in comparable yields. Clearly, the formation and persistence of VO⁺ as a major product ion in this system can be attributed to its very strong bond dissociation energy (5.98 ± 0.10 eV, or 577 ± 10 kJ/mol) (27).

One of the key questions of the study is to determine the mechanism for formation of V⁺. There are three reasonable alternatives, which are presented pictorially in Figure 5. Pathway 1 is the "ordinary" Type B mechanism characteristic of most organometallic complexes, that is, scission of all the metal-ligand bonds followed by ionization of the neutral metal atom. The customary signature of this mechanism is the observation of strong atomic resonances in the REMPI spectrum. However, no resonances were observed in this study for either the VO⁺ or V⁺ ion formation channels in the regions 430-440 nm and 380-390 nm, despite the existence of many possible transitions (28,29).

Pathways 2 and 3 correspond to unimolecular dissociation of VO⁺, the former aided by the absorption of additional laser photons. These two pathways can be distinguished by determining the dependence of V⁺ and VO⁺ ion signals on the laser intensity. The intensity dependence was condensed to a single parameter, n, by finding the best linear fit to the empirical equation

$$\ln \text{(ion signal)} = n \ln \text{(laser pulse energy)}. \tag{1}$$

The value of n was thereby determined at 7 different wavelengths in the 430-440 nm wavelength region, as shown in Table III. The actual value of n has no physical interpretation in this experiment because of uncertainties in the spatial and temporal profiles of the laser beam, but the comparison of n for two different ion channels under identical optical conditions is a valid one. At each wavelength, the power dependence

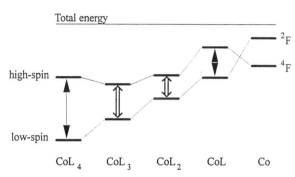

Figure 4. Qualitative energy level diagram for low-spin and high-spin photo-dissociation pathways leading from the saturated cobalt complex to the bare Co atom. Vertical arrows signify that spin-changing transitions in the early intermediates are facile due to the relatively high overall state densities and small electronic excitation energies. The spin-state populations become "locked in" prior to the crossing of high-spin and low-spin pathways, where the overall state density is relatively low and spin-changing transitions not as likely. This scenario provides a plausible mechanism for selective formation of excited doublet states of bare Co atoms.

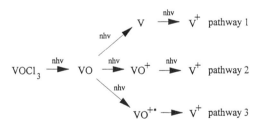

Figure 5. Scheme depicting possible pathways for formation of V^+ and VO^+ from MPD/MPI of $VOCl_3$.

Table III. Laser Power Dependence[a] of Ion Signals from $VOCl_3$

Wavelength (nm)	V^+ exponent[b]	VO^+ exponent[c]
431.710	3.7	3.7
432.340	8.9	4.2
433.145	4.4	3.2
434.290	4.0	2.7
436.340	8.8	2.5
438.950	3.3	2.0
439.500	7.4	5.3

[a] Exponent n in equation 1.
[b] Typical error for the value of the V^+ exponent is ±0.9.
[c] Typical error for the value of the VO^+ exponent is ±0.6.

for the V^+ channel was found to be greater than or equal to that of VO^+, showing that absorption of additional photons was required to generate the atomic ion by photodissociation of the metal oxide ion.

To our knowledge, this is the first clear-cut example of MPD/MPI for a transition metal complex in which the presence of a strong metal-ligand bond gives rise to a mechanism that favors photoionization (to VO^+) in preference to photodissociation (to V + O) in the pathway leading to the atomic metal ion (V^+). This result provides experimental support for the long-standing belief that a major source of differences in the MPD/MPI mechanisms of organic and organometallic compounds lies in the relative strengths of the bonds which are broken in forming the ionic fragments.

Conclusions

Pulsed laser studies of photodissociation and photoionization processes in organometallic complexes have yielded new insights into the fundamental photochemistry and photophysics of this interesting class of molecules. Interpretation of the experimental results can be clouded by the presence of multiple pathways in the MPD/MPI mechanisms. However, with the aid of carefully conceived experimental studies, considerable progress has been made towards understanding these processes at a molecular level.

One of the most fruitful areas of our research program has been to study the effects of "spectator" ligand substitution on free ligand vibrational and rotational state distributions. In the case of the series of cobalt nitrosyl complexes described in this review, we find strong evidence for statistical partitioning of available energy in the secondary and tertiary fragmentation of metal ligand bonds.

Our most surprising result is that the Co atom spin state distributions formed by MPD of $Co(CO)_3NO$ exhibit a high selectivity for formation of the first excited doublet term. This selectivity can be enhanced by lowering the laser intensity or by substituting CO with larger trialkylphosphine ligands. Comparison of these results to metal atom spin-state distributions from MPD of other transition metal complexes suggests that crossings to potential energy surfaces of higher spin multiplicity may generally occur during intermediate steps in the overall MPD process, but in the case of $Co(CO)_3NO$ this crossing may occur so late that the low-spin pathway is followed throughout.

Finally, we have recently shown that branching between photodissociation and photoionization channels during the course of laser MPD/MPI can be dramatically altered by the presence of strong metal-ligand bonds (e.g., the metal-oxide bond in $VOCl_3$). This has led us to the first clear-cut example of a mixed Type A/Type B mechanism for formation of an atomic metal ion from a transition metal complex.

Acknowledgments

The considerable contributions of Stanley Niles, Savas Georgiou, and Dr. Douglas Prinslow in carrying out this research are gratefully acknowledged. We also thank the reviewer for thought-provoking comments that significantly improved the final version of his paper. Financial support for this work was provided by the Office of Naval Research through the University of Utah Laser Institute (Contract No. N00014-91-C-0104), by a fellowship from the Alfred P. Sloan Foundation (C.A.W), and by a grant to P.B.A. from the National Science Foundation (CHE-8917980).

Literature Cited

1.	Hollingsworth, W. E.; Vaida, V. *J. Phys. Chem.* **1986**, *86*, 1235.
2.	Gerrity, D. P.; Rothberg, L. J.; Vaida, V. *J. Phys. Chem.* **1983**, *87*, 2222.
3.	Yardley, J. T.; Gitlin, B.; Nathanson, G.; Rosan, A. M. *J. Chem. Phys.* **1981**, *74*, 370.
4.	Ryther, R. J.; Weitz, E. *J. Phys. Chem.* **1992**, *96*, 2561.
5.	Hossenlopp, J. M.; Samoriski, B.; Rooney, D.; Chaiken, *J. Chem. Phys.* **1986**, *85*, 3331.
6.	Georgiou, S.; Wight, C. A. *Chem. Phys. Letters* **1986**, *132*, 511.
7.	Waller, I. M.; Davis, H. F.; Hepburn, J. W. *J. Phys. Chem.* **1987**, *91*, 506.
8.	Georgiou, S.; Wight, C. A. *J. Chem. Phys.* **1988**, *88*, 7418.
9.	Waller, I. M.; Hepburn, J. W. *J. Chem. Phys.* **1988**, *88*, 6658.
10.	Ray, U.; Brandow, S. L.; Bandukwalla, G.; Venkataraman, B. K.; Zhang, Z.; Vernon, M. *J. Chem. Phys.* **1988**, *89*, 4092.
11.	Venkataraman, B. K.; Bandukwalla, G.; Zhang, Z.; Vernon, M. *J. Chem. Phys.* **1989**, *90*, 5510.
12.	Georgiou, S.; Wight, C. A. *J. Chem. Phys.* **1989**, *90*, 1694.
13.	Although sequential bond dissociation energies of these cobalt complexes are not well established, the first two metal-carbonyl bond dissociation energies of $Fe(CO)_5$ are about 26 kcal/mol (see reference *14*) whereas typical metal-nitrosyl

bond strengths are thought to be generally in the range 40-50 kcal/mol (see reference *15*).

14. Schultz, R. H.; Crellin, K. C.; Armentrout, P. B. *J. Am. Chem. Soc.* **1991**, *113*, 8590.
15. Connor, J.A. *Top. Curr. Chem.* **1977**, *71*, 68.
16. Crichton, O.; Rest, A. J. *J. Chem. Soc. Dalton Trans.* **1977**, 536.
17. Rothberg, L. J.; Cooper, N. J.; Peters, K. S.; Vaida, V. *J. Am. Chem. Soc.* **1982**, *104*, 3546.
18. Martin, J. L.; Mingus, A.; Poyart, C.; Lecarpenter, Y.; Astier, R.; Antonetti, A. *Proc. Natl. Acad. Sci. U.S.A.* **1983**, *80*, 173.
19. Georgiou, S.; Wight, C. A. *J. Phys. Chem.* **1990**, *94*, 4935.
20. Prinslow, D. A.; Niles, S.; Wight, C. A.; Armentrout, P. B. *Chem. Phys. Letters* **1990**, *168*, 482.
21. Niles, S.; Prinslow, D. A.; Wight, C. A.; Armentrout, P. A. *J. Chem. Phys.* **1992**, *97*, 3115.
22. Niles, S.; Prinslow, D. A.; Wight, C. A.; Armentrout, P. B. *J. Chem. Phys.* **1990**, *93*, 6186.
23. Barton, T. J.; Grinter, R.; Thompson, A. J.; Davies, B.; Poliakoff, M. *J. Chem. Soc., Chem. Commun.* **1977**, 841.
24. Barnes, L. A.; Rosi, M.; Bauschlicher, C. W., Jr. *J. Chem. Phys.* **1991**, *94*, 2031.
25. Gedankin, A.; Robin, M. B.; Kuebler, N. A. *J. Phys. Chem.* **1982**, *86*, 4096.
26. Niles, S.; Armentrout, P. B.; Wight, C. A. *Chem. Phys.* **1992**, *165*, 143.
27. Clemmer, D. E.; Elkind, J. L.; Aristov, N.; Armentrout, P. B. *J. Chem. Phys.* **1991**, *95*, 3387.
28. It should be noted that atomic vanadium resonances were reported in a previous study of $VOCl_3$ (see reference *29*). There are some problems with the spectral assignments made in that paper, that are discussed in reference *(26)*, along with some possible reasons for the differences in experimental observations between our most recent study and the previous results.
29. Georgiadis, R.; Armentrout, P. B. *Chem. Phys. Letters* **1987**, *137*, 144.

RECEIVED January 19, 1993

Chapter 5

Tin Ester Photodissociation Processes

Excitation of the Ligand-Centered $^1(n,\pi^*)$ Chromophore

Jeanne M. Hossenlopp, Terrance R. Viegut, and Julie A. Mueller

Department of Chemistry, Marquette University, Milwaukee, WI 53202

Infrared absorption spectroscopy was used to probe photofragments from the 230 nm excitation of the ligand-centered $^1(n,\pi^*)$ chromophore of di-n-butyl tin diacetate. Production of CO_2, with a total absolute quantum yield of 0.2%, was observed with no evidence for photolytic formation of CO or acetic acid. Comparison of the photo-decarboxylation yield with that obtained for organic esters demonstrates the effect of a metalloid on fragmentation patterns. The total ion multiphoton ionization spectrum was obtained for di-n-butyl tin diacetate using the dissociative $^1(n,\pi^*)$ excitation as the resonant first step. Observed spectral features include a tin atomic resonance as well as evidence of a small molecular fragment.

Tin ester compounds such as di-n-butyl tin di-esters, $Sn(butyl)_2(O(CO)R)_2$, are commonly used as industrial catalysts and PVC stabilizers.(1) Di-n-butyl tin diacetate (DBTDA) is an effective esterification catalyst where the mechanism is believed to be activation by complexation of reactant oxygen atoms to the tin atom.(2) The efficiency of this catalyst is presumed to be due to its ability to activate both the carbonyl and alkoxyl group of the starting materials.(2) Photochemical initiation of new reaction pathways is a topic which is largely unexplored for this class of compounds. One- or multiphoton excitation holds promise for generation of in-situ catalysts for systems which require oxygen activation, as well as for new routes for deposition of tin or tin oxide films. In contrast to transition metal containing organometallics, virtually nothing is known about the primary photophysical properties of metalloid esters.

One feature of DBTDA which makes it a particularly interesting subject for photodissociation studies is the presence of the ligand carbonyl chromophore. West has predicted from a simple molecular orbital model that the ligand $^1(n,\pi^*)$ transition, which is localized on the carbonyl, should not shift in energy in metalloid esters from that observed in organic esters.(3) A number of reported absorption

0097–6156/93/0530–0075$06.00/0

spectra support this argument and a review of the early literature on this subject is provided by Ramsey.(4) Still open is the question of whether or not the presence of the tin atom will influence the subsequent photofragmentation pathways following the $^1(n,\pi^*)$ excitation. Use of this dissociative, ligand-centered transition as the resonant first step in a multiphoton excitation scheme is also of interest in understanding fundamental interactions of inorganic coordination compounds with lasers as well as a possible initiation step for new chemical reactions and thin film deposition.

In order to characterize the nature of tin ester photodissociation pathways, comparisons can be made with organic esters. The photodissociation dynamics of organic carbonyl compounds are the subject of a great deal of current interest.(5) The primary photodissociation channel in small carbonyls is cleavage adjacent to the carbonyl group. This α-cleavage process is known as Norrish Type I photochemistry.(6) Compounds large enough to have a hydrogen atom at the γ-position relative to the carbonyl are known to also undergo an intramolecular abstraction, via a six-center transition state, of the γ-hydrogen by the carbonyl. The result of this Norrish Type II process in esters is commonly molecular elimination of an alkene and an acid.(6) While there is only limited fundamental data on esters,(7) analogies have been made between ester photochemical processes and those of the better-understood ketones.(6) Figure 1 illustrates possible Norrish Type I and Type II channels for DBTDA. The work discussed here involving characterization of DBTDA photofragmentation patterns is one part of a series of systematic investigations into the relationship between ester molecular structure and photodissociation dynamics.(8)

The basic experimental approach involves excitation of the $^1(n,\pi^*)$ ligand-centered chromophore in DBTDA followed by observation of photoproducts by infrared absorption or multiphoton ionization spectroscopy. Comparisons of the observed one-photon pathways with those observed under identical conditions for organic compounds provide the initial picture of photo-induced decomposition of this compound. Multiphoton excitation provides the basis for comparison with other inorganic coordination compounds.

Experimental

The experimental apparatus used in this work is shown in Figure 2. Gas phase samples of DBTDA were excited in the $^1(n,\pi^*)$ band in the wavelength region of 225 -245 nm. A Questek 2520 vβ excimer laser was used to pump a Lambda Physik Fl3002 dye laser, operating with Coumarin 470 dye. The dye laser output was frequency-doubled with a BBO crystal housed in an InRad Autotracker. The fundamental was separated from the harmonic beam using a four prism beam separator and blocked. Tunable infrared diode lasers were used to provide an in-situ probe of stable photoproducts. For the infrared experiments, the pulsed UV photolysis laser and the continuous wave infrared probe laser were copropagated through a 275 cm long single-pass absorption cell. The UV beam was blocked at the output of the cell while the IR beam was passed through a 1/4 meter monochromator. The monochromator served to shield the HgCdTe detector from stray light and to also filter out any unwanted diode laser second mode emission.

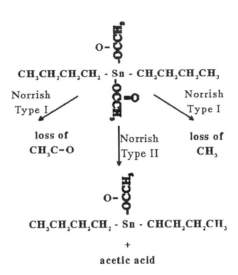

Figure 1. Possible di-n-butyl tin diacetate photolyis channels. Following excitation of the $^1(n,\pi^*)$ transition, Norrish Type I α-cleavage pathways can result in either the loss of CH_3 or CH_3CO from the acetate ligand. A possible Norrish Type II pathway involves intramolecular abstraction of a hydrogen from the n-butyl group, followed by elimination of acetic acid.

Multiphoton ionization spectra were obtained using parallel plate electrodes and were collected without the use of any external buffer gas for DBTDA. For the MPI experiments, the UV laser was focused with a 20 cm focal length lens. The ion detection plate voltage was 125 volts with a one centimeter plate separation. Data acquisition was performed using either a LeCroy 9410 digital oscilloscope or a computer-controlled gated integrator built from Evans Electronics boards.

Commercial samples of DBTDA (Pfaltz and Bauer, 95%) were opened under an inert, dry atmosphere and all vacuum pumps used in sample preparation and experimental measurements utilize liquid nitrogen traps. GC/Mass Spectroscopy was used to examine the contents of the samples used. Only one minor contaminant was observed, a less volatile, higher molecular weight tin complex. The mass spectral pattern of the larger molecular weight component is consistent with the monomeric unit of a hydrolysis product, $[(OH)R_2SnOSnR_2(OH)]_2$.(9) Carbon-13 and proton NMR spectra were obtained which confirm the presence of a minor contaminant which is roughly on the order of 3-5% of the sample. DBTDA, a liquid at room temperature, is reported (2) to have a boiling point of 145° C. at 10 torr. We attempted to distill our samples and found that while the sample boiled under the appropriate conditions, we obtained solely organic decomposition products.

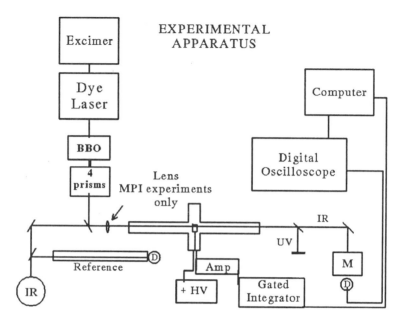

Figure 2. Experimental apparatus. Details are provided in the text of the experimental section. HgCdTe infrared detectors are represented as D and M indicates a 1/4 monochromator used for filtering stray light and unwanted second mode emission of the diode lasers.

Since the only detectable contaminant was less volatile than DBTDA, samples were prepared by several freeze-pump-thaw cycles with liquid nitrogen, followed by pumping at $0°$ C. and also briefly at room temperature to remove any residual volatile contaminants. Prior to any experimental measurements, the clean vacuum system is exposed to a flowing sample of DBTDA for several minutes.

Quantum yield measurements for the production of CO_2 were made under static cell conditions. The long-path absorption cell was filled with a gas-phase samples of the DBTDA at several pressures in the range of 10 - 100 mtorr. The UV absorption coefficient was determined using a Molectron JD-50 joule meter to measure laser power before and after the sample cell. After correcting for window absorption, an absorption coefficient of 1.6(+/- 0.2) cm^{-1} $torr^{-1}$ was obtained for excitation at 230 nm. This was then used to calculate the total number of DBTDA molecules excited during the photolysis run.

The production of CO_2 was determined by use of infrared absorption. Prior to the photolysis, the spectral profile of a single diode laser mode was determined by averaging 100 modulation cycles on the LeCroy. This provides a measure of the transmitted IR intensity across an approximately 2 cm^{-1} range. Static samples of DBTDA were photolyzed for 15 minutes with a 5 Hz photolysis repetition rate. We have varied the length of the photolysis as well as the repetition rate in order to find the best conditions for obtaining reproducible quantum yield data. The diode laser signal is again averaged and stored along with the pre-photolysis baseline measurement. Production of CO_2 is immediately apparent by growth of peaks in the spectrum. The assignment of the exact diode laser wavelength region and identities of the IR rovibrational transitions were made using the AFGL database.(10) The 10^00 P22 $^{12}C^{16}O_2$ transition at 2307.991 cm^{-1} was used to calculate pressure of CO_2 in the cell. Knowing the cell volume, this can be converted to a total number of CO_2 molecules produced and the quantum yield is defined as this number divided by the number of DBTDA molecules excited.

Multiphoton ionization spectra were obtained by averaging the integrated total ion signal over 25 laser shots per wavelength. Spectra were collected by stepping the dye laser in 0.005 nm increments (visible wavelengths). For comparison, both CO and Sn(II) diacetate were also studied in the wavelength regions where there was significant structure in the DBTDA spectrum. The CO spectra were measured with pressures ranging from 50 mtorr to one torr. Sn(II) diacetate was placed directly in the cell and spectra were measured at both its room temperature vapor pressure (< 1 mtorr) and also using 1 torr of CO_2, which is structureless in this wavelength region, as a signal multiplying buffer gas. Cu(II)diacetate monohydrate has also been studying using the same experimental conditions as described for Sn(II)diacetate.

Results

One-Photon Excitation. The first objective of our work was to compare the fragmentation patterns of DBTDA to that of organic esters also excited via the $^1(n,\pi^*)$ transition. The specific stable photofragments which were searched for via IR probing are CO_2, CO, and acetic acid. Acetic acid is a possible Norrish Type II molecular elimination product and CO_2 and CO are possible products resulting from Norrish Type I α-bond cleavages.

Acetic acid was probed in the 1780 cm^{-1} region. A gas-phase sample of acetic acid was used as a reference. Samples were photolyzed at 5 Hz for periods up to 30 minutes and there was no clear evidence of production of acetic acid. This indicates that acetic acid is not a major photoproduct. Loss of the acid via secondary photolysis (5f) is not sufficient to account for the lack of observed product under our experimental conditions. We cannot conclusively rule out small yields of this Norrish Type II product due to the presence of a number of broad infrared transitions of DBTDA in this wavelength region which overlap with a number of the acetic acid lines. There was no observed production of CO, probed in the 2100 cm^{-1} where there was no problem with spectral interference of the DBTDA.

Photolytic production of CO_2 was observed. With the photolysis laser set at 230 nm, a total absolute quantum yield of 2 (+/- 1) x 10^{-3} (0.2%) was obtained. The values for quantum yields obtained in this manner are total absolute yields and reflect contributions from secondary photolysis processes and side reactions as well as the one-photon production of CO_2 via the Norrish Type I process. For a stable photoproduct such as CO_2, this places an upper limit on the relative importance of the three body dissociation process which can be initiated by the initial loss of CH_3 as shown in Figure 1.

Multiphoton Excitation. Figures 3 and 4 illustrate portions of the total ion signal spectrum of DBTDA in the one photon wavelength region 230 - 240 nm.

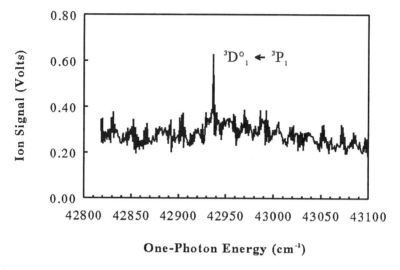

One-Photon Energy (cm⁻¹)

Figure 3. Di-n-butyl tin diacetate MPI spectrum. Total ion signal versus one-photon energy is shown for a selected portion of the spectrum, along with the assignment of the observed tin atomic transition.

There are two different types of structure observed in this wavelength region. The sharp atomic line shown in Figure 3 can be assigned to the Sn atomic transition $^3D^{\circ}_1 \leftarrow {}^3P_1$. The $^3D^{\circ}_j \leftarrow {}^3P_1$ transitions are the only atomic resonances expected in the wavelength region scanned in these experiments due to the use of one-color UV excitation. The $^3P_1$ component of the tin ground electronic state lies 1691.8 cm⁻¹ above the lowest $^3P_0$ level.(11)

The second type of structure is shown in Figure 4. The spacing between the features varies slightly as a function of wavelength. This broad structure is characteristic of molecular, or molecular fragment, resonances. No such resonance features are observed in the excitation of Sn(II) diacetate or Cu(II) diacetate. In the case of Sn(II) diacetate, the sharp atomic feature shown in Figure 1 is also not observed, while a number of Cu atomic resonances are observed in the spectrum of Cu(II) diacetate. (Wilkinson, H.A.; Hossenlopp, J.M., unpublished results.)

One possible source of the molecular structure is CO via a 1+1 MPI scheme. Excitation of the (17,0) band of the $A^1\Pi \leftarrow X^1\Sigma^+$ transition is known to be resonant with the one-photon wavelengths used in these experiments. One more photon

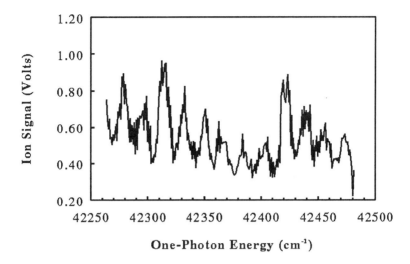

One-Photon Energy (cm^{-1})

Figure 4. Di-n-butyl tin diacetate MPI spectrum. The portion of the the total ion signal spectrum where broad molecular (fragment) resonances are observed. Some of the variation in peak intensity is due to fluctuations in laser power across the wavelength region scanned.

would be required for ionization.(12) Rotational assignments have been made for the (17,0) transition from emission experiments.(12) In order to explore this possibility, MPI spectra were run for pure CO at a range of pressures and no resonances were observed. Measuring the MPI spectrum at room temperature does not eliminate the possibility of hot-band transitions. However, due to the differences in vibrational frequency and anharmonicity constants of the electronic states in CO,(12) there are no assignments from vibrationally excited ground state CO which are consistent with structure in the observed region.

Discussion

One Photon Fragmentation Patterns. The primary motivation for the one-photon excitation experiments was to explore molecular structure effects on ester photodissociation processes. The ^{1}(n,π*) ligand-centered chromophore does not shift significantly in wavelength but there is a major effect on the observed fragmentation patterns. The most striking difference is the low quantum yield for CO_2 production.

Following excitation of the ^{1}(n,π*) transition, alkyl organic esters exhibit a variety of competing photochemical decomposition channels. Bond cleavage adjacent to the carbonyl group, known as Norrish Type I photochemistry, is observed. There is conflicting evidence in the literature concerning the relative importance of dissociation involving each of the two possible α-positions.(6) One possiblity for the general case of RO(CO)R' dissociation is to break the O-C bond, leading to RO and R'CO fragments. The R'CO radical may further decompose into R' + CO. The barrier for this second step where R' is an alkyl group is approximately 15 kcal/mole.(13) The other possible Norrish Type I channel involves breaking the bond between the R' group and the carbonyl carbon leading to RCO_2 and R' radicals. If R is an alkyl group there is only a 1-2 kcal/mole barrier for decarboxylation, leading to R and CO_2 as the ultimate products.(14)

With simple alkyl esters such as methyl acetate and ethyl formate, we find quantum yields for CO_2 production to be as much as two orders of magnitude greater than that observed with DBTDA. One explanation for this would be a decreased efficiency in the initial α-cleavage rate, possibly favoring the alternate α-site or the Norrish Type II channel. A second explanation would be a higher activation barrier for the decarboxylation step, leading to a decrease in free CO_2 yields in the static cell experiments. The possibility of a stabilizing interaction between the carbonyl oxygen and the tin atom would be a likely source for an increased barrier for the second step in the CO_2 production channel. Also possible in this tin-containing system would be crossing onto a potential surface which leads to loss of one of the n-butyl ligands.

Calculations of RRKM rates for Norrish Type I processes have been used with some limited sucess in modeling molecular structure effects in phenyl ketones and predict lower α-cleavage rates for larger molecules.(15) However, it should be noted that the excitation of the $^1(n,\pi^*)$ transition in organic esters does not lead to fragmentation patterns which are properly modeled with RRKM theory due to the preferential cleavage at the α-position and the possibility for photoproduct evolution from the ground electronic state, lowest triplet state, and the initially prepared $^1(n,\pi^*)$ state.(12) Attempts at direct observation of CH_3 and/or CH_3O radicals will be necessary to unambiguously determine the reason for lower CO_2 yields. Manipulation of the structure of the alkyl groups, possibly by subsitution of phenyl ligands, may be an important step in understanding the role of photo-induced decarboxylation in the degradation of these compounds. In the case of tri-organo tin carboxylates, the structure of the organic ligands is known to play an important role in the relative efficiency of thermal decarboxylation, with trialkyl compounds exhibiting lower yields of decarboxylation than do triphenyl compounds.(9)

The absence of any clear indication of CO or acetic acid production can be explained based on simple energetic arguments. The production of CO in a two-step process requires decomposition of a photo-generated CH_3CO radical. The 15 kcal/mole barrier (13) for the second step, coupled with the greater initial energy required for this α-cleavage compared to the site which leads to decarboxylation, leads to no observable CO production in any of the organic or organometalloid esters which we have investigated. Generally, the lower energy α-cleavage process is observed to dominate following the $^1(n,\pi^*)$ excitation.(16) This is due to the correlation between the stability of the radical being produced and the height of the triplet barrier for the cleavage due to an avoided crossing of the nπ* and ππ* states.(16) Loss of the methyl group rather than CH_3CO would be consistent with this general trend for α-site preference.

Acetic acid production, via the Norrish Type II split, would require abstraction of a hydrogen atom from the n-butyl group followed by elimation of the molecular product. Norrish Type II processes most frequently occur via a 6-center transition state.(6) Relative rates of such processes as a function of molecular structure are modeled (17) by considering the barrier for formation of the transition state and also the type of hydrogen atom being abstracted. Secondary hydrogen atoms exhibit intermediate reactivity compared to primary and tertiary positions. For DBTDA, the important factor in controlling Norrish Type II yields should be the barrier for transition state formation. Long alkyl chains tend to diminish the Norrish Type II yields in experiments. This has been observed in traditional photochemical measurements on product branching ratios for alkyl esters,(7) as well as in recent work in our laboratory on the influence of alkyl chain lengths on benzoate ester phosphorescence yields.(8)

Multiphoton Excitation. The role of molecular structure in influencing UV multiphoton dissociation pathways is a subject of on-going investigation in a number of groups. The dissociation followed by ionization behavior of metal carbonyls has been well-documented.(18) Photofragmentation patterns resulting from one- and multiphoton excitation have been characterized for a few alkyl metal complexes.(19) Tetramethyl tin is known to exhibit dissociation followed by ionization, leading to the observation of tin atomic resonances in the MPI spectrum.(19c) Also notable with respect to our work on tin ester compounds is that of Mikami, et al., on first row transition metal acetylacetonato complexes.(20) In their study of nine different acetylacetonato compounds, only the Cu and Cr analogues were observed to decompose via the route of dissociation followed by metal atom ionization. The metal atom dependent behavior was attributed (20) to correlation of the molecular super-excited state with the ground state electron configuration of the bare metal atom. Cu and Cr compounds, due to the s^1 configurations, therefore were expected to exhibit different MPD patterns than the other transition metal acetylacetonato complexes.

Comparisons between our system and the work of Mikami, et al. (20) requires consideration of the structure of the molecules as well as the ligand identity. The Cu(II) acetate and Sn(II) acetate are analogous to the acetylacetonato compounds due to the bidentate bridging bonding of the ligand. Observation of Cu, but not Sn, atomic resonances is consistent with the observation that the metal atom identity can make a difference in the observed multiphoton decomposition pathways. Additional work needs to be done in order to fully understand the source of the molecular structure dependence in these types of molecules.

Acetate ligands can also be found with unidentate bonding, such as in DBTDA. Here there is evidence for the production of Sn atoms as well as a small molecular fragment in the MPI results. Using the dissociative, ligand-centered $^1(n,\pi^*)$ excitation as the resonant first step of the multiphoton excitation scheme leads to fragmentation patterns which are not observed for the case of the bridging, bidentate ligand structure of Sn(II) diacetate. Similar to tetramethyl tin,(19c) formation of atomic Sn is clearly evident in our MPI spectrum. Unambiguous determination of the identity of the source of the molecular structure will require mass resolution. Carbon monoxide and carbon dioxide can be eliminated as possibilities due to the lack of observed structure in their spectra when measured in our apparatus under identical conditions. Another possibility is SnO, excited via the one-photon resonant E ← X transition.(21) Isotopic splittings, due to the large number of naturally abundant tin isotopes,(22) would be sufficient to produce the structure observed in Figure 4. For example, the (17-2) and (19-3) bands of this transition would be predicted to overlap the first two broad features in Figure 4. Rydberg transitions of CH are also known (23) to occur in this wavelength region; this product could be produced from fragmentation of either the acetate or the n-butyl ligand. Current efforts in our laboratory involve use of both fluorescence excitation spectroscopy and dispersed fluorescence to characterize the source of the molecular structure observed in these MPI experiments.

Conclusions

Excitation of the ligand-centered $^1(n,\pi^*)$ transition in di-n-butyl tin diacetate results in the production of CO_2, with yields much lower than those observed for organic esters. Using diode laser infrared absorption spectroscopy, no evidence for photolytic production of CO or acetic acid was obtained. Multiphoton excitation, using the ligand-centered transition as the resonant first step, leads to observation of atomic tin and as well as molecular features in the the MPI spectrum. Acetate ligand bonding structure plays an important role in determining the multiphoton decomposition pathways in tin esters.

Acknowledgments

Support for this work has been provided by a Camille and Henry Dreyfus Foundation New Faculty Award (JMH), the Donors of the Petroleum Research Fund, administered by the American Chemical Society, and the Marquette University Committee on Research. TRV acknowledges the support of the Department of Education National Needs Fellowship Program (P200A-90035-91). We are grateful to the National Science Foundation (CHE-8905465) for partial funding of the 300 MHz NMR spectrometer used in this research.

Literature Cited

1.   Evans, C.J.; Karpel, S. *Organotin Compounds in Modern Technology*; Elsevier: Amsterdam, 1985.
2.   Omae, I. *Organotin Chemistry*; Elsevier: Amsterdam, 1989.
3.   West, R. *J. Organomet. Chem.* **1965**, *3*, 314.
4.   Ramsey, B.G. *Electronic Transitions in Organometalloids*; Academic Press: New York, NY, 1969.
5.   See, for example, (a) Person, M.D.; Kash, P.W.; Butler, L.J. *J. Phys. Chem.* **1992**, *96*, 2021. (b) Hunnicut, S.S.; Waits, L.D.; Guest, J.A. *J. Phys. Chem.* **1991**, *95*, 562. (c) Trentelman, K.A.; Kable, S.H.; Moss, D.B.; Houston, P.L. *J. Chem. Phys.* **1989**, *91*, 7498. (d) Carleton, K.L.; Butenhoff, T.J.; Moore, C.B. *J. Chem. Phys.* **1990**, *93*, 3907. (e) Brouard, M.; Martinez, M.T.; O'Mahoney, J.; Simons, J.P. *Molec. Phys.* **1990**, *69*, 65. (f) Singleton, D.L.; Paraskevopoulos, G.; Irwin, R.S. *J. Phys. Chem.* **1990**, *94*, 695.

6. Calvert, J.G.; Pitts, J.W. *Photochemistry*; Wiley: New York, NY, 1966.
7. (a) Ausloos, P. *Can. J. Chem.* **1958**, *36*, 383. (b) Ausloos, P.; Rebbert, R.E. *J. Phys. Chem.* **1963**, *67*, 163.
8. Viegut, T.R.; Pisano, P.J.; Mueller, J.A.; Kenney, M.J.; Hossenlopp, J.M. *Chem. Phys. Lett.*, in press.
9. Okawara, R.; Ohara, M. In *Organotin Compounds*; Sawyer, A.K., Ed.; Marcel Dekker: New York, NY, 1971, Vol. 2.
10. AFGL HITRAN Database, 1986.
11. Tilford, S.G.; Simmons, J.D. *J. Phys. Chem. Ref. Data* **1972**, *1*, 147.
12. Moore, C.E. *Atomic Energy Levels*: Natl. Stand. Ref. Data Ser., (U.S.) Natl. Bur. Stand., 1971; Vol. 3.
13. Benson, S.W.; O'Neal, H.E. *Kinetic Data on Gas Phase Unimolecular Reactions*; NSRDS-NBS, 1970, Vol. 21.
14. Gray, P.; Thynne, J.C.J. *Nature* **1961**, *191*, 1357.
15. Rennert, A.R.; Steel, C. *Chem. Phys. Lett.* **1981**, *78*, 36.
16. Reinsch, M.; Klessinger, M. *J. Phys. Org. Chem.* **1990**, *3*, 81.
17. Sengupta, D.; Sumathi, R.; Chandra, A.K. *J. Photochem. Photobiol. A* **1991**, *60*, 149.
18. See, for example, (a) Kearney, Z.; Naaman, R.; Zare, R.N. *Chem. Phys. Lett.* **1978**, *59*, 33. (b) Gerrity, D.P.; Rothberg, L.T.; Vaida, V. *Chem. Phys. Lett.* **1980**, *74*, 1. (c) Leutwyler, S.; Even, U.; Jortner, J. *Chem. Phys. Lett.* **1980**, *74*, 11. (d) Fisanick, G.J.; Gedanken, A.; Eichelberger, T.S.; Kuebler, N.A.; Robin, M.B. *J. Chem. Phys.* **1981**, *75*, 5215. (e) Whetten, R.L.; Fu, K.-J.; Grant, E.R. *J. Chem. Phys.* **1983**, *79*, 4899. (f) Nagano, Y.; Achiba, Y.; Kimura, K. *J. Phys. Chem.* **1986**, *90*, 1288. (g) Hossenlopp, J.M., Samoriski, B.; Rooney, D.; Chaiken, J. *J. Chem. Phys.* **1986**, *85*, 3331. (h) Tyndall, G.W.; Jackson, R.L. *J. Amer. Chem. Soc.* **1987**, *109*, 582. (i) Belbruno, J.J. *Chem. Phys. Lett.* **1989**, 267. (j) Gergio, S.; Wight, C.A. *J. Chem. Phys.* **1989**, *90*, 1694.
19. (a) Mitchell, S.A.; Hackett, P.A.; Rayner, D.M. *J. Chem. Phys.* **1985**, *83*, 5028. (b) Jackson, R.L. *J. Chem. Phys.* **1992**, *96*, 5938 and references therein. (c) Robin, M.B. In *Advances in Laser Spectroscopy;* Garetz, B.A., Lombardi, J.R., Eds.; Wiley: New York, NY, 1986, Vol. 3; pp. 147-157.
20. Mikami, N.; Ohki, R.; Kido, H. *Chem. Phys.* **1990**, *141*, 431.
21. *Spectroscopic Data;* Suchard, S.N., Ed.; Plenum: New York, NY, 1975, Vol. 1, Part B.
22. Rai, S.B.; Sing, J. *Spec. Lett.* **1972**, *5*, 155.
23. Huber, K.P.; Herzberg, G. *Molecular Spectra and Molecular Structure;* Van Nostrand Reinhold Company: New York, NY, 1979, Vol. 4.

RECEIVED January 19, 1993

Chapter 6

Laser Photoelectron Spectroscopic Study of Gas-Phase Organometallic Molecules

Katsumi Kimura[1]

Institute for Molecular Science, Okazaki 444, Japan

Multiphoton ionization (MPI) photoelectron spectroscopy of
jet-cooled molecules using a ns laser source provides
spectroscopic information about molecular excited states
including non-radiative states as well as about their
ionic states. This article consists of two parts. One is
an MPI photoelectron study of some organic iron complexes,
which determines electronic states of photoinduced iron
atoms. The other is concerned with 'cm^{-1}-resolution'
threshold photoelectron spectroscopy which will be
powerful for future cation spectroscopy of organometallic
molecules.

During the last decade, 'resonantly enhanced multiphoton ionization'
(REMPI) with a nanosecond UV/VIS laser system has been combined with
a photoelectron spectroscopic technique, providing excited-state
photoelectron spectroscopy or REMPI photoelectron spectroscopy (1).
In connection with laser chemistry of organometallics, one of
interesting applications of this technique is to study electronic-
state population of metal atoms which are produced by laser
photofragmentation of organometallic molecules. We demonstrate a few
examples in which many low-lying non-radiative electronic states of
photofragment metal atoms can be identified from photoelectron energy
analysis. Recently, in this laboratory we have developed a compact
threshold photoelectron analyzer for high-resolution (1-2 cm^{-1})
cation spectroscopy with a two-color REMPI technique. In this

[1]Current address: Japan Advanced Institute of Science and Technology, Tatsunokuchi,
Ishikawa 923–12, Japan

0097–6156/93/0530–0086$06.00/0

article, therefore we also mention that this technique is very useful for cation spectroscopy of organometallic molecules.

Photodissociation of Some Iron Complexes

A molecule irradiated by a laser pulse undergoes two typical types of nonlinear photochemistry: ionization followed by fragmentation, and fragmentation followed by ionization (2). Laser excitation of gaseous $Fe(CO)_5$, for example, in the region around 280 nm produces Fe^+ ions with almost unit efficiency at relatively mild laser fluence, as earlier indicated by Duncan et al. (3).

In UV/visible MPI studies of metal carbonyls, Nagano et al. (4-7) have first assigned many MPI ion-current peaks to resonant ionizations of the ground-state metal atoms. Several prominent MPI peaks of $Fe(CO)_5$ in the visible region 447-466 nm are attributed to three-photon resonant ionizations of the ground-state Fe atoms. A total of thirteen low-lying electronic states of Fe atoms have been identified in the photodissociation of $Fe(CO)_5$ from further photoelectron spectroscopic studies (5-7). A similar photoelectron study has recently been reported by Niles *et al.* (8), who have studied photodissociation of $Fe(CO)_5$ in a more detail.

When metal atoms are produced in various electronic states, it is often difficult to identify the electronic states from only MPI ion-current spectra, since there are a number of possible resonant intermediates. From photoelectron energy analysis, it is possible to identify the initial atomic states of the MPI processes. This situation is schematically shown in Fig. 1. Broad bands and background signals appearing in MPI ion-current spectra for iron complexes have been studied by photoelectron spectra (6,7). Without photoelectron energy data, it is especially difficult to interpret the broad feature and background signals of MPI ion-current spectra.

Photoelectron Determination of Electronic States

A typical MPI ion-current spectrum observed for $Fe(CO)_5$ in the laser wavelength 447-450 nm is shown in Fig. 2, indicating many weak and strong peaks (5). From the photoelectron energy analysis, all these peaks have been interpreted in terms of the three-photon resonant ionizations originating from the lowest four electronic states of Fe atoms. A typical photoelectron spectrum due to Fe^+ ions are shown in Fig. 3, obtained at the strongest MPI peak at 447.65 nm (Fig. 2). The photoelectron energy K in the three-photon ionization of Fe atom is given by

$$K = (\text{three-photon energy}) - I + E(i") - E(j^+) \qquad (1)$$

where I is the first ionization potential (7.90 eV), and $E(i")$ and $E(j^+)$ are the energies of the ith initial neutral electronic state and the jth final ionic state, respectively. The electronic terms of

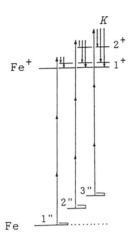

Figure 1. Schematic energy diagram of the Fe atom and Fe^+ ion, showing the photoelectron determination of initial states.

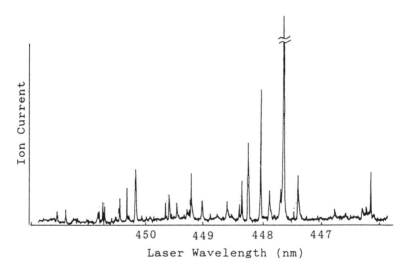

Laser Wavelength (nm)

Figure 2. MPI ion-current spectrum due to Fe atoms, obtained from $Fe(CO)_5$. (Reproduced with permission from reference 5. Copyright 1986 American Institute of Physics.)

Fe and Fe^+ are hereafter abbreviated as i"(1", 2", etc.) and j^+ (1^+, 2^+, etc.), respectively. Formation of the lowest ionic state gives rise to the maximum photoelectron energy which is given by

$$K_{max} = \text{(three-photon energy)} - I + E(i") \qquad (2)$$

since $E(1^+) = 0$. From Eq. (2), the maximum photoelectron energies for Fe atoms populated at the lowest four electronic states have been evaluated to be $K_{max} = 0.33$-0.56, 1.19-1.45, 1.82-2.05, and 2.51-1.66 eV (5). When Fe atoms are produced in many different electronic states, maximum photoelectron energies are quite helpful for interpreting MPI ion-current peaks.

The photoelectron spectrum shown in Fig. 3 consists of two bands 1^+ and 2^+ which both originate from 1". The 2^+ band is resolved into a few J levels, since the energy resolution in this region is better than 15 meV. The 2^+ state is not observed in HeI photoelectron spectra of Fe atoms (13).

From the photoelectron analysis, it has been indicated that the lowest four electronic states of Fe atoms are responsible for all the MPI ion-current peaks shown in Fig. 2, although resonant intermediate states in the three-photon resonant ionizations of 3" and 4"(5)" have not been identified yet. In general, such photoelectron spectroscopic determination of the initial electronic states should be important for studying electronic states of photodissociation products.

Ligand Effect on Electronic-State Population

In their photoelectron study, Nagano *et al.* (6) have also indicated that the electronic-state population of Fe atoms largely depends on the ligand. The MPI ion-current spectra observed for three different iron complexes are shown in Fig. 4; (a) ferrocene $Fe(Cp)_2$, (b) iron tris (acetylacetonate) $Fe(Acac)_3$, and (c) iron pentacarbonyl $Fe(CO)_5$. All these spectra, showing remarkable differences in spectral pattern, have been attributed to two- or three-photon ionizations of Fe atoms from the photoelectron energy analysis (6). Such remarkable differences in the MPI ion-current spectra are due to significant changes in the electronic-state population of Fe atoms.

Spectrum (a) in Fig. 4 is the simplest among the three spectra, mainly consisting of four sharp peaks, while spectrum (b) shows several additional peaks and some weak broad bands. Spectrum (c) showing many broad bands is striking contrast to spectra (a) and (b) in Fig. 4.

Ferrocene and iron pentacarbonyl are two extreme cases, yielding remarkably different patterns in the MPI ion-current spectra. In the one extreme case (ferrocene), ground-state Fe atoms are dominantly produced. However, in the other extreme case (iron pentacarbonyl), Fe atoms are broadly populated among the excited states up to 13". In the middle case, the excited-state population is in between the two extreme cases, as shown by spectrum (b) in Fig. 5. Many other Fe complexes are probably in between the two extreme cases.

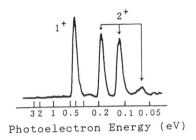

Photoelectron Energy (eV)

Figure 3. Photoelectron spectrum of the Fe atom in the ground state. (Reproduced with permission from reference 5. Copyright 1986 American Institute of Physics.)

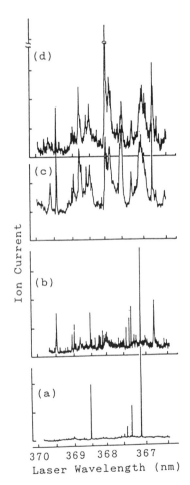

Figure 4. MPI ion-current spectrum observed for a, ferrocene; b, iron tris(acetylacetonate); c, $Fe(CO)_5$; and d, $FeCl_3$. (Reproduced from reference 7. Copyright 1986 American Chemical Society.)

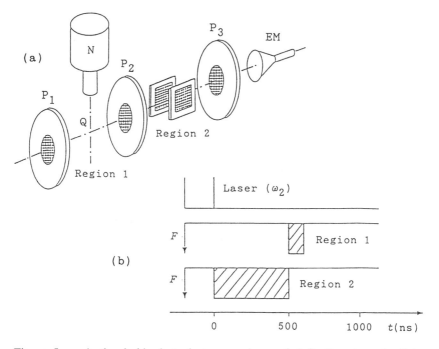

Figure 5. a, A threshold photoelectron analyzer of deflection type. b, Pulse profiles of the laser and the electric fields in regions 1 and 2.

The excitation energy of the parent iron complexes is considered to rapidly transfer to the low-frequency vibrational modes of the ligands. Roughly speaking, the density of the vibrational states increases in the order of $Fe(CO)_5$, $Fe(Acac)_3$ and $Fe(Cp)_2$. The ratios of the number of the ligand vibrational modes to the number of the metal-ligand stretching modes in the Fe complexes are 5.4 for $Fe(CO)_5$, 22.0 for $Fe(Acac)_3$, and 28.5 for $Fe(Cp)_2$ (13). These differences probably give rise to the different population of excited-state Fe atoms.

Compact cm^{-1} Resolution Threshold Photoelectron Analyzer

Recently, in this laboratory we have developed a compact photoelectron analyzer capable of high-resolution (1-2 cm^{-1}) cation spectroscopy of molecules in two-color REMPI experiments (9-13). Our compact deflector-type threshold photoelectron analyzer is schematically shown in Fig. 5(a). Photoelectrons with energies lower than a few cm^{-1} are collected as a function of the second (ionization) laser wavelength by applying a pulsed electric field at 500 ns at each laser shot, while energetic photoelectrons are removed by angular- and time-resolved discrimination. In connection with two-color REMPI photoelectron spectroscopy, Müller-Dethlefth *et al.*(14) have developed a zero kinetic energy (ZEKE) photoelectron technique to perform very high resolution photoelectron spectroscopy.

Two-color (n+1') REMPI experiments are especially important, since the second ionization source can be employed independently of the first excitation source. The wavelength of the first laser is scanned while keeping the second laser wavelength constant. The advantage of the two-color ionization is that only a special molecule among a mixture of analogous molecular species can be selectively ionized to provide its cation spectroscopy.

Let us consider a sphere of 10 mm in diameter surrounding the ionization point (Q). Electrons with energies lower than a few cm^{-1} should remain in the sphere for as long as 500 ns after each laser shot. In other words, only threshold photoelectrons can be collected typically at 500 ns after each laser shot. Such very low energy electrons can be therefore collected with a small-size compact analyzer by applying a pulsed electric field (a few V/cm) across P_1 and P_2. The deflection-type analyzer has a pair of deflection plates (D). Most energetic electrons can be removed by the deflector before reaching the detector. Figure 5(b) shows schematically how to apply the electric pulses in the regions 1 and 2 with respect to the ns laser pulses (first and second lasers).

A threshold photoelectron spectrum due to NO^+, which was obtained by two-color (1+1') REMPI via the A^2R^+ state ($v'=0$, $N'=7$), has indicated well-resolved rotational peaks ascribed to $N = N^+ - N' = 0$, ±1, ±2, and ±3 transitions (9). Although the energy resolution (fwhm) depends on the delay time of the electric field as well as on the field strength, we have evaluated the energy resolution to be 2 cm^{-1} (fwhm) from the threshold photoelectron spectrum.

Examples of High-Resolution Threshold Photoelectron Spectra

The van der Waals vibrations of the aniline-Ar_n complexes (n=1,2) in the S_1 states have been studied by Bieske *et al.* (15) by using a mass-selected REMPI technique. The geometrical structure of the neutral aniline-Ar complex (n=1) has been studied by a LIF rotational analysis by Yamanouchi *et al.*(16).

Figure 6 show threshold photoelectron (ZEKE photoelectron) spectra of aniline and its Ar complexes (n=1,2), obtained by (1+1') REMPI *via* S_1 (12). Each cation spectrum consists of several vibrational bands in the region below 1300 cm^{-1}. Upon complex formation, each [aniline]$^+$ vibrational band is split into several sub-bands. The [aniline]$^+$ vibrational frequencies are little perturbed by the Ar atoms. The band splittings observed for the van der Waals cations are due to combination of the [aniline]$^+$ vibrational mode with some low-frequency van der Waals vibrational modes.

From the photoelectron bands (0^{+0}) in Fig. 6, the adiabatic ionization potentials (I_a) have been determined as

I_a(aniline) = 62268 ± 4 cm^{-1},
I_a(aniline-Ar) = 62157 ± 4 cm^{-1},
I_a(aniline-Ar_2) = 62049 ± 4 cm^{-1}.

The I_a shift of the 1:2 complex (219 cm^{-1}) is almost twice as much as that of the 1:1 complex (111 cm^{-1}), this fact suggesting that the [aniline]$^+$ component is planar. This is also supported by the fact that the photoelectron intensity due to the [aniline-Ar]$^+$ inversion mode is very weak. The molecular geometries of the amino groups in [aniline-Ar_n]$^+$ are almost the same as those in the neutral S_1 states. The geometry of aniline in the S_1 state has been indicated to be planar from an analysis of the inversion potential function by Hollas *et al.* (17). As a result, we may conclude that the [aniline]$^+$ component has a planar geometry.

From the ZEKE spectra of the aniline-Ar_n complexes (n=1,2), the low-frequency vibrational frequencies have been determined to be 16 cm^{-1} (n=1) and 11 cm^{-1} (n=2). From Franck-Condon calculations, these low-frequency vibrational modes have been assigned to the van der Waals 'symmetric bending' and 'in-phase bending' mode in the [aniline-Ar]$^+$ and [aniline-Ar_2]$^+$ cations, respectively (12).

Concluding Remarks

As demonstrated in the cases of the organometallic molecules in this article, MPI photoelectron spectroscopy is useful to study various electronic levels including non-radiative states from photoelectron energy analysis.

Our compact analyzer useful for two-color REMPI threshold

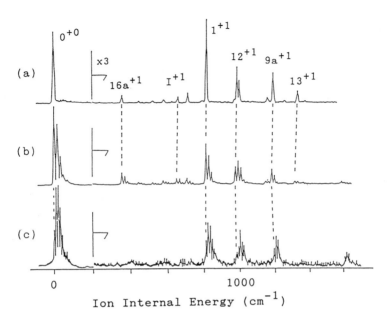

Figure 6. The (1+1') threshold photoelectron spectra due to a, [analine]+; b, [analine-Ar]+; and c, [analine-Ar$_2$]+. (Reproduced with permission from reference 12. Copyright 1992 American Institute of Physics.)

photoelectron measurements with time-resolved discrimination is excellent for carrying out threshold photoelectron spectroscopy with a few cm^{-1}. Such a very high resolution photoelectron technique has enormous potential for 'cation spectroscopy' and it is applicable for any molecules and molecular complexes in supersonic jets. It should also be mentioned that individual molecular species or molecular complexes can be selectively ionized in REMPI experiments.

This compact threshold photoelectron analyzer makes it possible even to measure rotational structure of simple molecular cations such as NO+. If we have an appropriate VUV laser system, it is also possible to study higher excited electronic states of molecular cations from threshold photoelectron measurements. This is an advantage of photoelectron spectroscopy.

Literature Cited

1 (a) K. Kimura, Adv. Chem. Phys., 60, 161 (1985); (b) K. Kimura, Intern. Rev. Phys. Chem., 6, 195 (1987); (c) S. T. Pratt, P. M. Dehmer and J. L. Dehmer, in S. H. Lin (Ed.), Advances in Multi-Photon Processes and Spectroscopy, World Scientific, Singapore, 1988, Vol. 4, p 69; (d) R. N. Compton and J. C. Miller, in D. K. Evans (Ed.), Laser Applications in Physical Chemistry, Marcel Dekker, New York, 1989, p. 221.

2 A. Gedanken, M. B. Bobin, and N. A. Kuebler, *J. Phys. Chem.*, 84, 4096 (1982).

3 M. A. Duncan, T. G. Dietz, and R. E. Smalley, *Chem. Phys.*, 44, 415 (1979).

4 Y. Nagano, Y. Achiba, K. Sato, and K. Kimura, *Chem. Phys. Lett.*, 93, 510 (1982).

5 Y. Nagano, Y. Achiba, and K. Kimura, *J. Chem. Phys.*, 84, 1063 (1986).

6 Y. Nagano, Y. Achiba, and K. Kimura, *J. Phys. Chem.*, 90, 615 (1986).

7 Y. Nagano, Y. Achiba, and K. Kimura, *J. Phys. Chem.*, 90, 1288 (1986).

8 S. Niles, D. A. Prinslow, C. A. Wight, and P. B. Armentrout, *J. Chem. Phys.*, 93, 6186 (1990).

9 M. Takahashi, H. Ozeki, and K. Kimura, *Chem. Phys. Lett.*, 181, 255 (1991).

10 M. Takahashi and K. Kimura, *J. Chem. Phys.*, 97, 2920 (1992).

11 K. Okuyama, M. C. R. Cockett, and K. Kimura, *J. Chem. Phys.*, 97, 1649 (1992).

12 M. Takahashi, H. Ozeki, and K. Kimura, *J. Chem. Phys.*, 96, 6399 (1992).

13 K. Kimura and M. Takahashi, *"High-Brightness cm^{-1}-Resolution Threshold Photoelectron Spectroscopic Technique"*, in Proceedings of SPIE (Vol. 1638) on *"Optical Methods for Time- and State-Resolved Chemistry"*, Ed. by C.-Y. Ng (1992).

14 K. Müller-Dethlefs, M. Sander, and E. W. Schlag, *Chem. Phys. Lett.*, 112, 291 (1984).

15 (a) E. J. Bieske, M. W. Rainbird, 1. M. Atkinson, and A. E. W. Knight, *J. Chem. Phys.*, 91, 752 (1989); (b) E. J. Bieske, M. W. Rainbird, and A. E. W. Knight, *J. Chem. Phys.*, 94, 7019 (1991).

16 K. Yamanouchi, S. Isogai, S. Tsuchiya, and K. Kuchitsu, *Chem. Phys. Lett.*, 116, 123 (1987).

17 J. M. Hollas, M. R. Howson, and T. Ridley, *Chem. Phys. Lett.*, 98, 611 (1983).

RECEIVED February 10, 1993

Chapter 7

Photodissociation Dynamics of Metal Carbonyls
Branching Ratios and Bond Dissociation Energies

David M. Rayner, Yo-ichi Ishikawa[1], Carl E. Brown[2], and Peter A. Hackett

Steacie Institute for Molecular Sciences, National Research Council of Canada, 100 Sussex Drive, Ottawa, Ontario K1A 0R6, Canada

Recent progress in the study of the single photon photodissociation dynamics of transition metal carbonyls is reviewed with emphasis on the information regarding individual bond dissociation energies which can be obtained from fragment branching ratios. Approaches which determine CO fragment energy distributions directly are also discussed. Statistical models for the prediction of metal carbonyl fragment internal energy distributions are outlined along with a kinetic model which uses this information to predict product branching ratios. Techniques used to obtain experimental fragment yields are discussed and the application of the models to extract information on individual bond energies is demonstrated using experimental data from time-resolved infrared spectroscopic studies on group 6 metal carbonyl photolysis. Finally the role of excited electronic states in metal carbonyl photodissociation and its effect on the modeling are considered.

The photodissociation dynamics of transition metal carbonyls, $M(CO)_n$, is of intrinsic interest because it is representative of a class of reaction in which absorption of a single photon by a polyatomic molecule deposits sufficient energy to release more than one leaving group. This field of study is also of interest because it opens methods to approach the measurement of individual bond dissociation energies in both stable and coordinatively unsaturated organometallic species. Such thermochemical information is relatively scarce but much needed to quantify the organometallic chemistry of important areas such as gas-phase catalysis, metal refining and metal deposition. Here we review recent progress in understanding the photodissociation dynamics of metal carbonyls and present an assessment of the

[1]Current address: Kyoto Institute of Technology, Matsugasaki, Sakyoku, Kyoto, 606 Japan
[2]Current address: Environment Canada, 3439 River Road, Ottawa, Ontario K1A 0H3, Canada

0097–6156/93/0530–0096$06.00/0
Published 1993 American Chemical Society

extent to which thermochemical information can presently be obtained from the study of the ultraviolet photolysis of transition metal coordination compounds, particularly from branching-ratio measurements.

Approaches to the Photodissociation Dynamics of Metal Carbonyls

The study of the photodissociation dynamics of transition metal carbonyls has followed two complementary approaches involving either photofragment yield measurements or direct energy disposal measurements. Chemical-trapping measurements on $Fe(CO)_5$ *(1,2)* gave the initial indications that $M(CO)_n$ photolysis follows a sequential, CO elimination mechanism. Subsequently a series of time-resolved infrared spectroscopic (TRIS) studies which include $Fe(CO)_5$ *(3,4)*, $Cr(CO)_6$ *(5-7)*, $Mo(CO)_6$ *(7,8)*, $W(CO)_6$ *(7,9)*, $V(CO)_6$ *(10)*, $Ru(CO)_5$ *(11)* and $Os(CO)_5$ *(12)* has confirmed this and has led to the following general mechanism:

$$M(CO)_n + h\nu \rightarrow M(CO)_{n-1}{}^{**} + CO \tag{1a}$$

$$M(CO)_{n-1}{}^{**} \rightarrow M(CO)_{n-2}{}^{**} + CO \tag{1b}$$

$$M(CO)_{n-1}{}^{**} + Q \rightarrow M(CO)_{n-1} + Q \tag{1c}$$

.
.
.

$$M(CO)_{x+1}{}^{**} \rightarrow M(CO)_x{}^{*} + CO \tag{1d}$$

$$M(CO)_{x+1}{}^{**} + Q - M(CO)_{x+1} + Q \tag{1e}$$

$$M(CO)_x{}^{*} + Q - M(CO)_x + Q \tag{1f}$$

where Q is the buffer gas and $M(CO)_x{}^{**}$ and $M(CO)_x{}^{*}$ respectively indicate $M(CO)_x$ internally excited above and below its dissociation limit.

In all cases, the extent of fragmentation, its increase with photon energy and the spectroscopic observation of internally excited nascent $M(CO)_x$ fragments lead to the conclusion that the dynamics of each elimination step is largely determined by statistical considerations. Given the above mechanism it is clear that more detailed measurements of photofragment branching ratios could lead to further understanding of the dynamics. In particular the buffer gas pressure and temperature dependence of the ratios should offer a way to test models which describe the dynamics and kinetics and provide estimates for bond dissociation energies. Such a buffer gas pressure dependence was first observed for the ratio $[Cr(CO)_4]/[Cr(CO)_5]$ measured by TRIS in the 308 nm photolysis of $Cr(CO)_6$ *(7)*. Subsequently this study was expanded to include temperature as an experimental variable and was applied to additional examples in group 6 carbonyl photolysis *(13)*.

Details of the dynamics and kinetics modelling used in this work and its results are discussed below.

Direct energy disposal measurements in $M(CO)_n$ photolysis are an extension of the laser, molecular beam and mass-spectrometry techniques developed to probe the molecular dynamics of small molecules. Detailed information available from state-to-state and vector correlation measurements is not available for polyatomics of this size. What can be measured are $CO(v,J)$ and $CO(E_T)$ energy distributions from VUV LIF (14-16) and diode laser (17) experiments, and translational energy distributions in $M(CO)_x$ from molecular beam photofragment mass-spectroscopy (18). The measured distributions are necessarily a weighted superposition of the individual distributions associated with each operative elimination step in reaction 1. The extraction of dynamical information from these results can again only be approached by modelling. Branching ratios must be obtained in order to determine weighting factors. It should be noted that $CO(v,J)$ modes carry only a small fraction of the excess energy. Subtle effects in the exit channels may distort their distributions significantly from full statistical predictions without markedly altering the distribution in the major energy carrier, the internal modes of the polyatomic fragment. To this extent the two types of experiment emphasize different features of the dynamics, the branching ratios being determined by energy retention in the fragment and reflecting the grosser features of the dissociation whilst the direct measurements may be more sensitive to subtler features of the process.

Modelling $M(CO)_n$ Photodissociation

In the presence or absence of collisions energy disposal is a central factor in determining the extent to which the sequence of reactions 1 proceeds. Purely impulsive dynamical models, where large fractions of available energy are partitioned to translation are neither consistent with TRIS product fragment distributions (3-12) nor with direct CO LIF Doppler-width translational energy measurements (14, 16). Statistical models of some form are required to approach energy disposal in the photodissociation of $M(CO)_n$.

Statistical Models of the Dynamics of CO Elimination. The full statistical model in which equal probability is given to all product quantum states allowed by energy conservation is our starting point. This theory has been formulated for $CO(v,J)$ distributions from $M(CO)_n$ photolysis by several groups (14-17). Our interest in branching ratios has led us to reframe the model to predict directly the distribution of internal energy retained in $M(CO)_x$. Although this information is central to modelling the energy disposal in sequential reactions, it had previously only been dealt with using approximations.

Taking energy conservation restraints into consideration and allowing for angular momentum constraints by allowing two rotational modes of the reactant to be inactive, it is found (13) that, for any elimination step in the sequence, the probability, $G(E_I,E)$, of observing a given internal energy, E_I, in a fragment , at a total available energy, E, is given by

$$G(E_I,E) \propto N_I(E_I) \sum_{v=0}^{v_{max}} (E - E_I - E_v)^2 \,. \tag{2}$$

Here $N_I(E_I)$ is the density of states of the polyatomic fragment, v is the vibrational quantum number for CO, $E_v = vhv$, and v_{max} is the maximum value of v allowed under the requirement $E_v \leq E - E_I$. For completeness, although not required for analyzing the yield results, the CO rovibrational distributions are given by

$$G(v,J,E) \propto (2J + 1) \int_{E_I=0}^{E-E_v-E_J} N(E_I) (E - E_v - E_J - E_I) \, dE_I \,, \tag{3}$$

where J is the rotational quantum number for CO, and $E_J = B_{CO}J(J+1)$. The translational energy distribution is given by

$$G(E_T,E) \propto E_T^{1/2} \sum_{v=0}^{v_{max}} \sum_{J=0}^{J_{max}} (2J + 1) \int_{E_I=0}^{E-E_T-E_v-E_J} N(E_I) (E - E_T - E_v - E_J - E_I)^{-1/2} \, dE_I \tag{4}$$

where J_{max} is the maximum value J is allowed under the constraint $E_J \leq E - E_T - E_v$.

Kinetic Model. Aspects of the kinetics of this system are similar in many ways to those of chemical activation (*19,20*). $M(CO)_x{}^*$ starts with an internal energy distribution, $P_x(E)$, determined in the previous step. To find the complete nascent distribution in the product $M(CO)_{x-1}{}^*$, $P_{x-1}(E)$, we require the distribution of *reacting* molecules, $F_x(E)$ and $G_{x-1}(E_I,E)$, the distribution expected in $M(CO)_{x-1}{}^*$ for a fixed available energy E as given in equation 2. $P_{x-1}(E)$ is a convolution of these two functions,

$$P_{x-1}(E) = \int_E^{\infty} F_x(E') \, G_{x-1}(E,E') \, dE', \tag{5}$$

where E' is the total available energy in the reacting fragment. Under collision free conditions, or when $E \gg D_x$, the CO bond dissociation energy, reaction dominates deactivation,. Then $F_x(E)$ is simply $P_x(E)$ for $E > D_x$ and 0 elsewhere. In this case the yield of stabilized $M(CO)_x$, ϕ_x, is given simply by

$$\phi_x = \int_0^{D_x} P_x(E) \, dE. \tag{6}$$

In other cases calculation of $F_x(E)$ requires a knowledge of the energy dependent rate constant for CO elimination, $k(E)$, and a model to describe collisional deactivation. $k(E)$ can be calculated from RRKM theory which gives

$$k(E) = L W^+(E - D_x) / h N(E) \qquad (7)$$

where L is the reaction path degeneracy, $W^+(E - D_x)$ is the sum of states in the transition state up to an internal energy $(E - D_x)$ and $N(E)$ is the density of states in the reactant. Collisional activation is adequately described (13) by a step ladder model for weak collisions as developed for chemical activation (19, 20). The large internal energy allows up-transitions to be neglected. Taking down-transitions to have a probability of one for the transfer of a fixed energy $<E>$, and 0 elsewhere, $F_x(E)$ is given by

$$F_x(E) = \frac{k(E)}{k(E) + \omega} \int_{E'}^{\infty} P_x(E') \prod_{i=E/<E>}^{0} \frac{\omega}{k(i<E>) + \omega} dE', \qquad (8)$$

where ω is the collision frequency. The yield of stabilized $M(CO)_x$, ϕ_x, is then

$$\phi_x = 1 - \int_0^{\infty} F_x(E) \, dE. \qquad (9)$$

Sequential computational evaluation of equation 8 followed by equation 5 allows distributions and yields to be followed down the reaction sequence. $P_n(E)$, the starting distribution in the parent carbonyl, is taken as the thermal Boltzmann distribution shifted up by the energy of the absorbed photon, E_λ.

Input Parameters. The outcome of this model depends on a set of input parameters which reduces finally to: the photon energy, E_λ, and the temperature, T, through $P_n(E)$; the dissociation energies, D_x, and the associated A factors A_x through $k(E)$; the buffer gas pressure, p, through $\omega = Zp$, where Z is the collision number; and the vibrational frequencies used to compute the sums and densities of states required for $G(E_I, E)$ and $k(E)$. Vibrational frequencies for the parent carbonyls are available in the literature and provide adequate estimates for the fragment frequencies. Empirically, $k(E)$ is determined largely by the parameters D_x and A_x for this type of bond dissociation reaction and is virtually independent of details of the transition state. A_x is conveniently adjusted by tuning the energies of low vibrational and/or internal rotations in the transition complex. E_λ, p and T then form a set of independent experimental variables with which we can test the model against measured product branching ratios.

Branching Ratios

Figure 1 shows the predicted yields of $Fe(CO)_x$ fragments as a function of photolysis wavelength in the dissociation of $Fe(CO)_5$ using the full statistical model for the dynamics and a representative set of the input parameters D_x and A_x. The general features of metal carbonyl photodissociation are reproduced. The increase in fragmentation as E_λ is increased is well demonstrated. Large spreads in the extent of fragmentation at a particular E_λ are not predicted and this is in accord with almost all TRIS results.

Given photolysis wavelengths limited to those of excimer lasers the opportunity to observe branching is restricted. However in those cases where branching can be observed, the effect of varying p and T on the relative yields tests aspects of the model including its sensitivity to D_x and A_x and the energy disposal models. Depending where our confidence lies in prior knowledge of these factors, the results add to our understanding of the thermochemistry and/or dissociation dynamics of metal carbonyls.

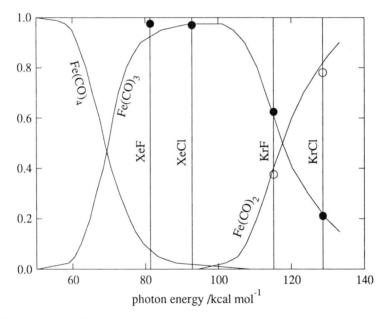

Figure 1. $Fe(CO)_x$ fragment yields as a function of photon energy. The solid lines are predicted yields, the circles show experimental yields for $Fe(CO)_3$, •,and $Fe(CO)_2$, ∘, estimated from TRIS studies (*3*, Ishikawa Y., Hackett P.A. and Rayner D.M., National Research Council Canada, unpublished data). Vertical lines mark excimer laser photon energies.

Measurement of Branching Ratios. Experimentally, branching ratios have been measured by chemical trapping (*1*), TRIS (*7,13*) and by molecular beam mass spectroscopy, MBMS (*21,22*). The chemical trapping work on $Fe(CO)_5$ reported wider fragmentation patterns than later found by TRIS (*3,4*). The problems of high trapping gas pressures, secondary photolysis and secondary reactions make it difficult to obtain reliable branching ratios using chemical trapping. TRIS does not suffer from these difficulties. The measurement of branching ratios from TRIS does require confidence in the assignment of infrared absorption features to particular $M(CO)_x$ species. Comparison with low temperature matrix spectra and "bootstrap" kinetic measurements with added CO allow this to be done with some confidence (*23*). To go from qualitative fragmentation patterns to quantitative branching ratios requires estimates for relative absorption coefficients. In most cases this can be achieved by comparing spectra taken under photolysis conditions which favour the production of the single fragments. MBMS is a third route to measuring fragment yields but the difficulties of electron impact fragmentation must be dealt with adequately. There is agreement between MBMS and TRIS studies in the 248 nm photolysis of $Cr(CO)_6$ (*21*), but not for $Mo(CO)_6$ photolysis at 248 nm (*22*). We report $Mo(CO)_4$ as the major product with some $Mo(CO)_3$ formation (*7,13*). The relative yield of $Mo(CO)_3$ goes up as the buffer gas pressure is decreased. We assigned absorption previously attributed to $Mo(CO)_5$ in an earlier TRIS study (*8*) to hot CO, based mainly on the absence of absorption attributable to $Mo(CO)_5$ in the 308 nm photolysis. Diode laser TRIS, which can distinguish hot CO absorptions (*4*) is required to confirm our assignment. Until then yields obtained by MBMS must be treated with some caution.

Bond Dissociation Energies from Branching Ratios. Figure 2 demonstrates the ability of the model to describe experimental $M(CO)_x$ fragment branching ratios and their dependence on buffer gas pressure and temperature. Using the statistical model for energy disposal, satisfyingly good agreement can be obtained by adjusting the parameters D_x and A_x. In addition a given set of D_x and A_x which reproduces the pressure dependence at one temperature also fits the higher temperature data without alteration. A reduced-statistical model which treats the eliminated CO as a quasi-atom fails to reproduce the experimental behavior.

   Sensitivity analysis shows that the branching ratios unfortunately do not define a unique set of D_x and A_x . However, if all but one of these parameters are known or can be reasonably estimated by other methods then the remaining parameter is established with reasonable precision (within $\pm$ 0.2 kcal mol^{-1} in some cases). As a starting point D_n, the dissociation energy for CO loss from the parent carbonyls, has been measured by infrared laser sensitized pyrolysis in several cases. A_x can be estimated by comparison with other diatomic-releasing bond-dissociation reactions and transition state theory (*24*). Taking D_6, A_6 and A_5 from this literature for the group 6 carbonyls we have estimated D_5 and D_4 from branching ratio measurements (*13*). The errors involved are systematic and depend on the reliability of both the input data and the statistical model. Despite thesequalifications the results do allow the evaluation of trends in individual bond dissociation energies in the group. For example, in the case of group 6 carbonyls it is found that for Mo and W individual binding energies are close to the average

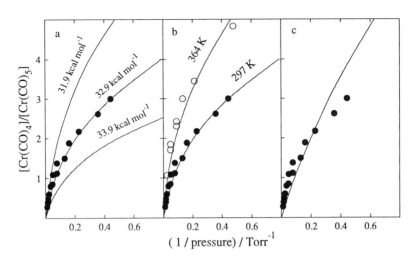

Figure 2. $[Cr(CO)_4]/[Cr(CO)_5]$ branching ratios in the 308 nm photolysis of $Cr(CO)_6$, showing the sensitivity of fit to: a) D_5; b) temperature; c) the photodissociation dynamics model.

values but for Cr the first three ligands are significantly less strongly bound than the last three. In conjunction with recent theoretical advances such information is helping extend our understanding of binding in metal carbonyls.

The Role of Excited States in $M(CO)_n$ Photodissociation

An area of uncertainty in $M(CO)_n$ photodissociation is the role of the initially excited state and the identity of the reacting state involved in the elimination of the first CO. It is common in the photodissociation of organometallics in solution to identify particular transitions as labilizing (25). Theoretical studies often draw correlations between individual reactant excited states and reaction pathways. In contrast many of the gas phase studies cited here point to a common reacting state, independent of the initially populated state. Wherever they have been measured CO translational and rotational distributions are not consistent with the impulsive elimination of CO from a dissociative excited state (14-17). Group 6 metal carbonyl fragmentation patterns show no correlation with the initially excited state (7). The dynamics experiments on $Fe(CO)_5$ (14), $Ni(CO)_4$ (15), $Cr(CO)_6$ (13) and $Mo(CO)_6$ (13,16) all require that statistical dynamics dominates energy disposal in the first elimination step as well as the subsequent ones. In this respect the reacting state has the character of a highly internally excited ground state such as would be formed by internal conversion. The reacting state does not necessarily correlate directly with the fragment ground state. In the case of $Fe(CO)_5$ (4) and $Ni(CO)_4$ (15) there is now evidence that excited electronic states of fragments can be involved. The participation of an excited electronic state of $Ni(CO)_3$ in $Ni(CO)_4$ photolysis is required to predict $CO(v,J)$ distributions using the statistical model (15). Also diode laser TRIS has now revealed that electronically excited $Fe(CO)_3$

and $Fe(CO)_4$ are produced in the 248 and 351 nm photolysis, respectively of $Fe(CO)_5$ (4). Both excited fragments are rapidly converted to the ground state by collisions with buffer gas. It is not clear from the present results if the excited electronic state pathway is the only route or if branching between excited and ground state pathways occurs as has been suggested for $Ni(CO)_4$. Either pathway can be supported by the statistical model in the absence of firm information on D_x and the energies of the accessible electronic states of $Fe(CO)_x$ species.

Conclusions

We have reviewed a large body of evidence which points to the dominance of statistical dynamics in the photodissociation of transition metal carbonyls. This is strongly reinforced by statistical modeling of product fragment yields obtained in the bulk gas-phase. These measurements are well matched in information content to our present level of understanding of the dynamics of metal carbonyl photodissociation. Subtler aspects of the dynamics and potential energy surfaces are revealed by direct $CO(v,J)$ measurements but will require much further work to understand in detail. Acceptance of the statistical approach opens the way to extracting information on individual bond dissociation energies from branching ratio measurements. These measurements describe a surface of acceptably-defined molecular parameters, including the bond dissociation energies and related A factors. Confidence in one particular parameter extracted from this information depends on our confidence in the rest of the set. At the present time this confidence is sufficient to allow us to evaluate trends in CO binding as a function of the degree of unsaturation and the identity of the central metal atom.

Literature Cited

1. Nathanson, G.; Gitlin, B.; Rosan, A. M.; Yardley, J. T. *J. Chem. Phys.* **1981**, *74*, 361.
2. Yardley, J.T.; Gitlin, B.; Nathanson, G.; Rosan, A. M. *J. Chem. Phys.* **1981**, *74*, 370.
3. Seder, T. A.; Ouderkirk, A. J.; Weitz, E. *J. Chem. Phys.* **1986**, *85*, 1977.
4. Ryther, R. J.; Weitz E. *J. Phys. Chem.* **1991**, *95*, 9841.
5. Seder, T. A.; Church, S. P.; Weitz, E. *J. Am. Chem. Soc.* **1986**, *108*, 4721.
6. Fletcher, T. R.; Rosenfeld, R.N. *J. Am. Chem. Soc.* **1985**, *107*, 2203.
7. Ishikawa, Y.; Brown, C. E.; Hackett, P. A.; Rayner, D. M. *J. Phys. Chem.* **1990**, *94*, 2404.
8. Ganske, J. A.; Rosenfeld, R. N.; *J. Phys. Chem.* **1989**, *93*, 1959.
9. Ishikawa, Y.; Hackett, P. A.; Rayner D. M. *J. Phys. Chem.* **1988**, *92*, 3863.
10. Ishikawa, Y.; Hackett, P. A.; Rayner D. M. *J. Am. Chem. Soc.* **1987**, *109*, 6644.
11. Bogdan, P. L.; Weitz, E. *J. Am. Chem. Soc.* **1989**, *111*, 3163.
12. Bogdan, P. L.; Weitz, E. *J. Am. Chem. Soc.* **1990**, *112*, 639.
13. Rayner D. M.; Ishikawa, Y.; Brown, C. E.; Hackett, P. A. *J. Chem. Phys.* **1991**, *94*, 5471.

14. Waller, I. M.; Hepburn, J. W. *J. Chem. Phys.* **1988**, *88*, 6658.
15. Schlenker, F. J.; Bouchard, F.; Waller, I. M.; Hepburn, J. W. *J. Chem. Phys.* **1990**, *93*, 7110.
16. Buntin, S. A.; Cavanagh, R. R.; Richter, L. J.; King, D. S. *J. Chem. Phys.* **1991**, *94*, 7937.
17. Holland, J. P.; Rosenfeld, R. N. *J. Chem. Phys.* **1988**, *89*, 7217.
18. Venkataraman, B.; Hou, H.; Zhang, Z.; Chen, S.; Bandukwalla, G.; Vernon, M. *J. Chem. Phys.* **1990**, *92*, 5338.
19. Robinson, P. J.; Holbrook, K. A. *Unimolecular Reactions*; Wiley-Interscience: New York, 1972.
20. Forst, W. *Theory of Unimolecular Reactions*; Academic: New York, 1973.
21. Tyndall, G. W.; Jackson, R. L.; *J. Chem. Phys.* **1989**, *91*, 2881.
22. Tyndall, G. W.; Jackson, R. L.; *J. Phys. Chem.* **1991**, *95*, 687.
23. Weitz, E. *J. Phys. Chem.* **1987**, *91*, 3945.
24. Lewis, K. E.; Golden, D. M.; Smith, G. P. *J. Am. Chem. Soc.* **1984**, *106*, 3905.
25. Geoffroy, G. L.; Wrighton, M. S. *Organometallic Photochemistry*; Academic: New York, 1979.

RECEIVED January 13, 1993

Chapter 8

Site-Specific Laser Chemistry Relevant to III–V Semiconductor Growth

Xiaodong Xu, Jeffrey L. Brum, Zhongrui Wang, Subhash Deshmukh, Yu-Fong Yen, and Brent Koplitz

Department of Chemistry, Tulane University, New Orleans, LA 70118

Despite its toxicity, arsine remains a "workhorse" in the field of III-V semiconductor research because it provides a relatively clean (i.e. carbon-free) means of delivering atomic As to a growth surface. However, the feasibility of using alternative arsenic sources is currently being investigated by the semiconductor community. Two liquid metalorganic alternatives, triethylarsenic and monoethylarsine, contain carbon that can appear as a significant impurity in the semiconductor material itself. We present results on the laser-induced photochemistry of the above two compounds as well as monoethylamine, a second row analog. Our efforts are focused on using lasers to generate and detect atomic hydrogen, a species that is known to be a good radical scavenger in many semiconductor growth environments. Experiments on selectively-deuterated group V compounds are used to identify primary photolytic pathways for H-atom production.

This presentation is concerned with the laser-induced photochemistry of several metalorganic precursor compounds associated with the fabrication of III-V semiconductor materials. In particular, we are interested in discovering and controlling processes that lead to the generation of atomic hydrogen. From a fundamental perspective, the goal is to determine the site(s) that result in H-atom production, and selectively-deuterated compounds allow us to investigate this site-specific photochemistry. In practice, H atoms can be important as alkyl radical scavengers in various semiconductor growth environments. Of particular interest in our studies are ethyl-containing metalorganic compounds such as triethylarsenic (TEAs) and monoethylarsine (MEAs). Experimentally, we ask the following questions. In what manner does excimer laser radiation (193, 222, or 248 nm) interact with TEAs or MEAs? Can we identify and quantify the important processes involved, especially with respect to H-atom production? Can we alter or influence the various photolysis pathways, either by manipulating the photolysis laser or introducing some additional radiation (e.g., another excimer laser beam)? To be of direct relevance, it is important to perform these studies using convenient excimer lines, but excitation at other wavelengths may also prove beneficial. Two main goals of our work are (**1**) to measure the extent to which primary 193 nm excimer laser photolysis results in significant H-atom production and (**2**) to determine the degree of

0097–6156/93/0530–0106$06.00/0
© 1993 American Chemical Society

enhancement achievable if 248 nm radiation is introduced in the form of a second photolysis laser pulse. Ultimately, we hope to correlate our findings with changes in the properties of III-V semiconductor films.

Practical Motivation

Metalorganic compounds are routinely used to grow III-V semiconductor materials by methods such as metalorganic chemical vapor deposition (MOCVD) or metalorganic molecular beam epitaxy (MOMBE), also called chemical beam epitaxy (1-7). Unfortunately, perpetual problems are encountered (e.g. intolerable carbon incorporation, lattice mismatches, slow growth rates, etc.) that have prompted many researchers to seek alternative ways of fabricating these materials. In the field of photo-assisted growth, an active area of research has involved using excimer lasers to alter growth conditions, hopefully "steering" the deposition process in a favorable direction (8-14). Mechanistically speaking, chemical change can be brought about either photochemically or pyrolytically, and progress has been made toward understanding the role of laser radiation in growth environments (8-14). However, many fundamental *and* practical questions remain. It is apparent that understanding the gas-phase photochemistry of the compounds of interest can guide one toward selecting the laser conditions that will produce the maximum benefits for materials growth.

Because they are liquids and not high pressure gases, compounds such as TEAs, MEAs, and tertiarybutylarsine (TBAs) have emerged as replacement candidates for arsine (15-19). However, the presence of the alkyl group can still create problems, for example the introduction of unwanted carbon impurites into the bulk material (17). Consider TEAs as an arsine alternative used to grow GaAs. If ethyl radicals become trapped on the GaAs growth surface, they can eventually appear as unwanted C atoms within the bulk material. One mechanism by which this probability is reduced involves the scavenging of ethyl radicals by neutral H atoms to form ethane, a molecule which should have a lower sticking coefficient than the ethyl radical. As an approach to alleviate the carbon problem, one might use a laser to dissociate TEAs in the gas phase. Is it possible to photolytically transform a C_2H_5 photoproduct into $C_2H_4 + H$? Presumably, ethylene would be less likely to stick to the surface in the first place, resulting in a "cleaner" growth environment. Moreover, H atoms would be available to scavenge alkyl radicals. A depiction of this process is shown in Figure 1. Here, a simplistic representation is shown that contrasts the differences for semiconductor growth when using unphotolyzed TEAs or photolyzed TEAs. Ideally, the goal would be to convert ethyl radicals to ethylene and atomic hydrogen *before* they impinge upon a growth surface.

Experimental Approach

A schematic diagram of the experimental setup is shown in Figure 2. Photolysis of the molecule of interest is achieved using an excimer laser (operational at a variety of wavelengths), while an excimer-pumped dye laser serves as the probe. In addition to tunable visible and ultraviolet radiation, frequency-tripling of the dye laser output in a cell containing krypton allows tunable vacuum ultraviolet (VUV) radiation to be produced. A second excimer laser is available as an additional source of photolysis photons. For detection, a time-of-flight mass spectrometer (TOFMS) is used. Samples are introduced via molecular leak in the current studies.

Several features should be stressed that make this experimental system well-suited for investigating the photolysis of metalorganic compounds. First of all, the

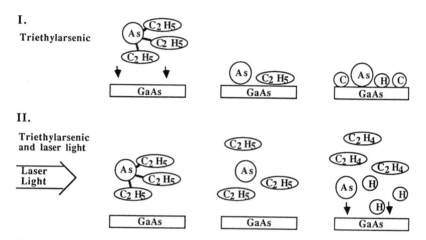

Figure 1. Possible positive effects of photolysis on the carbon incorporation problem in III-V semiconductor growth. Case I shows the incorporation of carbon impurities resulting from the use of TEAs, without laser irradiation. Case II shows the effects of laser photolysis of TEAs, leading to the formation of ethylene and atomic hydrogen.

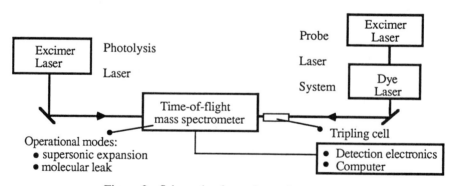

Figure 2. Schematic of experimental apparatus.

availability of tunable VUV radiation means that H atoms (as well as D atoms) can be readily detected through two-frequency, two-photon ionization via Lyman-α (*20,21*). As will be shown, having the ability to detect atomic hydrogen is essential to our research. Secondly, the output of an excimer laser often produces extensive fragmentation and/or ionization, and the presence of a mass spectrometer provides a convenient means of identifying a particular fragment. If ions are formed (either the parent or fragments), their detection by the TOFMS is straightforward.

Results and Discussion

As we have shown previously, 193 nm photolysis of TEAs will produce a substantial H-atom signal, but the photolysis power "window" for maximal H-atom production is rather narrow (*22*). Too little photolysis power will not induce sufficient chemistry, yet too much photolysis power will result in two-photon ionization of the parent molecule. This ionization process comes at the expense of neutral fragmentation pathways as measured by H-atom production. For neutral fragmentation, the initial step will presumably involve ejection of an ethyl radical via the reaction

(I) $As(C_2H_5)_3 + h\nu$ (193 nm) $\rightarrow As(C_2H_5)_2 + C_2H_5$

although another possible route involves the concerted mechanism in which a β-hydrogen atom is transferred to the As atom and ethylene is ejected:

(II) $As(C_2H_5)_3 + h\nu$ (193 nm) $\rightarrow AsH(C_2H_5)_2 + C_2H_4$

Reaction (II) is thought to play a role in the pyrolytic decomposition of TEAs (*23*), but mechanistically, if excitation is followed *promptly* by dissociation, reaction (I) will dominate. In general, one of two reactive events can occur for a nascent ethyl radical: dissociation or photon absorption. Depending on the photolysis conditions and the nature of the precursor molecule, both events are possible. To examine the interplay between these two processes, we have studied the photolysis of simplified ethyl precursors, specifically haloethanes.

Site-Specific Haloethane Photochemistry. Although the primary focus of this work is on the photochemistry of ethyl-containing metalorganic compounds, investigating haloethanes can provide additional insight into the photochemistry of the ethyl radical itself. When studying H-atom production in ethyl-containing systems, it seems reasonable to ask the following simple question. Which site(s) produces the H atoms? In the case of the 193 nm dissociation of ICD_2CH_3, there is no obvious evidence for dissociation occurring at a particular carbon atom (*24*). The observed ratio (1.6:1) is consistent with the initial H:D ratio of 3:2 present in the parent molecule. However, when the wavelength for ICD_2CH_3 photolysis is switched to 248 nm, the situation is remarkably different. In this case, the H/D ratio is ~11:1, clearly indicating a preference for bond cleavage at the β carbon (*24*). Even if one chooses to adjust the ratio by a factor of 1.5 for simple statistical bias, the ratio is still 7:1. By using ICH_2CD_3, i.e. the H and D positions have been reversed, we have also demonstrated that the dissociation process is indeed carbon-atom specific and not due solely to some type of H-atom dynamics (*24*).

H-atom power dependence studies on ICH_2CH_3 at 248 nm yield a log-log plot that is clearly nonlinear, the slope being ~1.7 (25). This finding is consistent with the argument that given various energy considerations (e.g. the $I-CH_2CH_3$ bond energy and the $I-CH_2CH_3$ dissociation energetics), absorption of a single 248 nm photon by iodoethane will not sequentially break both a C-I and a C-H bond in most cases. For the majority of C-I dissociation events, there simply will not be enough internal energy deposited in the ethyl radical to overcome the barrier (~1.7 eV) to H-atom production (25). The logical candidate for secondary absorption is the ethyl radical, which possesses a reasonable absorption cross section (2×10^{-18} cm^2) at 248 nm (26).

Two-color studies on chloroethane photolysis can probe the importance of photon absorption by the ethyl radical in producing site-specific behavior. In contrast to ICH_2CH_3, $ClCH_2CH_3$ has no appreciable absorption cross section at 248 nm (27). However, 193 nm light *will* access the first electronic state of $ClCH_2CH_3$. (Note that this electronic state is directly dissociative, analogous to the A-state transition observed at 248 nm in ICH_2CH_3 and other iodoalkanes) (27). The ethyl radical is thought to be a major product in the 193 nm photolysis of chloroethane (28). If our understanding of the overall mechanism for H-atom generation in these haloethane systems is correct for 248 nm photolysis, one should be able to generate CH_2CH_3 by photolyzing $ClCH_2CH_3$ with 193 nm radiation and use a second laser operating at 248 nm to enhance H-atom production via CH_2CH_3 photolysis. Since CH_2CH_3 has a reasonable absorption cross section at 248 nm as stated above, H-atom enhancement is certainly a logical expectation. Such a two-color experiment has been conducted in our laboratory (21). An initial 193 nm photolysis pulse results in cleavage of the C-Cl bond in $ClCH_2CH_3$, while a second pulse at 248 nm does, in fact, dramatically enhance H-atom production. Furthermore, experiments on $ClCH_2CD_3$ and $ClCD_2CH_3$ clearly show that this enhancement occurs preferentially via bond cleavage at the β carbon site, and it is apparent that 248 nm photon absorption by the ethyl radical is an important step in the overall process.

In a related area, the experiments on iodoethane can be extended to 1-iodopropane. Is the site preference exhibited by iodoethane at 248 nm a β carbon effect, or is it a terminal carbon effect? To address this question, the photolysis of three selectively-deuterated 1-iodopropanes has been studied (29). In each case, a specific carbon site has been labelled α, β, or γ with respect to the I atom. As is the case with iodoethane, atomic hydrogen production resulting from the 193 nm photolysis of 1-iodopropane displays no dominant site preference. In contrast, 248 nm excitation clearly produces a propensity for carbon-hydrogen bond fission at the β position, not the terminal (γ) site. The computed site-specific branching ratios in 1-iodopropane are 0.2, 0.7, and 0.1 for the α, β, and γ sites, respectively. Once again, it is apparent that photolysis of an intermediate (presumably the propyl radical) is important (29).

Group V Photochemistry. In order to advance our understanding of group V photochemistry, we have been studying $C_2H_5NH_2$ and $C_2H_5AsH_2$ photolysis with respect to H-atom production (30,31). Preliminary results on the 193 nm photolysis of both $C_2H_5NH_2$ and $C_2H_5AsH_2$ are very encouraging. Our work on 193 nm $C_2H_5NH_2$ photolysis has shown that a large H-atom signal can be readily induced (30). Which site(s) was producing the H atom? (Note that the N-H bond is ≥ 90 kcal/mol, while the C-N bond is only 72 kcal/mol) (32). The experimental procedure was simply to photolyze $C_2H_5ND_2$, for example, and look for H versus D atom formation. At lower laser powers ($\leq$ several MW/cm^2), virtually all atomic hydrogen results from N-D bond cleavage *(30)*. With sufficient laser power, one can see the C-H channel "turn on." Power dependence studies indicate that a two-photon process is responsible for H-atom production, while N-D bond cleavage is produced through a one-photon route. In the sense of absolute H-atom production referenced to NH_3 (where the quantum yield is ~1.0), H-atom generation from $C_2H_5NH_2$ is ~30-50% that resulting from NH_3. (Note that differences in absorption cross sections have been taken into account.) Consequently, one can conclude that N-H bond cleavage is fairly competitive with C-N bond cleavage, despite the difference in bond strength *(32)*.

Results on the 193 nm photolysis of $C_2H_5AsH_2$ are shown in Figure 3, and it is clear from the broadening of the H-atom Doppler profile that a second H-atom channel is "turning on" dramatically at higher laser powers *(31)*. Since the H-atom profile is broadening significantly, it is likely that the important intermediate photolysis step involves irradiation of AsH_2. Studies are currently being planned that will involve the selectively-deuterated compounds $C_2D_5AsH_2$ and $C_2H_5AsD_2$. In this case, *positive* identification of the reactive site is possible. It must also be noted that the absolute amount of H-atom production resulting from MEAs photolysis is comparable to that for 193 NH_3 photolysis, where the H-atom quantum yield is ~1.0. Consequently, the prospects for inducing macroscopic H-atom production in MEAs growth environments are encouraging.

Enhancing H-atom Production. To influence the overall photochemistry of $C_2H_5NH_2$, we propose to use two photolysis lasers operating at 193 and 248 nm. By using this additional 248 nm photon source, one can alter the 193-nm induced H/D ratio that results from $C_2H_5ND_2$ photolysis *(33)*. As shown in our recent two-color experiment on $ClCH_2CH_3$, H-atom production from the ethyl radical can be enhanced by using 248 nm radiation *(21)*. By using this second photolysis source operating at 248 nm, we can also photolytically increase H-atom production from a C_2H_5 radical produced via the 193 nm photolysis of $C_2H_5ND_2$ *(33)*.

Significant H-atom enhancement is also possible with TEAs and MEAs *(33)*. Figure 4a shows H-atom Doppler profiles at Lyman-α produced via TEAs photolysis using three different photolysis combinations. Clearly, H-atom formation arising from a combination of 193 and 248 nm laser pulses is enhanced significantly when compared with H-atom generation resulting from the sum of the two photolysis lasers used separately. Since the absorption cross sections for TEAs at 193 nm and 248 nm are reasonable (1.76 x 10^{-17} cm^2 and 1.2 x 10^{-18} cm^2, respectively) *(34)*, it is not surprising that H-atom formation is observed in each individual case. However, the net H-atom signal is enhanced by a factor of 5 when compared with the H-atom signal resulting from the sum of the two lasers acting alone. We submit that in the case of TEAs, the desired reaction ($C_2H_5 \rightarrow C_2H_4 + H$) is being photolytically enhanced by the 248 nm laser pulse.

In contrast to TEAs, MEAs has direct H-atom formation channels available. Nonetheless, we still expect that 193 nm excitation of MEAs will induce significant As-C dissociation in addition to As-H bond cleavage, since the As-C bond is significantly weaker than the As-H bond *(35)*. If As-C bond cleavage does in fact occur, then ethyl radicals will be produced. Will 248 nm radiation enhance H-atom formation following 193 nm MEAs photolysis in a fashion similar to the case of 193 nm TEAs photolysis? The results of such an experiment can be found in Figure 4b. When compared with the TEAs experiment, the main difference for MEAs is the fact

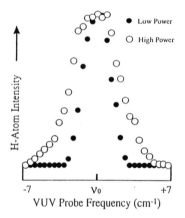

Figure 3. H-atom Doppler profiles resulting from the 193 nm photolysis of $C_2H_5AsH_2$ at two photolysis powers. The larger photolysis power has clearly activated a second H-atom channel, as evidenced by the Doppler broadening. Here, $v_0 = 82{,}259.1$ cm^{-1}.

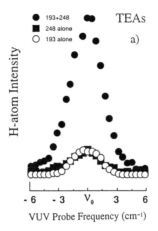

Figure 4a. H-atom Doppler profiles resulting from $(C_2H_5)_3As$ photolysis using three different excitation schemes: 193 nm alone, 248 nm alone, and 193 nm followed by 248 nm. It is clear that significant enhancement of the H-atom signal occurs when both lasers are used. Here, v_0 is equal to Lyman-α (82,259 cm^{-1}).

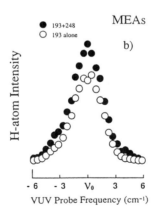

Figure 4b. H-atom Doppler profiles resulting from $C_2H_5AsH_2$ photolysis using two different excitation schemes: 193 nm alone and 193 nm followed by 248 nm. (When only 248 nm radiation was used, there was no discernible H-atom signal.) It is clear that some enhancement of the H-atom signal occurs when both lasers are used. Here, v_0 is equal to Lyman-α (82,259 cm^{-1}).

that 248 nm radiation alone produced no discernible H-atom signal, probably due to a low absorption cross section at this wavelength. However, the H-atom signal is enhanced (albeit slightly) when 193 nm and 248 nm pulses are used together as opposed to 193 nm radiation acting alone. The actual enhancement is by ~50%. If this enhancement is occurring via the ethyl radical, then we are achieving the desired photochemistry. Once again, experiments with selectively-deuterated MEAs compounds should be particularly useful for addressing this issue. Finally, we re-emphasize that understanding the photochemistry of group V ethyl-containing compounds may ultimately prove important in understanding and/or enhancing III-V semiconductor growth processes that utilize metalorganic compounds. In this area, MEAs is a particularly attractive candidate since it contains only one ethyl group.

Acknowledgments

Acknowledgment is made to the Donors of the Petroleum Research Fund, administered by the American Chemical Society, for partial support of this research. We also acknowledge support by the Louisiana Board of Regents, the Center for Bioenvironmental Research at Tulane University, and the Department of Energy via its NIGEC program. Special thanks go to Shannon Koplitz for presentation assistance.

Literature Cited

1. Manasevit, H.M. *J. Cryst. Growth* **1981**, *55*, 1.
2. Dapkus P.D.; Manasevit H.M.; Hess K.L.; Low T.S.; Stillman G.E. *J. Cryst. Growth* **1981**, *55*, 10.
3. Tsang W.T. *Appl. Phys. Lett.* **1984**, *45*, 1234.
4. Tsang W.T.; Chiu T.H.; Cunningham J.E.; Robertson A. *Appl. Phys. Lett.* **1987**, *50*, 1376.
5. Tsang W.T.; Chiu T.H.; Cunningham J.E.; Robertson A. *J. Appl. Phys.* **1987**, *62*, 2302.
6. Robertson A.; Chiu T.H.; Tsang W.T.; Cunningham J.E. *J. Appl. Phys.* **1988**, *64*, 87.
7. Putz N.; Heinecke H.; Heyen M.; Balk P.; Weyers M.; Luth H. *J. Cryst. Growth* **1986**, *74*, 292.
8. Kukimoto H.; Ban Y.; Komatsu H.; Takechi M.; Ishizaki M. *J. Cryst. Growth* **1986**, *77*, 223.
9. Nishizawa J.; Kurabayashi T.; Hoshina J. *J. Electrochem. Soc.* **1987**, 502.
10. McCaulley J.A.; McCrary V.R.; Donnelly V.M. *J. Phys. Chem.* **1989**, *93*, 1148.
11. Donnelly V.M.; McCrary V.R.; Appelbaum A.; Brasen D.; Lowe W.P. *J. Appl. Phys.* **1987**, *61*, 1410.
12. Tu C.W.; Donnelly V.M.; Beggy J.C.; Baiocchi F.A.; McCrary V.R.; Harris T.D.; Lamont M.G. *Appl. Phys. Lett.* **1988**, *52*, 966.
13. Donnelly V.M.; Brasen D.; Appelbaum A.; Geva M. *J. Appl. Phys.* **1985**, *58*, 2022.
14. Nishizawa J.; Kurabayashi T.; Abe H.; Sakurai N. *J. Vac. Sci. Technol. A* **1987**, *5*, 1572.
15. Chen C.H.; Larsen C.A.; Stringfellow G.B. *Appl. Phys. Lett.* **1987**, *50*, 219.
16. Lum R.M.; Klingert J.K.; Lamont M.G. *Appl. Phys. Lett.* **1987**, *50*, 284.
17. Lum R.M.; Klingert J.K.; Wynn A.S.; Lamont M.G. *Appl. Phys. Lett.* **1988**, *52*, 1475.
18. Hummel S.G.; Zou Y.; Beyler C.A.; Grodzinski P.; Dapkus P.D.; McManus

J.V.; Zhang Y.; Skromme B.J.; Lee W.I. *Appl. Phys. Lett.* **1992**, *60*, 1483.
19. Speckman D.M.; Wendt J.P. *Appl. Phys. Lett.* **1990**, *56*, 1134.
20. Zacharias H.; Rottke H.; Danon J.; Welge K.H. *Opt. Commun.* **1981**, *37*, 15.
21. Brum J.L.; Deshmukh S.; Koplitz B. *J. Chem. Phys.* **1991**, *95*, 2200.
22. Xu X.; Deshmukh S.; Brum J.L.; Koplitz B. *Appl. Phys. Lett.* **1991**, *58*, 2309.
23. DenBaars S.P.; Maa B.Y.; Dapkus P.D.; Melas A. *J. Electrochem. Soc.* **1990**, *138*,2068.
24. Brum J.L.; Deshmukh S.; Koplitz B. *J. Chem. Phys.* **1990**, *93*, 7504.
25. Deshmukh S.; Brum J.L.; Koplitz B. *Chem. Phys. Lett.* **1991**, *176*, 198.
26. Arthur N. *J. Chem. Soc. Faraday. Trans.* **1986**, *82*, 1057.
27. Robin M.B. *Higher Excited States of Polyatomic Molecules*; Academic Press: New York, NY, 1974; Vol. 1.
28. Kawasaki M.; Kasatani K.; Sato H.; Shinohara H.; Nishi N. *Chem. Phys.* **1984**, *88*, 135.
29. Brum J.L.; Deshmukh S.; Koplitz B. *J. Phys. Chem.* **1991**, *95*, 8676.
30. Xu X.; Deshmukh S.; Brum J.L.; Koplitz B. *Chem. Phys. Lett.* **1992**, *188*, 32.
31. Xu X.; Brum J.L.; Deshmukh S.; Koplitz B. *J. Phys. Chem.* **1992**, *96*, 2924.
32. Sanderson R.T. *Chemical Bonds and Bond Energy*; Academic Press: New York, NY, 1971, 153.
33. Xu X.; Wang Z.; Brum J.L.; Yen Y.-F.; Deshmukh S.; Koplitz B. *J. Phys. Chem.* **1992**, *96*, 7048.
34. McCrary V.R.; Donnelly V.M. *J. Cryst. Growth* **1987**, *84*, 253.
35. Berkowitz J. *J. Chem. Phys.* **1988**, *89*, 7065.

RECEIVED January 6, 1993

Chapter 9

UV Photodissociation and Energy-Selective Ionization of Organometallic Compounds in a Molecular Beam

Jeffrey A. Bartz, Terence M. Barnhart, Douglas B. Galloway,
L. Gregory Huey, Thomas Glenewinkel-Meyer, Robert J. McMahon,
and F. Fleming Crim

Department of Chemistry, University of Wisconsin—Madison,
Madison, WI 53706

This chapter introduces the application of molecular beams, ultraviolet lasers, energy-selective ionization, and time-of-flight mass spectrometry to the general study of the Laser Chemistry of Organometallics. It briefly contrasts energy-selective vacuum ultraviolet ionization and time-of-flight mass spectrometry with other laser methods used in investigating the photodecomposition of organometallic molecules. After a description of the experimental apparatus, a summary of results from photodissociation studies of dicarbonyl(η^5-cyclopentadienyl)(1-propyl)iron appears.

LASERS AND ORGANOMETALLIC MOLECULES

Lasers can provide useful information about the primary photodissociation pathways of organometallic compounds. Techniques such as chemical trapping, transient infrared spectroscopy, and mass spectrometry explore the photolyses of organometallic molecules in a "solvent-free" environment. These gas phase investigations generate information on energy disposal and fragmentation pathways for volatile organometallic systems.

A notable example of early work on the gas phase photochemistry of organometallic molecules is that of Yardley and coworkers ($1,2$). They used PF_3 to trap the fragments from the 248-nm photodissociation of $Fe(CO)_5$. They found a distribution of coordinatively unsaturated iron carbonyl fragments, but, because collisions between different iron carbonyl fragments randomize the extent of unsaturation, they did not unambiguously determine the detailed distribution of the initial fragments. Direct spectroscopic methods can also detect the primary photodissociation fragments from organometallic compounds. Weitz and coworkers studied the photodissociation of $Fe(CO)_5$ by time-resolved infrared spectroscopy (TRIS) ($3,4$) and observed the CO stretching frequencies of fragments formed by ultraviolet (UV) laser photolysis. They detected the same fragments trapped by Yardley and coworkers ($1,2$).

Another method for detecting products from the laser photodissociation of organometallic molecules is mass spectrometry using electron impact or multiphoton

0097–6156/93/0530–0116$06.00/0
© 1993 American Chemical Society

ionization. The ions observed in a mass spectrum help identify the primary photodissociation pathways of the molecules in a *collision-free* environment. As one example, Mikami et al. (*5*) used a molecular beam, multiphoton ionization (MPI), and time-of-flight mass spectrometry (TOFMS) to detect metal atoms and other fragments from the photodissociation of a series of metal acetylacetonates. Multiphoton ionization is a useful technique for identifying products, although many of the species come from secondary photolyses. Thus, the observed masses reflect more than just the primary photodissociation event. A more conventional ionization method, electron impact, can detect all products from a photodissociation of an organometallic compound. For example, Vernon and coworkers studied the 248-nm photodissociation of $Fe(CO)_5$ (*6*), $Zn(C_2H_5)_2$ (*7*) and $M(CO)_6$ (*8*) (M = Cr, Mo, W) by electron impact mass spectrometry. The typical electron energies used for mass spectrometry both ionize the primary photodissociation products and extensively fragment those same ions.

We have implemented an energy-selective method that avoids many of the complications of secondary dissociations in the ionization step of mass spectrometric detection. We use vacuum ultraviolet (VUV) ionization and TOFMS to identify the primary fragments from the photodissociation of a variety of molecules, including organometallic compounds. This method is energy-selective as it ionizes only the chemical species with ionization potentials of 9.9 eV or lower, which for most of them is near their ionization threshold. Because these ions have little excess energy from the VUV photon, they undergo secondary dissociations less extensively compared to those formed by MPI or electron impact ionization. Our approach is similar to the single photon ionization work of Van Bramer and Johnston (*9*), although we generate our VUV light differently. VUV ionization coupled with UV laser photodissociation and TOFMS is a valuable tool for determining decomposition pathways in a collision-free environment and is a means of identifying the important intermediates in chemical vapor deposition or in condensed phase photolysis.

METHOD

Figure 1 shows a portion of our experimental molecular beam apparatus, which has two differentially pumped chambers separated by a skimmer. The organometallic compound resides in a sample holder, inside the molecular beam source chamber, approximately 5 cm behind the nozzle of the pulsed valve. Thermal coaxial cable heats this holder, which has an inlet for the external carrier gas, He, and an outlet that connects to the nozzle. Typically, the source chamber operates at a temperature of 80 °C and the carrier gas pressure is about 300 torr. The nozzle pulses at 20 Hz and produces 1 μs-long bursts of gas. The valve sits in an aluminum holder, wrapped in thermal coaxial cable. Typically, the pulsed valve operates at 90 °C, a higher temperature than the sample holder, to reduce condensation or sublimation of the organometallic compound in the nozzle assembly. The seeded carrier gas travels out of the nozzle and expands, producing a molecular beam that passes through a skimmer and travels into the interaction region, about 3 cm from the nozzle. In the interaction region, 280-nm UV light crosses the beam of organometallic molecules and photolyzes it. After a short time delay (0-1000 ns), 125-nm VUV light ionize the resulting photofragments. An extraction field accelerates the ions into the field-free region of a TOFMS where they separate by mass. To remove any contributions from either the UV or VUV photons alone, the mass spectrum presented later is the result of subtracting the UV-only and VUV-only signal contributions from the signal acquired by the method described above. In general, these background contributions are negligible in the mass region presented.

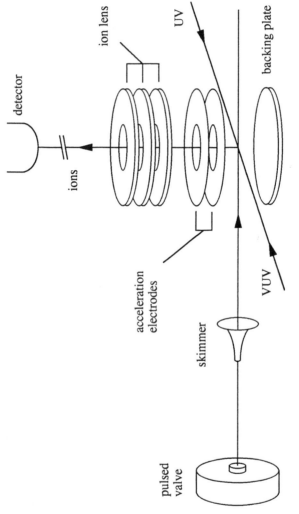

Figure 1. The experimental apparatus.

LASER WAVELENGTHS AND ENERGETIC CONSIDERATIONS

The apparatus uses two separate Nd:YAG-pumped dye laser systems. The first produces UV light for photolysis (280 nm) by frequency-doubling the dye laser light. The second generates VUV light by four-wave mixing in mercury. Light from the second dye laser (625 nm) passes through a KDP crystal where a portion of the beam doubles in frequency. The two wavelengths (625 nm + 312.5 nm) travel co-linearly through a lens and into a Hg heat pipe where the combination of two 312.5-nm UV photons and one 625-nm visible photon produces a single 125-nm (9.9 eV) VUV photon through resonant four-wave mixing. The details of the heat pipe design appear elsewhere (*10*). As with electron impact and multiphoton ionization, small ionization cross-sections and secondary dissociations can obscure the pathways we hope to observe. Furthermore, the energy of the VUV photons limits the fragments detected to those with ionization potentials less than 9.9 eV. Published ionization potentials (*11,12*) and previous results from the photodissociation of 1-nitropropane (*10*) in this laboratory guide our assignment of primary photodissociation fragments. In the photolysis of $(\eta^5\text{-}C_5H_5)Fe(CO)_2(CH_2CH_2CH_3)$ described below, 9.9-eV photons ionize all the products, excluding CO, which has an ionization potential (14.0 eV) (*11*) greater than the energy of the VUV ionization photons (9.9 eV).

$(\eta^5\text{-}C_5H_5)Fe(CO)_2(CH_2CH_2CH_3)$ RESULTS

Solution phase photodissociation of $(\eta^5\text{-}C_5H_5)Fe(CO)_2(\text{alkyl})$ **1** produces alkanes (*13*) and alkenes (*14*) (Scheme 1). The production of alkanes apparently

Scheme 1

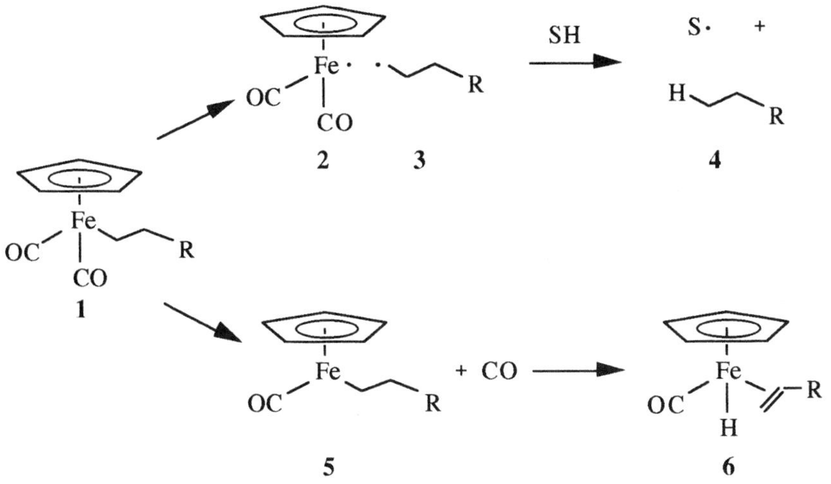

begins with Fe-alkyl bond homolysis, and the resulting alkyl radical **3** abstracts a hydrogen atom from solvent to yield an alkane **4**. The proposed primary step in the production of alkenes involves the photochemical generation of a vacant coordination site, either through CO loss to give **5** or through ring slippage. Subsequent insertion of the metal into a β-C-H bond (β-hydride insertion) produces a metal-alkene hydride complex **6**. Alkane production occurs most readily in fluid solution, where geminate

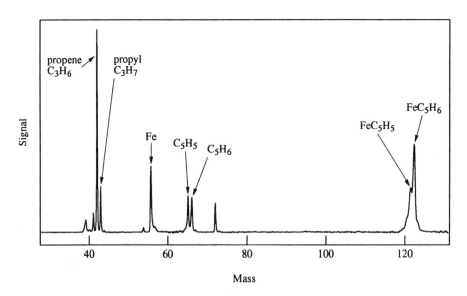

Figure 2. TOF mass spectrum of $(\eta^5\text{-}C_5H_5)Fe(CO)_2(CH_2CH_2CH_3)$ following 280-nm photolysis.

radical pairs diffuse apart rapidly (*13,15,16*), while alkene production dominates in matrices at low temperature (*14,17*). The partitioning between alkane and alkene pathways is unknown either in solution or in the gas phase. The measurements described here provide direct evidence for both pathways occurring in the gas phase.

Figure 2 shows the mass spectrum obtained upon photodissociation of (η^5-C_5H_5)Fe(CO)$_2$(CH$_2$CH$_2$CH$_3$) (**1**, R=CH$_3$) at 280 nm. The presence of propyl and propene as primary photoproducts, following the absorption of one UV photon, provides strong evidence for the two separate photodissociation pathways. The products containing iron, FeC$_5$H$_5$ from Fe-R bond homolysis and FeC$_5$H$_6$ from β-hydride elimination, also one-photon products, are an additional indication of two decomposition channels. Moreover, the products C$_5$H$_5$ and C$_5$H$_6$, which come from multiphoton dissociation, are also consistent with those photopathways. (To determine the number of photons required to produce any feature in the mass spetrum we measure the variation in product yield as a function of laser power.) The number of carbonyls attached to any Fe-containing intermediates following 280-nm photolysis remains unclear from our data. The detected FeC$_5$H$_5$ and FeC$_5$H$_6$ ions may not correspond to the residual Fe-containing species following alkyl loss or alkene production because of excess energy from the ionizing VUV photon. With the absorption of a 9.9-eV photon, secondary loss of CO may occur for any product containing Fe-CO bonds. For example, the parent molecule **1**, with an ionization potential between 7 and 8 eV (*12*), dissociates both carbonyls following VUV ionization. Despite these secondary dissociations, two distinct primary photodissociation pathways appear from the absorption of a 280-nm photon: Fe-alkyl bond homolysis and direct production of alkenes. These results agree qualitatively with those of condensed phase experiments (*18*) which produce alkenes (*14,17*) and, to a lesser extent, alkyl radicals (*14-16*). We are preparing a more detailed analysis of these data.

CONCLUSIONS

The laser chemistry of organometallic compounds in a molecular beam provides clues to the mechanisms for photodecomposition of metal-containing species. The combination of UV photolysis, single-photon VUV ionization, and TOFMS, can observe many species that may also appear in solution or in laser-assisted chemical vapor deposition. In the case of dicarbonyl(η^5-cyclopentadienyl)(1-propyl)iron, the measurements clearly show that both alkyl radical and alkene production are primary processes.

ACKNOWLEDGMENTS

J.A.B. wishes to thank the U.S. Department of Education for fellowship support, and T.G.M. gratefully acknowledges the support of the Alexander von Humboldt Foundation by a Feodor-Lynen Fellowship. The Army Research Office and the National Science Foundation support this work.

Literature Cited

(1)   Nathanson, G.; Gitlin, B.; Rosan, A. M.; Yardley, J. T. *J. Chem. Phys.* **1981**, *74*, 361-369.
(2)   Yardley, J. T.; Gitlin, B.; Nathanson, G.; Rosan, A. M. *J. Chem. Phys.* **1981**, *74*, 370-378.
(3)   Ouderkirk, A. J.; Weitz, E. *J. Chem. Phys.* **1983**, *79*, 1089-1091.
(4)   Ryther, R. J.; Weitz, E. *J. Phys. Chem.* **1992**, *96*, 2561-2567.
(5)   Mikami, N.; Ohki, R.; Kido, H. *Chem. Phys.* **1990**, *141*, 431-440.
(6)   Ray, U.; Brandow, S. L.; Bandukwalla, G.; Venkataraman, B. K.; Zhang, Z.; Vernon, M. *J. Chem. Phys.* **1988**, *89*, 4092-4101.
(7)   Hou, H.; Zhang, Z.; Ray, U.; Vernon, M. *J. Chem. Phys.* **1990**, *92*, 1728-1746.
(8)   Venkataraman, B.; Hou, H.; Zhang, Z.; Chen, S.; Bandukwalla, G.; Vernon, M. *J. Chem. Phys.* **1990**, *92*, 5338-5362.
(9)   Van Bramer, S. E.; Johnston, M. V. *Anal. Chem.* **1990**, *62*, 2639-2643.
(10)   Huey, L. G. Ph. D. Thesis, University of Wisconsin - Madison, 1992.
(11)   Lias, S. G.; Bartmess, J. E.; Liebman, J. F.; Holmes, J. L.; Levin, R. D.; Mallard, W. G. *Gas-Phase Ion and Neutral Thermochemistry;* Supplement 1 ed.; American Chemical Society and American Institute of Physics: New York, 1988; Vol. 17.
(12)   Lichtenberger, D. L.; Fenske, R. F. *J. Am. Chem. Soc.* **1976**, *98*, 50-63.
(13)   Alt, H. G.; Herberhold, M.; Rausch, M. D.; Edwards, B. H. *Z. Naturforsch.* **1979**, *34b*, 1070-1077.
(14)   Kazlauskas, R. J.; Wrighton, M. S. *Organometallics* **1982**, *1*, 602-611.
(15)   Blaha, J. P.; Wrighton, M. S. *J. Am. Chem. Soc.* **1985**, *107*, 2694-2702.
(16)   Hudson, A.; Lappert, M. F.; Lednor, P. W.; MacQuitty, J. J.; Nicholson, B. D. *J. Chem. Soc. Dalton Trans.* **1981**, 2159-2162.
(17)   Mahmoud, K. A.; Rest, A. J.; Alt, H. G. *J. Chem. Soc. Dalton. Trans.* **1985**, 1365.
(18)   Pourreau, D. B.; Geoffroy, G. L. In *Advances in Organometallic Chemistry*; F. G. A. Stone and R. West, Ed.; Academic Press: New York, 1985; Vol. 24; pp 249-352.

RECEIVED January 25, 1993

Chapter 10

Buffer-Gas Quenching as a Dynamical Probe of Organometallic Photodissociation Processes

Rong Zhang[1] and Robert L. Jackson

IBM Research Division, Almaden Research Center, 650 Harry Road, San Jose, CA 95120

We have developed an indirect method for probing the photodissociation dynamics of organometallic molecules. This method relies on the sequential multiple-bond-scission processes that typically occur in the photodissociation of organometallics, even upon absorption of a single photon. Photodissociation is carried out in the presence of a buffer gas and the yields of both the primary and secondary products are determined as a function of buffer-gas pressure. The resulting "quenching curve" gives the pressure-dependent probability that either the primary or secondary dissociation processes are quenched via collisional stabilization of the reactant. The quenching data are then fit using a time-dependent master equation to describe the competition between reaction and stabilization. From the fitted data, we can readily determine if the primary photodissociation process occurs in a statistical manner, and if vibrational energy is partitioned statistically to the primary photoproducts. The quenching technique is illustrated for the photodissociation of two distinctly different classes of organometallic molecules: the dialkyl zincs and ferrocene.

Experimental studies of photodissociation dynamics have contributed extensively to our understanding of chemical reactions at the molecular level (1). The development of key experimental techniques over the years, including molecular beams, laser-induced fluorescence (LIF), and resonance-enhanced multiphoton ionization (REMPI), has enabled researchers to determine the translational, rotational, and vibrational energy distributions of products formed upon photodissociation of small and intermediate-sized molecules (1). By comparing such distributions with theoretical predictions, it is possible to learn a great deal about the dynamics of many different classes of photoinitiated reaction processes.

Studies of the photodissociation dynamics of large molecules, including organic and organometallic species, have not kept pace with studies of small molecules, primarily because of experimental difficulties. For example, it is difficult to resolve

---

[1]Current address: SRI International, Menlo Park, CA 94025

0097–6156/93/0530–0122$06.00/0
© 1993 American Chemical Society

and interpret spectroscopic detail at the rotational level, and sometimes even the vibrational level, for molecules composed of more than a few atoms, so experimental techniques such as LIF and REMPI that seek to determine product state distributions spectroscopically are rarely applicable to large product molecules. Also, organometallics present a unique problem in molecular-beam experiments because of their tendency to undergo multiple bond-cleavages upon absorption of only a single photon (2). Using a crossed laser-molecular beam apparatus, one can determine photoproduct translational energy distributions by monitoring the time required for products to travel from the point of creation to the detector, which is usually a mass spectrometer. If the products spontaneously dissociate on the way to the detector, however, there is no unique point of product origin, making it difficult to determine product translational energy distributions uniquely (3). Also, because organometallics tend to fragment extensively upon electron impact, it can be difficult to tell whether a given product ion results from photodissociation, electron-impact-induced fragmentation in the mass spectrometer, or both (3,4).

Because of these experimental limitations, we have begun to apply an indirect method to examine the photoreaction dynamics of organometallics in the gas phase (5-7). We photodissociate an organometallic molecule with a pulsed laser and determine the yields of the various products, including those formed in spontaneous secondary fragmentation reactions. Photodissociation is performed in the presence of a buffer gas in order to stabilize the parent molecule or the primary photoproducts before they can undergo secondary dissociation. The probability of stabilization vs. dissociation is determined at several pressures. The resulting quenching data are then fit using a time-dependent master equation (8), treating the spontaneous secondary dissociation processes, or the primary photodissociation process itself, as statistically-driven reactions. By comparing the calculated and experimental quenching curves, we can learn about the dynamics of the primary photodissociation process without directly measuring product state distributions. While our method is suitable only for photoprocesses where the primary photoproducts undergo spontaneous secondary dissociation, this is rarely a limitation for organometallics, since such secondary dissociation processes are the rule, rather than the exception. We demonstrate how the quenching technique can be used to probe the photodissociaton dynamics of organometallic molecules, presenting results obtained upon photodissociation of two distinctly different types of organometallic compounds: the dialkyl zincs (5,6) and ferrocene (7).

Experimental

Details of the experimental apparatus and product detection methods have been described elsewhere (6), so we will only provide a brief outline here. A schematic of the experimental appratus is shown in Fig. 1. Photodissociation was conducted in a stainless steel flow cell using a pulsed excimer laser as a light source. Product detection was accomplished via LIF driven by a counter-propagating Nd:YAG-pumped dye laser tuned to specific product bands. Fluorescence was collected at normal incidence, dispersed with a monochromator, and detected using a photomultplier tube or diode array. The fluorescence intensities and relative yields are corrected for variations in the intensity of the excimer and dye laser beams during a photolysis run, as well as for pressure effects on spectroscopic linewidths and fluorescence lifetimes.

The organometallic compound was contained within a stainless steel vessel and was delivered into the cell using He as a carrier gas. A separate carrier gas line was also attached to the cell, allowing the organometallic reactant and He partial pressures to be controlled separately. Flow rates were controlled accurately using mass flow controllers. *Caution: the dialkyl zinc compounds flame spontaneously upon exposure to the atmosphere.*

Photodissociation of the Dialkyl Zincs

We begin by briefly reviewing prior photochemical and spectroscopic work on the dialkyl zincs as well as the dialkyls of the other Group 12 metals, Cd and Hg. The UV spectra of these metal alkyls show two broad, overlapping bands in the deep UV (9,10). The lower energy band has little or no discernible structure, while the higher energy band shows diffuse vibrational structure. The linewidths are consistent with a photodissociation lifetime on the order of 50 fs (10). The photodissociation process yields metal atoms via a stepwise process (11):

$$R_2M \xrightarrow{\text{h}\nu} RM^\dagger + R \qquad (1)$$
$$RM^\dagger \longrightarrow R + M \qquad (2)$$

where R denotes an alkyl radical and † denotes vibrational excitation. For M=Zn, the vibrationally hot monoalkyl intermediate, $RZn^\dagger$, must retain about half of the energy partitioned to products in the 248-nm photodissociation process to undergo spontaneous secondary dissociation to Zn atom (see Table I).

Table I: Bond dissociation energies [a]

Molecule	DH^0_1 (kcal/mole)	DH^0_2 (kcal/mole)
$(CH_3)_2Zn$[b]	63.7 ± 1.5	24.5 ± 4.0
$(C_2H_5)_2Zn$[b]	52.4 ± 2.0	22.0 ± 4.2
$(n\text{-}C_3H_7)_2Zn$[c]	52.4 ± 2.0	22.0 ± 4.2
$FeCp_2$	91 ± 3[d]	67 ± 5[e]

SOURCE: Adapted from Ref. 6.

[a] DH^0_1 represents the dissociation energy of the first metal-ligand bond, while DH^0_2 represents the dissociation energy of the second metal-ligand bond. [b] Jackson, R.L. *Chem. Phys. Lett.* **1989**, *163*, 315. [c] Bond dissociation energies in $(n\text{-}C_3H_7)_2Zn$ are assumed to be equal to those in $(C_2H_5)_2Zn$. See Benson, S.W.; Francis, J.T.; Tsokis T.T. *J. Phys. Chem.* **1988**, *92*, 4515. [d] Ref. 30. [e] Puttemans, J.P.; Smith, G.P.; Golden, D.M. *J. Phys. Chem.* **1990**, *94*, 3226.

In our work (5,6), we detect the yield of Zn atom and of the monoalkyl zinc intermediate via LIF upon photodissociation of the dialkyl zinc in the presence of He buffer gas. The yields of Zn and CH_3Zn vs. He pressure are shown in Fig. 2 for photodissociation of $(CH_3)_2Zn$. The yield of Zn falls, while the yield of CH_3Zn rises with increasing buffer-gas pressure because the secondary spontaneous dissociation step [Eqn (2)] is quenched with increasing efficiency as the buffer gas pressure rises. The yield of Zn falls from ~95% at very low pressure to ~50% at 400 torr. The yields shown in Fig. 2 are determined by normalizing the LIF data to a total yield (Zn + CH_3Zn) of unity at all pressures.

Similar results are obtained upon photodissociation of $(C_2H_5)_2Zn$ and $(n\text{-}C_3H_7)_2Zn$, as shown in Fig. 3. Again, the yield of Zn falls with increasing He pressure. Unfortunately, we could not obtain complementary data on the yield of the

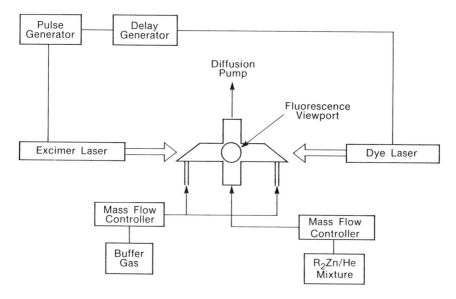

Figure 1: Experimental apparatus (Reproduced with permission from Ref. 6. Copyright 1992 American Institute of Physics).

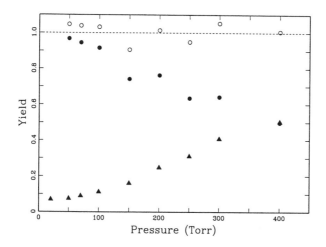

Figure 2: Yield of Zn (▲) and CH₃Zn (●) obtained upon photodissociation of $(CH_3)_2Zn$ vs. He buffer gas pressure. The data are normalized to an average total yield of 1.0 (○), represented by the dotted line (Reproduced with permission from Ref. 6. Copyright 1992 American Institute of Physics).

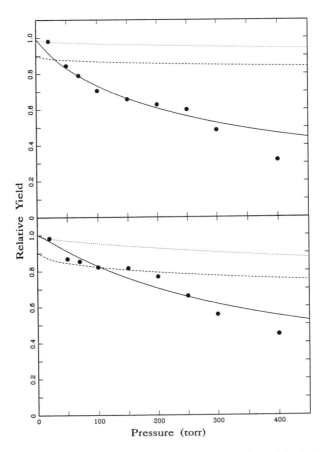

Figure 3:   Yield (●) of Zn obtained upon photodissociation of $(C_2H_5)_2Zn$ (top) and $(n\text{-}C_3H_7)_2Zn$ (bottom) vs. He pressure.   The data are normalized to an extrapolated (Stern-Volmer) yield of unity at zero pressure.   The solid line represents the best fit obtained from master equation calculations (see text). The dotted line represents the fit obtained assuming that the vibrational energy distribution is a Gaussian function with a width (FWHM) of 3000 cm^{-1}. The dashed line represents the fit obtained assuming that the vibrational energy distribution is given by phase-space theory (Reproduced with permission from Ref. 6. Copyright 1992 American Institute of Physics).

monoalkyl zinc for $(C_2H_5)_2Zn$ and $(n-C_3H_7)_2Zn$, since no fluorescence signals attributable to C_2H_5Zn and $n-C_3H_7Zn$ were observed. As discussed in Ref. 6, the lack of fluorescence from these species can be understood from symmetry arguments. Still, the declining yield of Zn with increasing He pressure confirms the two-step photodissociation mechanism for $(C_2H_5)_2Zn$ and $(n-C_3H_7)_2Zn$, and confirms that the C_2H_5Zn and $n-C_3H_7Zn$ intermediates are stabilized with increasing probability as the He pressure rises.

Time-Dependent Master Equation Calculations. To determine the nascent vibrational energy distribution of the monoalkyl zinc intermediate from the quenching data shown in Figs. 2 and 3, we use a fitting procedure based on a time-dependent master equation formalism. The following is a brief outline of this procedure, since it has been described in detail elsewhere (8). Denoting the monoalkyl zinc species as A, the appropriate form of the time-dependent master equation is:

$$\frac{dA(E,t)}{dt} = A(0)\phi(E)dE + \omega \int_0^\infty P(E,E')A(E',t)dE' - \omega A(E,t) - k(E)A(E,t). \quad (3)$$

The first term on the right-hand side of Eqn (3) defines the concentration and vibrational energy distribution of A created by photodissociation at t=0, the second and third terms define the rate of collisional energy transfer between the monoalkyl zinc and the buffer gas, while the fourth term defines the rate of spontaneous dissociation of the monoalkyl zinc. The meaning of the variables and their values are given below:

$\phi(E)dE$ is the probability that photodissociation yields a monoalkyl zinc molecule with internal energy between E and E + dE. This function is the differential form of the nascent vibrational energy distribution for the monoalkyl zinc photoproduct, and is treated as an adjustable function in order to fit the quenching data.

ω is the rate of collisions between A and the buffer gas. We employed Lennard-Jones collision rates, which were determined as described in Ref. 6.

$k(E)$ is the J-averaged (12) RRKM rate constant for dissociation of the monoalkyl zinc intermediate. The RRKM calculations are described in detail in Ref. 6.

$P(E',E)$ is the normalized probability that a single collision with He transforms A(E) into A(E'). Theories of collisional vibrational energy transfer are not yet sufficiently well developed for polyatomic molecules to yield a general functional form for $P(E',E)$, so a suitable mathematical model must be specified. A common model, and the one we used in this work, is the exponential-down model (13):

$$P(E',E) = \frac{1}{N(E)} \exp\left(-\frac{E-E'}{\alpha(E)}\right) \quad E \geq E', \quad (4)$$

where $N(E)$ is a normalization factor and $\alpha(E)$ is a parameter with units of energy that denotes the average quantity of vibrational energy transferred per collision. The value of $\alpha(E)$ can be determined from direct measurements, as have been performed on several polyatomic molecules (14), but such data are not available for the monoalkyl zincs. Semi-quantitative data on the rate of thermalization of the hot CH_3Zn photoproduct are available from our LIF experiments, however, as described in Ref. 6. The value of $\alpha(E)$ for each calculation was chosen to be consistent with the measured rate of thermalization. Analogous data are not available for C_2H_5Zn and $n-C_3H_7Zn$, but recent studies indicate that the average quantity of vibrational energy removed per collision by a given buffer gas does not vary greatly from one molecule to the next (15). Hence, we used the thermalization rate data for CH_3Zn to determine $\alpha(E)$ for all the monoalkyl zincs.

The goal of our fitting procedure is to determine the function $\phi(E)$ that provides the best fit to the quenching data for all three dialkyl zinc compounds. It is not possible to determine $\phi(E)$ uniquely, since many functions could conceivably fit the data. We restricted the set of trial $\phi(E)$ functions to continuous functions, such as a Gaussian, Boltzmann, or Poisson distribution. Use of these model functions will give a reasonably accurate picture of the peak energy and the width of the nascent internal energy distribution for the monoalkyl zinc photoproducts.

Fitting Results for $(C_2H_5)_2Zn$ and $(n-C_3H_7)_2Zn$. We found that good fits to the quenching data for $(C_2H_5)_2Zn$ and $(n-C_3H_7)_2Zn$ could be obtained only by assuming that the nascent vibrational energy distribution of the monoalkyl zinc photoproduct was narrow and quite hot, i.e. highly non-statistical. For $(C_2H_5)_2Zn$, a good fit to the quenching data is obtained by representing $\phi(E)$ as a Gaussian function with a full-width at half-maximum of 500 cm^{-1} centered at 7600 cm^{-1}. For $(n-C_3H_7)_2Zn$, a good fit to the quenching data is obtained by representing $\phi(E)$ as a Gaussian function with a full-width at half-maximum of 1250 cm^{-1} centered at 9200 cm^{-1}. Figure 3 shows the fits to the quenching data obtained with these choices of $\phi(E)$.

The hot, narrow vibrational distribution functions that provide the best fit to the quenching data for $(C_2H_5)_2Zn$ and $(n-C_3H_7)_2Zn$ may be compared with the distributions derived from phase-space theory (PST) (16) or separate statistical ensemble (SSE) theory (17). Each theory has been used successfully to compute photoproduct rotational and vibrational distributions for photodissociation processes that follow a statistical dissociation pathway. The vibrational energy distribution functions predicted by each theory for the monoalkyl zinc photoproducts are approximately Gaussian in shape, but are quite broad, as shown in Table II. Fig. 3 shows that these broad distribution functions provide a very poor fit to the quenching data for both $(C_2H_5)_2Zn$ and $(n-C_3H_7)_2Zn$.

Table II. Energy maximum and width (FWHM) of the vibrational energy distributions computed via PST and SSE theory for the monoalkyl zincs

Molecule	PST		SSE	
	Maximum (cm^{-1})	Width (cm^{-1})	Maximum (cm^{-1})	Width (cm^{-1})
CH$_3$Zn	7800	8200	9200	7500
C$_2$H$_5$Zn	10800	7600	11200	7800
n-C$_3$H$_7$Zn	11100	7000	9000	7600

SOURCE: Reprinted with permission from Ref. 6.

To ensure that our fitting results are accurate, we adjusted the values of $k(E)$ and $\omega P(E',E)$ in order to force a fit using the monoalkyl zinc vibrational energy distributions predicted by phase-space theory. A reasonable fit can be obtained, but only by making huge changes to one or both rate constants. For $(C_2H_5)_2Zn$, the RRKM rate constants would have to be reduced by a factor of ~100 in order to force a reasonable fit. Alternatively, the calculated collisional energy transfer rates would have to be increased by a factor of ~100. For $(n-C_3H_7)_2Zn$, the errors would also have to be quite large (~35 ×). There is no reason to believe that our calculations have such large errors.

To rationalize the hot, narrow vibrational energy distribution found for the C_2H_5Zn and n-C_3H_7Zn photoproducts, we examine the photodissociation pathways that are available in the dialkyl zincs. Fig. 4 shows a one-dimensional cross-section, taken along the breaking C-Zn bond, of the potential energy surfaces involved. The symmetry species assigned to each state are based on the highest molecular symmetry (C_s) preserved throughout the course of the bond-breaking process. Excitation of the dialkyl zinc at 248 nm populates the lower-energy $^1A'$ excited state, but this state cannot dissociate directly to ground-state products without crossing onto the repulsive $^3A'$ surface or internally converting onto the ground-state potential surface. We can readily eliminate the internal conversion mechanism, since it is well-known that dissociation processes of this kind follow a statistical pathway (*17-20*). Hence, we conclude that photodissociation proceeds via crossover from the optically prepared $^1A'$ excited state to the repulsive $^3A'$ state. Because bond-breaking occurs on the timescale of a single vibration, however, it is probably more precise to consider the optically prepared state to be an admixture of the $^1A'$ excited state and the repulsive $^3A'$ state, with the $^1A'$ state carrying most of the oscillator strength.

Based on this picture of the potential surfaces involved in the photodissociation process, we propose the following explanation for the hot, narrow vibrational energy distribution of the monoalkyl zinc photoproduct. The dialkyl zinc molecule is placed on the repulsive $^3A'$ potential surface within a very narrow region of phase space defined by several factors, including the wavelength of the photon source, the Franck-Condon matrix elements connecting the ground state and the $^1A'$ excited state, and the spin-orbit matrix elements connecting the $^1A'$ excited state with the repulsive $^3A'$ state. Because the accessible phase space is limited, the product vibrational energy distributions are much narrower than those expected from statistical theories. Also, both the alkyl radical and monoalkyl zinc fragments are born with a geometry nearly identical to their geometries within the ground-state dialkyl zinc. As discussed in Ref. 6, these geometries differ significantly from the equilibrium geometries for the separated fragment, so both fragments are formed with a relatively hot vibrational energy distribution.

Fitting Results for $(CH_3)_2Zn$. In contrast to the good fits obtained for $(C_2H_5)_2Zn$ and (n-$C_3H_7)_2$, a good fit could not be obtained for $(CH_3)_2Zn$. Neither a narrow Gaussian vibrational distribution function, such as those used successfully for $(C_2H_5)_2Zn$ and (n-$C_3H_7)_2Zn$, nor the broader vibrational distribution functions predicted by PST or SSE could fit the quenching data. For all reasonable distribution functions, the effect of He on the spontaneous secondary dissociation probability for the CH_3Zn photoproduct was much stronger than predicted by our calculations.

To understand why the master equation procedure fails for $(CH_3)_2Zn$, we need only look at the RRKM calculations on the dissociation rate of the CH_3Zn photoproduct. CH_3Zn is a small molecule with a low barrier to dissociation. As a result, the vibrational state density of CH_3Zn in the vicinity of the dissociation barrier is extremely low, ~$10/cm^{-1}$. Experiments by Bauer and coworkers (*21*) indicate that RRKM theory fails to predict reaction rates in cases where the state density of the reactant is so low. The reason for this failure is that the rate of internal vibrational energy redistribution (IVR) is not fast enough at very low state densities to ensure that the reaction follows a purely statistical pathway. RRKM theory will thus predict a reaction rate that is much faster than the actual reaction rate, translating into calculated quenching probabilities that are much smaller than observed.

Photodissociation of Ferrocene

We now turn to a discussion of our work on the photodissociation of ferrocene (*7*) beginning with a brief review of prior work. Ferrocene has been irradiated numerous

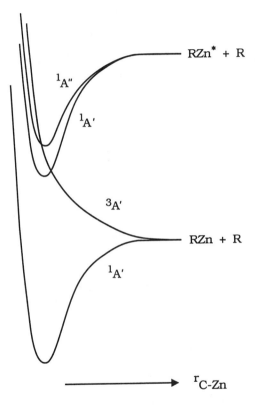

Figure 4: Cross section taken along the breaking C-Zn bond of the potential energy surfaces involved in photodissociation of the dialkyl zincs (Reproduced with permission from Ref. 6. Copyright 1992 American Institute of Physics).

times in condensed phases, and it is clear from published results that the quantum yield for photoreaction is immeasurably low (*22*). Excitation of ferrocene in the UV is apparently followed immediately by internal conversion to the ground state, since no fluorescence is observed and the phosphorescence quantum yield is very low (*23*). Internal conversion in condensed phases is of course followed very rapidly by energy transfer to the surrounding medium. Since the first Fe-Cp bond dissociation energy is quite large (see Table I), dissociation cannot compete with relaxation following absorption of a single photon.

While the photostability of ferrocene in condensed phases is well-documented, it has been known since the 1956 flash photolysis work of Thrush (*24*) that photodissociation of ferrocene in the gas phase yields measurable quantities of cyclopentadienyl radical and iron atoms. Recent laser studies by several groups show that dissociation is a multiphoton process (*25-28*). Gas-phase multiphoton dissociation of ferrocene proceeds stepwise (*25-28*), with the monocyclopentadienyl iron species formed as an intermediate:

$$Cp_2Fe \xrightarrow{\ nh\nu\ } CpFe^+ + Cp \qquad (5)$$

$$CpFe^+ \longrightarrow Cp + Fe \qquad (6)$$

There is disagreement over the dynamics of the primary dissociation process [Eqn (5)], however. Engelking and coworkers concluded from multiphoton ionization studies of the iron atom photoproduct that multiphoton dissociation of ferrocene at wavelengths ≤ 248 nm proceeds via a repulsive potential surface (*27*). The primary photodissociation step is thus a highly non-statistical process. In contrast, Vernon and coworkers concluded from molecular-beam measurements of product translational energy distributions that two-photon dissociation of ferrocene at 193 nm follows a statistical pathway, where internal conversion precedes dissociation and reaction occurs from the ground-state potential surface (*28*). Vernon and coworkers concluded that two-photon dissociation of ferrocene at 248 nm also follows a statistical pathway, although the translational energy distribution data obtained at 248 nm was not quantitative.

Using our quenching technique, together with a time-dependent master equation fit of the quenching data, it is easy for us to determine if the primary photodissociation step in ferrocene is a statistical process. If dissociation exhibits statistical behavior, we should observe significant quenching of the primary photodissociation process (see below). If dissociation occurs via a repulsive potential surface, no quenching will be possible.

Upon photodissociation of ferrocene at 248 nm, we observed both cyclopentadienyl radical and iron atom as products, consistent with the original flashlamp experiments of Thrush (*24*). The intensity of the LIF signal for both products scales approximately as I^2, where I is the excimer laser intensity, indicating that photodissociation is a two-photon process. We saw no indication of a one-photon reaction.

Fig. 5 shows the relative yield of cyclopentadienyl radical and iron atom vs. He pressure. The yield of cyclopentadienyl is largely unaffected by increasing pressure (except for an experimental artifact at low He pressure; see below), while the yield of iron atom falls off significantly as the He pressure increases. Note that the y-axis in Fig. 5 is a relative yield scale. We did not normalize to an absolute yield scale, as in Fig. 2. Note also that the falloff in the yield of iron atom with increasing pressure is not accompanied by a falloff in the yield of cyclopentadienyl radical. This can occur only if nearly all of the cyclopentadienyl radical is formed in the primary photprocess, Eqn (5), i.e. only if the secondary step that creates iron atom, Eqn (6), is a very minor process. We can thus conclude immediately from Fig. 5 that the major product in the two-photon dissociation of ferrocene at 248 nm is the monocyclopentadienyl iron species, FeCp.

Actually, there is a noticeable falloff in the yield of cyclopentadienyl radical as the He pressure *decreases*, but we do not believe that this falloff is real for the following reason. In our LIF process, we detect cyclopentadienyl radicals by exciting the 0_0 band of the $^2A_2" \leftarrow {}^2E_1"$ transition (*29*), so only those molecules that are in the vibrational ground state are detected. Cyclopentadienyl is formed vibrationally hot, however, so we must allow time for collisional thermalization of the cyclopentadienyl radical population before firing the dye laser. Otherwise, the LIF signal will not have reached maximum intensity. Because complete thermalization is slower at low pressures, cyclopentadienyl radicals begin to escape the detection zone before collisional thermalization is complete. We could not correct for this effect quantitatively, but it appears that the low-pressure falloff is entirely due to the competition between escape and thermalization, and that the yield of cyclopentadienyl does not actually change more than a few per cent over the entire pressure range examined.

Time-Dependent Master Equation Calculations. Calculations on ferrocene are performed using Eqn (3), just as they were for the dialkyl zincs. In the ferrocene work, our goal is to determine the quenching behavior expected for the cyclopentadienyl radical product assuming that the primary photodissociation process is statistical. Hence, we are not trying to obtain the best fit by varying an adjustable function; we are simply calculating the expected quenching curve to determine if it reproduces the experimental data. The values of the variables that must be defined to solve Eqn (3) are given below:

$\phi(E)dE$ - In our ferrocene calculations $\phi(E)dE$ represents the internal energy distribution of the excited ferrocene reactant, A. The distribution function may be represented with reasonable accuracy by adding the photon energy to the thermal internal energy distribution of ground-state ferrocene prior to excitation. $\phi(E)dE$ is thus very readily calculated by standard means.

ω – Lennard-Jones collision rates were determined as described in Ref. 7.

$k(E)$ - RRKM rate constants for statistical dissociation of ferrocene were computed as described in Ref. 7. The calculated rate constants reproduce the temperature-dependence and the high-pressure A-factor for the thermal dissociation of ferrocene, as measured by Lewis and Smith (*30*).

$P(E',E)$ - We used the exponential-down model to define the probability of energy transfer between ferrocene and He. Collisional vibrational energy transfer data for ferrocene are unavailable, so we were forced to estimate the quantity of energy transferred per collision [$\alpha(E)$ in Eqn (4)] from values measured for related molecules. We chose the energy-transfer data obtained for azulene by Troe and coworkers (*15*). Azulene is a C_{10} hydrocarbon comparable in size to ferrocene. While it is impossible to know the accuracy of this estimate with certainty, the quantity of energy transferred per collision is most likely underestimated by this procedure, particularly at the high excitation energies reached in a two-photon process at 248 nm (>80000 cm^{-1}). Quenching probabilities calculated at a given pressure are thus expected to be a lower limit.

Experimental data on the relative yield of the cyclopentadienyl radical vs. He pressure are compared to the results of our master equation calculation in Fig. 6. It can be readily seen that the calculated yield of cyclopentadienyl radical falls off much more quickly with increasing pressure than the yield measured experimentally. Since our master equation most likely *underestimates* the quenching probability, we can readily conclude from Fig. 6 that the primary photodissociation step in the two-photon dissociation of ferrocene is not a statistical process.

We can force the calculated quenching curve to fit the experimental data by making very large changes in either $k(E)$ or $P(E',E)$, but the changes are well outside any reasonable error. To force a fit, $k(E)$ would have to be increased by seven-fold or $P(E',E)$ would have to be decreased by the same factor. Given the excellent fit of our RRKM rate constants to the thermal dissociation data for ferrocene, such a large error

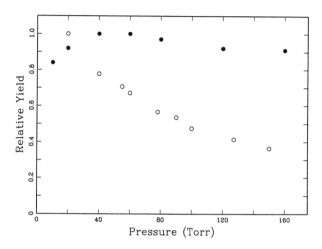

Figure 5: Relative yields as a function of He pressure of cyclopentadienyl radical (●) and iron atom (○) produced in the two-photon dissociation of ferrocene at 248 nm.

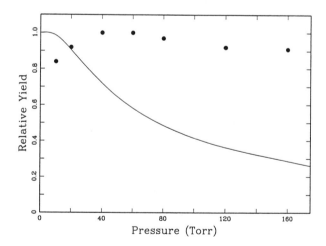

Figure 6: Fit of the cyclopentadienyl yield as a function of He pressure (●) to the yield calculated (solid line) using Eqn (3), assuming that the primary photodissociation process is statistical.

in k(E) is not possible. Our estimating procedure for P(E',E) is undoubtedly subject to some error, but since we believe that our estimated value lies below than the actual value, it is extremely unlikely that P(E',E) could actually be overestimated by a factor of seven.

Our calculations do not reveal the actual mechanism for the primary photodissociation step, but they show conclusively that it occurs much faster than the statistically predicted rate. Such results would be expected for a photodissociation process that involves predissociation or direct excitation to a repulsive potential surface. Our results thus support the conclusions of Engelking and coworkers (27) that dissociation occurs via the direct process.

Literature Cited

1. For a general discussion, see "*Advances in Gas-Phase Photochemistry and Kinetics, Molecular Photodissociation Dynamics*", Ashfold, M.N.R.; Baggott, J.E., Eds.; Royal Society: London, 1987.
2. For example, see Weitz, E. *J. Phys. Chem.* **1987**, *91*, 3945.
3. Venkataraman, B.; Hou, H.; Zhang, Z.; Chen, S.; Bandukwalla, G.; Vernon, M. *J. Chem. Phys.* **1990**, *92*, 5338.
4. Ray, U.; Brandow, S.L.; Bandukwalla, G.; Venkataraman, B.K.; Zhang, Z.; Vernon, M. *J. Chem. Phys.* **1988**, *89*, 4092.
5. Jackson, R.L. *J. Chem. Phys.* **1990**, *92*, 807.
6. Jackson, R.L. *J. Chem. Phys.* **1992**, *95*, 5938.
7. Zhang, R.; Jackson, R. L., manuscript in preparation.
8. Jackson, R.L. *Chem. Phys.* **1991**, *157*, 315.
9. Chen, C.J.; Osgood, R.M. *J. Chem. Phys.* **1984**, *81*, 327.
10. Amirav, A.; Penner, A.; Bersohn, R. *J. Chem. Phys.* **1989**, *90*, 5232.
11. Jonah, C.; chandra, P.; Bersohn, R. *J. Chem. Phys.* **1971**, *55*, 1903.
12. Smith, S.C.; Gilbert, R.G. *Int. J. Chem. Kinet.* **1988**, *20*, 307.
13. For example, see Gilbert, R.G.; Luther, K.; Troe, J. *Ber. Bunsenges. Phys. Chem.* **1983**, *87*, 169.
14. Oref, I.; Tardy, D.C. *Chem. Rev.* **1990**, *90*, 1407.
15. Damm, M.; Deckert, F.; Hippler, H.; Troe, J. *J. Phys. Chem.* **1991**, *95*, 2005.
16. Pechukas, P.; Light, J.C. *J. Chem. Phys.* **1965**, *42*, 3281.
17. Wittig, C.; Nadler, I.; Reisler, H.; Noble, M.; Catanzarite, J.; Radhakrishnan, G. *J. Chem. Phys.* **1985**, *83*, 5581.
18. Hippler, H.; Luther, K.; Troe, J.; Wendelken, H.J. *J. Chem. Phys.* **1983**, *79*, 239.
19. Klippenstein, S.J. *Chem. Phys. Lett.* **1990**, *170*, 71.
20. Chen, I.; Green, Jr., W.H.; Moore, C.B. *J. Chem. Phys.* **1988**, *89*, 314.
21. Borchardt, D.B.; Bauer, S.H. *J. Chem. Phys.* **1986**, *85*, 4980 and references cited therein.
22. For example, see Borrell, P.; Henderson, E. *Inorg. Chim. Acta* **1975**, *12*, 215.
23. Smith, J.J.; Meyer, B. *J. Chem. Phys.* **1968**, *48*, 5436.
24. Thrush, B.A. *Nature* **1956**, *178*, 155.
25. Leutwyler, S.; Even, U.; Jortner, J. *J. Phys. Chem.* **1981**, *85*, 3026.
26. Liou, H.T.; Ono, Y.; Engelking, P.C.; Moseley, J.T. *J. Phys. Chem.* **1986**, *90*, 2888.
27. Liou, H.T.; Engelking, P.C.; Ono, Y.; Moseley, J.T. *J. Phys. Chem.* **1986**, *90*, 2892.
28. Ray, U.; Hou, H.Q.; Zhang, Z.; Schwarz, W.; Vernon, M. *J. Chem. Phys.* **1989**, *90*, 4248.
29. Nelson, H.H.; Pasternack, L.; McDonald, J.R. *Chem. Phys.* **1983**, *74*, 227.
30. Lewis, K.E.; Smith, G.P. *J. Am. Chem. Soc.* **1984**, *106*, 4650.

RECEIVED February 10, 1993

LASER-INITIATED BIMOLECULAR CHEMISTRY

Chapter 11

Photoionization and Photodissociation Dynamics for Small Clusters of Refractory Metals
Time-Resolved Thermionic Emission

Andreas Amrein[1], Bruce A. Collings, David M. Rayner, and Peter A. Hackett

Steacie Institute for Molecular Sciences, National Research Council of Canada, 100 Sussex Drive, Ottawa, Ontario K1A 0R6, Canada

The chemistry of "naked" metal clusters, as revealed by molecular beam studies, is reviewed in the general context of chemistry at metal atom sites. Special emphasis is given to the chemistry of niobium clusters and their derivatives. The recent observation of delayed ionization from multiphoton excited, strongly bound clusters is highlighted. Recently, we have used this novel form of molecular ionization process to develop a reliable and calibrated thermometer for the vibrational temperature of metal cluster beams. This advance has meant that temperature may be used as a well defined variable in metal cluster beam experiments. This will be helpful for the development of "molecular" surface science using derivitized metal cluster beams.

The agglomeration of metal particles is a common feature of the laser chemistry of organometallics. Such particles may be desirable intermediates in metal deposition or laser writing processes. They may also provide particularly effective supports or active media for heterogeneous catalysis. The physical or chemical properties, for instance the porosity, of these catalysts may be varied over wide ranges in materials which have well defined particle size distributions on the length scale of angstroms i.e. "cluster assembled" materials (1). Finally, small metal particles may be the undesired end products of too vigorous a laser photochemical reaction. The properties of the smallest of metal particles, the metal clusters, are as interesting as their occurrence is ubiquitous. A small fraction of these properties is reviewed in this contribution.

[1]Current address: Technikum Winterthur, Postfach 805, CH–8401, Switzerland

0097–6156/93/0530–0136$06.00/0
Published 1993 American Chemical Society

The Role of "Naked" Metal Clusters in Metal-Ligand Chemistry

The chemistry and photochemistry of "naked" metal clusters is an important area of the chemistry occurring at metal centres. General features of this chemistry are outlined in the matrix represented in Figure 1 in which the number of metal atoms and the number of ligands present in a complex M_xL_y provide a logical basis with which to categorize areas of chemistry ranging from surface science and organometallic chemistry through to metal atom chemistry. Within this context, the roles (i) of complexes, ML, between a single ligand and a metal atom, as prototypical organometallic systems, (ii) of coordinatively unsaturated organometallic species, ML_x, as reactive intermediates in organometallic catalytic cycles, and (iii) of single ligands on extended metal surfaces, $M_\infty L$, as reactive intermediates in surface catalyzed reactions, are linked with those of "naked" metal clusters, M_x. Moscovits discussed these links in his prescient review of the field in 1979, emphasizing the essential differences in the ability of metal atoms in these various environments to accommodate ligands through electronic or geometrical reorganization (2).

It is apparent that small metal clusters may act as of finite models for the extended phase of metallic materials since molecular processes observed in clusters may have analogues in bulk matter. By observing the dynamics of molecular processes in clusters one can gain a detailed understanding, using the language of molecular science, of processes occurring in the extended phase.

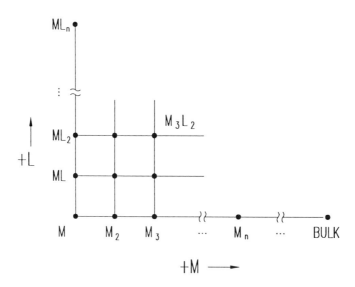

Figure 1: Chemistry at metal centres.

If we are resolved to use small derivitized clusters, M_xL, as simple models for examining chemistry occurring on extended metal surfaces, then we are required to define the properties, particularly the electronic and geometrical structural properties, of the "naked" metal cluster, M_x. A working definition of the term cluster is "a collection of atoms for which only the number of constituents is known". Once we know any molecular property of a cluster, the term cluster becomes inappropriate and the item becomes a molecule. A cluster of sixty carbon atoms was an interesting curiosity but aromatic C_{60}, with the soccer ball structure, is a far more compelling object. Our strategy is clear. We shall attempt to determine the molecular properties of small metal clusters and then we shall use the chemistry observed between these cluster molecules and ligands as simple molecular models for chemistry occurring in the extended state.

In this presentation, we shall review progress to date, using molecular beam techniques, on the chemistry of refractory metal clusters particularly niobium. We shall highlight the recent observation of a new molecular ionization process observed in these clusters. This molecular process is equivalent to the property of thermionic emission in bulk metals. We shall show how we have been able to use this process to explore the vibrational energy content of metal clusters on metal cluster beams. This work has allowed us to define the vibrational temperature of metal cluster beams for the first time and opens the way to the use of vibrational temperature as a well defined variable for future studies of reactions of "naked" metal clusters.

Chemistry of Niobium Clusters as Revealed by Molecular Beam Studies

The chemistry of niobium clusters is interwoven with the general development of metal cluster beam science as niobium is an ideal element with which to demonstrate aspects of metal cluster beam technology. Laser vaporization of niobium easily leads to clusters of high nuclearity. Niobium has only one naturally occurring isotope and it is therefore an ideal subject for mass spectroscopy.

Ionization Potentials for Niobium Clusters. Detailed threshold photoionization studies of niobium clusters have been made and ionization potentials for Nb_2 to Nb_{76} have been reported (3,4). The dependence of the ionization potential on cluster size is generally consistent with that expected from a consideration of image potential effects on the ionization energy of metal spheres (radius R) from which classical electrostatics would suggest equation 1 :

$$IP = W.F. + (1/2)e^2/R \tag{1.}$$

A smooth decrease in ionization potential with increasing nuclearity is indeed observed. Thus, it would seem that the evolution of this particular property from molecule to bulk is well understood. However, as Moscovits has pointed out (5), this interpretation is too simplistic. Firstly, equation 1 is not followed exactly. In this size range a coefficient of 0.2 rather than 0.5 fits the data. Secondly, the work function of bulk niobium is highly dependent on the choice of crystal face (6). Thirdly, there are large deviations from the smooth trend with cluster size for particular clusters. Finally, there is evidence that some clusters exist as two isomeric forms each with distinct ionization potential (*7*).

Clearly, the general agreement of the trend in ionization energies with the form of equation 1 should not be over interpreted to imply that bulk-like behavior has been achieved for clusters in the size range studied. Rather one should conclude that quantum effects and geometrical and electronic structure remain of importance in this size range.

Chemisorption on Niobium Clusters. Preliminary measurements of simple chemisorption reactions have been made on a variety of transition metal clusters. There is controversy over the roles of geometrical and electronic structural effects in determining the rates of these reactions. In this regard, it is unhelpful that the most compelling and widely accepted model in metal cluster chemistry and physics, the jellium model, is a shell model which ignores the position of the nuclei, assuming instead a spherically symmetric and uniform charge distribution (*8*).

Hydrogen Chemisorption. The first observation of significant cluster size-dependent effects on metal cluster reaction rates was made for the relative rates of dissociative chemisorption of deuterium molecules on niobium clusters (*9*). A highly structured pattern of reactivity was found as a function of cluster size. Highly structured reactivity patterns were also found for chemisorption of nitrogen (*10*); whereas, carbon monoxide was found to give a rate of nondissociative chemisorption which increased monotonically with cluster size (*10,11*).

Anti-correlation with Ionization Potential. When ionization potentials of size-selected niobium clusters were determined, it was found that the deuterium reactivity trends were anti-correlated with the electron binding energy of individual niobium clusters (*12*). This correlation is anticipated by a charge-transfer model for the dissociative chemisorption process in which an important initial interaction is charge donation from the metal "surface orbitals" into the $\sigma*$ orbital of the deuterium molecule. Such a correlation had previously been shown to exist for the reaction of iron clusters with deuterium (*13*). However, it was pointed out that the correlation may have its origin in other more subtle effects. Molecular and electronic structure are often correlated and the size-dependent variation in reaction rates may reflect an underlying geometrical factor (*12*). It has been suggested, for instance, that dissociative absorption may be favoured at outer shell nearest neighbors (*13*). Recently, it has been proposed that the less reactive niobium clusters have smooth surfaces and are convex at their edges (*14*).

The ionization potential changes following hydrogen chemisorption have been measured and are found to be consistent with the formation of two metal-hydride bonds per hydrogen molecule chemisorbed via net metal to hydrogen charge donating interactions (*15*).

Non-exponential reaction kinetics were observed for reaction of Nb_9, Nb_{11} and Nb_{12} with deuterium and nitrogen in a fast-flow reactor study (*16,17*). Two isomeric forms were proposed. Recently, the reaction with deuterium has been used to quench the reactive form and this has allowed the ionization potentials of the reactive and unreactive forms of Nb_9 and Nb_{12} to be determined (*13*). The strong anti-correlation between ionization potential and reaction rate was once again confirmed.

Chemisorption on Cluster Ions. Niobium cations and anions were the first to be used succesfully in the generation of cold metal cluster *ion* beams (*18*). Subsequently, specific niobium cations were stored in a Fourier transform ion cyclotron resonance trap and their relative reaction rate constants for dissociative chemisorption of hydrogen were determined (*19*). A reaction propensity curve similar to that obtained for neutral clusters was found. As well, it was shown that Nb_{19}^+ occurred in two forms, one with high and one with low reactivity. It was suggested that these two observations supported the critical importance of geometric structure in determining reactivity (*19*).

The reactivity of niobium cluster ions towards dissociative chemisorption of hydrogen has also been investigated using a fast-flow reactor (*20*). It was found that neutral, anionic and cationic clusters had similar reaction patterns of reactivity as a function of cluster size and that the overall pattern was unaffected by charge. Evidence for reactive and unreactive forms of Nb_9, Nb_{12} and Nb_{12}^+ was presented. At first sight, the absence of a charge effect on the reaction propensity curves strongly supports the contention that geometry is the more important rate determining factor. However, the authors of this work interpreted subtle changes in the relative rates in the range Nb_7 to Nb_{16}, as a function of charge state, as evidence for the importance of the electronic effect discussed above. We comment here that the fast-flow reactor used in most metal cluster reactivity research is far from ideal for kinetics studies. It has been noted that the temperature is not well defined in such reactors and may differ from study to study (*20*). In addition, supporting mechanistic experiments are difficult to perform in this environment. An alternative explanation for the similar reactivity trends observed for niobium clusters of different charge-state would be that the neutral, cationic and anionic clusters are maintained in an equilibrium mixture within the reactor, by charge exchange reactions for instance, and that reaction with hydrogen is determined by reaction with clusters of a specific charge.

Other Niobium Cluster Reactions. The reactions of niobium clusters with benzene have been examined (*21,22*). It was observed that Nb_3 to Nb_{15} may coordinate benzene and that Nb_5, Nb_6 and Nb_{11} have a special propensity to dehydrogenate the aromatic ring. When perdeuterobenzene was used a large kinetic isotope effect was claimed (*23*). Subsequently a similar reactivity pattern was observed for niobium cluster cations (*24*). Photochemical dehydrogenation of benzene on niobium clusters has been studied (*23,24*) as has the competition between bromine abstraction and dehydrogenation in the reaction of niobium clusters with various brominated hydrocarbons (*25,26*).

Determination of Bond Dissociation Energies for Niobium Clusters. The first photofragmentation study of specific metal cluster ions was made using Nb_2^+ (*27*). An unassigned vibronic transition of the dimer ion was observed near 18,000 cm^{-1}. It was suggested that binding energies for specific niobium cluster cations could be determined from photofragmentation thresholds. Bond dissociation energies of Nb_3^+

to Nb_{11}^+ were suggested to be less than 2.33 eV *(27)*. These estimates are in disagreement with the binding energies for Nb_2^+ to Nb_{11}^+ determined using collision-induced dissociation of niobium cluster beams in a guided ion beam mass spectrometer *(28-30)*. The primary dissociation pathway at low collision energies is atom loss; at higher energies sequential atom losses occur. The observation that the atom binding energies are in the range 5 to 6 eV is consistent with the fact that niobium metal possesses one of the largest cohesive energies of all elements. It was suggested that the photofragmentation experiments were in fact due to three-photon excitation and not single photon excitation as was previously claimed *(27)*.

As both the cluster ionization potentials and cluster ion binding energies have been measured, the binding energies of neutral clusters can be derived from the collision induced dissociation work. Binding energies in the range 4.1 to 6.5 eV range were reported *(29-31)* . As with the ionization energies, *vide supra*, the trend in bond dissociation energies with increasing cluster size, in this case an increase, is in accord with the expectation of a simple spherical droplet model which takes into account the increasing importance of the surface energy as size decreases *(32)*. However, as with the ionization energies, one should be cautioned that there are significant deviations from this general trend and that the trend should not be over interpreted to imply that quantum effects are unimportant in this size range.

Thermionic Emission from Niobium Clusters.

The combination of the fact that the ionization potential of bulk niobium , 4.3 eV, is less than its cohesive energy, 6.3 eV, with the general tendencies for the ionization potential to decrease and the bond dissociation energy to increase as the cluster size increases, leads to the interesting situation that at a certain cluster size the ionization potential becomes equal to the bond dissociation energy. Clusters larger than this will in fact have lower ionization potentials than bond dissociation energies. The implication of this observation is that thermally activated cluster molecules may prefer to "evaporate" an electron rather than an atom. We first reported such a thermally activated molecular ionization process for small clusters of niobium, tantalum and tungsten *(33)*. Clusters in the size range 6 to 10 showed ionization times in the order of microseconds when activated following two photon absorption of light in the 355 to 266 nm region. It was noted that this delayed ionization process had much in common with the thermionic emission of electrons from heated surfaces. Elements such as niobium, tantalum and tungsten are the materials of choice with which to construct electron emission filaments because of their high cohesive energy. By analogy, we have called the small clusters of these elements which undergo the delayed ionization process *nanofilaments*.

Mechanism of Delayed Ionization. The mechanism of delayed ionization of strongly bound metal clusters was postulated following a series of experiments investigating the laser wavelength, laser intensity and cluster size dependence of the ionization rate *(33,34)*. The mechanism described in equations 2 through 7 was proposed.

$$M_n \quad + \quad h\nu_L \quad \rightarrow \quad M_n^* \tag{2}$$

$$M_n^* \rightarrow M_n^\# \qquad (3)$$

$$M_n^* + h\nu_L \rightarrow M_n^+ + e^- \qquad (4)$$

$$M_n^\# + h\nu_L \rightarrow M_n^{\#*} \qquad (5)$$

$$M_n^{\#*} \rightarrow M_{n-1} + M \qquad (6)$$

$$M_n^{\#*} \rightarrow M_n^+ + e^- \qquad (7)$$

Here, M_n^*, represents an electronically excited cluster, $M_n^\#$ and $M_n^{\#*}$ represent clusters in which the excitation is distributed over all *vibronic* states acessible at the one and two photon energies respectively. Statistical calculations of the rates of reactions 6 and 7, based on assumed molecular properties of the relevant clusters, were in accord with the experimentally determined values (34). Calculations based upon size-corrected materials properties also gave reasonable agreement (34).

An interesting result of the delayed ionization experiments was the behavior of derivitized clusters in which carbon or oxygen were substituted for niobium. Niobium oxide clusters, $Nb_{x-n}O_n$, do not give delayed ionization and, because strongly bound NbO is an ideal leaving group, it was suggested that dissociation was favoured for these clusters. For niobium carbide clusters, $Nb_{x-n}C_n$, delayed ionization was observed. The ionization rate constant was independent of n but dependent on x. This intriguing result is explained by the fact that a statistical reaction rate constant is strongly dependent only upon the number of particles if the thermochemistry is similar, the total energy is equal and if all vibrational degrees of freedom are in their high temperature limit. These conditions are evidently satisfied by the niobium carbide system (34). These observations have recently been confirmed by *ab initio* calculations, using local density functional theory, of the bond dissociation and ionization energies of $Nb_{x-n}O_n$ and $Nb_{x-n}C_n$ (Salahub, D. and Erickson, L., University of Montreal, Personal communication, 1992).

The rate of reaction 3, $k_{relaxation}$, has recently been investigated using autocorrelation measurements of the delayed ionization yield following excitation by 50 ps pulses of the third harmonic of a Nd-YAG laser. A lower limit of 2×10^9 s^{-1} was found for $k_{relaxation}$ (Collings, B. A., Rayner D. M. and Hackett P. A., National Research Council of Canada, unpublished data). This observation supports the proposed mechanism in which it was assumed that $k_{relaxation}$ was fast.

Detailed studies of the rates of delayed ionization have now been reported for tungsten (35), tantalum (36) and carbon clusters (37).

The Temperature Dependence of the Thermionic Emission Rate and The Vibrational Temperature of Metal Cluster Beams.

Casual inspection of the mechanism outlined in equations 2-8 shows that the delayed ionization rate should depend upon the initial vibrational energy content of M_n. Recently experiments have been carried out which confirm this expectation

(Amrein, A., Collings, B. A., Rayner D. M. and Hackett, P. A., National Research Council of Canada, unpublished data). The experimental apparatus has been described previously (*34*). Briefly, metal clusters are formed in a laser vaporization source, flow along a 15 cm long by 2 mm diameter flow tube, and expand into a time-of-flight mass spectrometer equipped with laser ionization. Delayed ionization rates for particular clusters are measured as a function of both the photoionization laser wavelength and the temperature of the flow tube. The latter may be varied over the range 77 to 700 K.

Klotts has analyzed evaporative dissociation processes on a systematic basis (*38*) and has presented an analysis to determine statistical behavior (*39*). In brief, the procedure requires the construction of an "Arrhenius like plot" in which an effective temperature, $T_{internal}$, for the microcannonical ensemble is defined. Figure 2 is the result of an analysis on the basis of reference 39 and provides further evidence that the delayed ionization is a statistical process. The ionization rate is plotted against the $T_{internal}$ which is varied by varying the vibrational temperature of the metal clusters, via the temperature of the flow tube and the wavelength of the ionization laser.

As Figure 2 shows, the temperature of the flow tube has a marked influence on the delayed ionization decay rate. This is the first direct evidence for the involvement of the vibrational degrees of freedom in the ionization process and supports the mechanism presented above.

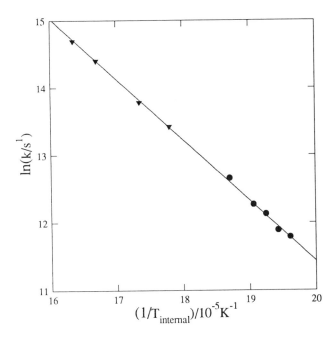

Figure 2. Thermionic Emission Rates for Nb_9 excited at 266 (circles) and 300 nm, (triangles).

A statistically determined rate constant is of course an excellent "thermometer" for the vibrational degrees of freedom of a molecule. Figure 2 may therefore be used to define the internal vibrational temperature of the niobium clusters prior to photoexcitation. Furthermore, by varying the photoexcitation wavelength, we change $T_{internal}$ in a precisely determined way. Hence, we may calibrate our "thermometer". When we make the assumption that the vibrational degrees of freedom of the clusters have the same temperature as the walls of the flow tube, both before and after the expansion into the time-of-flight mass spectrometer (as was done in producing Figure 2) we find that the data from different wavelengths lie on the same line. The implication of this observation is that the vibrational degrees of freedom of the metal clusters are not significantly cooled in our weak supersonic expansion. Therefore we have devised a simple and useful method of determining the vibrational temperature of metal clusters on metal cluster beams and have shown that vibrational temperature may be employed as a well defined variable in suitably designed experiments employing supersonic jet sources of metal clusters.

Conclusions

In this short contribution we have shown that much progress has been made in the use simple metal clusters as model systems for metal-ligand interactions. Futhermore, we have shown that metal clusters sometimes reveal novel molecular science. Much work remains to be done to establish the geometrical and electronic structures for metal clusters, particularly for transition series elements. However, there are already remarkable indications that geometrical structure does play a significant role in the chemistry of these species and that these molecules have a rich and selective chemistry. The challenges for future work in this field are many and varied. Among these are the central problems of the structure of the metal cluster and the structure of the ligand following interaction with the cluster. Progress in the interpretation of chemical reactivity of metal clusters may be aided if the experimental conditions can be controlled so that well characterized chemical measurements can be reported. In this regard, our measurements of the vibrational temperature of niobium cluster beams may be helpful.

Acknowledgements

This research was supported in part by the Network of Centres Excellence in Molecular and Interfacial Dynamics (CEMAID) in association with the National Research Council of Canada (NRC) and the Natural Sciences and Engineering Research Council of Canada (NSERCC).

Literature Cited

1. Andres, R. P.; Averback, R. S.; Brown, W. L.; Brus, L. E.; Goddard III, W. A.; Kaldor, A.; Louie, S. G.; Moscovits, M.; Peercy, P. S.; Riley, S. J.; Siegel, R. W.; Spaepen F.; Wang, Y. *J. Mater. Res.* **1989**, *4*, 704.
2. Moscovits, M. *Acc. Chem. Res.* **1979**, *12*, 229.

3. Whetten, R. L.; Zakin, M.R.; Cox, D. M.; Trevor, D. J.; Kaldor, A. *J. Chem. Phys.* **1986**, *85*, 1697.
4. Knickelbein M. B.; Yang, S. *J. Chem. Phys.* **1990**, *93*, 5760.
5. Moscovits, M. *Ann. Rev. Phys. Chem.* **1991**, *42*, 465.
6. Leblanc, R. P.; Vanbrugghe, B. C.; Girouard, F. E. *Can. J. Chem.* **1974**, *52*, 1589.
7. Knickelbein M. B.; Yang, S. *J. Chem. Phys.* **1990**, *93*, 1476.
8. Knight, W. D.; Clemenger, K.; deHeer, W. A.; Saunders, W. A.; Chou, M. Y.; Cohen, M. L. *Phys. Rev. Lett.* **1984**, *52*, 2141.
9. Geusic, M. E.; Morse, M. D.; Smalley, R. E. *J. Chem. Phys.* **1985**, *82*, 590.
10. Morse, M. D.; Geusic, M. E.; Heath, J. R.; Smalley, R. E. *J. Chem. Phys.* **1985**, *83*, 2293.
11. Cox, D. M.; Reichmann, K. C.; Trevor, D. J.; Kaldor, A. *J. Chem. Phys.* **1988**, *88*, 111.
12. Whetten, R. L.; Cox, D. M.; Trevor, D. J.; Kaldor, A. *Phys. Rev. Lett.* **1985**, *54*, 1494.
13. Phillips, J. C. *J. Chem. Phys.* 1986 **84**, 1951.
14. Kharas, K. C. C. *Chem. Phys. Lett.* **1989**, *161*, 339.
15. Zakin, M. R.; Cox, D. M.; Trevor, D. J.; Whetten, R. L.; Kaldor, A. *Chem. Phys. Lett.* **1987**, *133*, 223.
16. Hamrick, Y. M.; Taylor, S.; Lemire, G. W.; Fu, Z.-W.; Shui, J.-C.; Morse, M. D. *J. Chem. Phys.* **1988**, *88*, 4095.
17. Hamrick Y. M.; Morse, M. D. *J. Phys. Chem.* **1989**, *93*, 6494.
18. Zheng, L.-S.; Brucat, P. J.; Pettiette, C. L.; Yang, S.; Smalley, R. E. *J. Chem. Phys.* **1985**, *83*, 4273.
19. Elkind, J. L.; Weiss, F. D.; Alford, J. M.; Laaksonen, R. T.; Smalley, R. E. *J. Chem. Phys.* **1988**, *88*, 5215.
20. Zakin, M. R.; Brickman, R. O.; Cox, D. M.; Kaldor, A. *J. Chem. Phys.* **1988**, *88*, 3555.
21. St. Pierre, R. J.; Chronister, E. L.; El-Sayed, M. A. *SPIE* **1987**, *742*, 122.
22. St. Pierre R. J.; El-Sayed, M. A. *J. Phys. Chem.* **1987**, *91*, 763.
23. Zakin, M. R.; Cox, D. M.; Kaldor, A. *J. Phys. Chem.* **1987**, *91*, 5244.
24. Zakin, M.R.; Brickman, R. O.; Cox, D. M.; Kaldor, A. *J. Chem. Phys.* **1988**, *88*, 5943.
25. St. Pierre, R. J.; Chronister, E. L.; El-Sayed, M. A. *J. Phys. Chem.* **1987**, *91*, 5228.
26. Song L.; El-Sayed, M. A. *Chem. Phys. Lett.* **1988**, *152*, 281.
27. Brucat, P. J.; Zheng, L.-S.; Pettiette, C. L.; Yang, S.; Smalley, R. E. *J. Chem. Phys.* **1986**, *84*, 3078.
28. Loh, S. K.; Lian, L.; Hales, D.; Armentrout, P. B. *J. Chem. Phys.* **1988**, *89*, 3388.
29. Loh, S. K.; Lian, L.; Armentrout, P. B. *J. Am. Chem. Soc.* **1989**, *111*, 3167.
30. Loh, S. K.; Lian, L.; Armentrout, P. B. *J. Chem. Phys.* **1989**, *91*, 6148.
31. Hales, D. A.; Lian,L; Armentrout, P. B. *Int. J. Mass Spectrom. and Ion Proc.* **1990**, *102*, 269.
32 Miedma, A. R. *Faraday Symp. Chem. Soc.* **1980**, *14*, 136.

33. Amrein, A.; Simpson, R.; Hackett, P. A. *J. Chem. Phys.* **1991**, *94*, 4663.
34. Amrein, A.; Simpson, R.; Hackett, P. A. *J. Chem. Phys.* **1991**, *95*, 1781.
35. Leisner, T.; Athenassenas, K.; Echt, O.; Kandler, O.; Kreisle, D.; Recknagel, E. *Z. Phys .D.* **1991**, *20*, 127.
36. Leisner, T.; Athenassenas, K.; Kandler, O.; Kreisle, D.; Recknagel, E.; Echt, O. *Mater. Res. Soc. Proc. Symp.* **1991**, *206*, 259.
37. Wurz, P.; Lykke, K. R. *J. Chem. Phys.* **1991**, *95*, 7008.
38. Klotts, C. E. *Z. Phys. D.* **1991**, *20*, 105.
39. Klotts, C. E. *Chem. Phys. Lett.* **1991**, *186*, 73.

RECEIVED January 13, 1993

Chapter 12

Bond Energies and Reaction Kinetics of Coordinatively Unsaturated Metal Carbonyls

Eric Weitz, J. R. Wells, R. J. Ryther, and P. House

Department of Chemistry, Northwestern University, Evanston, IL 60208

Recent measurements of rate constants for reactions of coordinatively unsaturated iron carbonyls with $Fe(CO)_5$ obtained using time resolved infrared spectroscopy in the gas phase are presented and discussed. These measurements are consistent with the hypothesis that in these and related reactions, the spin states of the reactants and products can significantly influence the magnitude of the rate constant. Measurements of the rates of reaction of $Fe(CO)_4$ with O_2 and H_2 provide further support for this conclusion. However, variations in rate constants for spin allowed reactions of coordinatively unsaturated metal carbonyls indicate that other factors can significantly influence the magnitude of rate constants for association reactions. Also discussed is a technique for the determination of bond dissociation energies of organometallic compounds. This technique has been used to determine bond enthalpies in systems with weak bonds, such as the Group VI metal carbonyls bound to Xe. Use of time resolved fourier transform infrared (TRFTIR) spectroscopy to probe bond dissociation processes allows for extension of this technique to long time scales and more strongly bound systems. The bond enthalpy for N_2O bound to $W(CO)_5$ has been determined using TRFTIR as the kinetic probe. Implications of significant bond enthalpies for typical solvents interacting with coordinatively unsaturated metal carbonyls are discussed in the context of measurements of rate constants for association reactions in solution.

Over the last decade the photochemistry, spectroscopy, and reactivity of metal carbonyls has been the subject of intense interest (1-3). It has been found that metal carbonyls undergo a wide range of facile photochemical reactions. However, it is still the case that, in general, the pathways for these reactions have

0097–6156/93/0530–0147$06.00/0
© 1993 American Chemical Society

not been fully characterized. A motivation for study of these systems is that the coordinatively unsaturated metal carbonyls that are the products of many of these reactions can act as efficient catalysts. For example, $Fe(CO)_5$ can be used to photolytically generate efficient catalysts for olefin isomerization, hydrogenation and hydrosilation reactions (3-6). In many cases, even the initial photoproducts in these reactions have not been characterized. Studies of the product distribution that result from photolytic excitation of organometallic compounds and the iron carbonyls in particular have also been an area of current interest as has been the mechanism of ligand loss following such excitation (1,7-13).

Much of the difficulty in characterizing either photoproducts or reaction intermediates arises from their extreme reactivity. For example, it is generally agreed that $Cr(CO)_5$ coordinates even hydrocarbon solvents on a sub-nanosecond timescale (14,15). Techniques such as matrix isolation spectroscopy have provided considerable information about highly reactive species including coordinatively unsaturated organometallics. Much of what is known about the spectroscopy and structure of these species comes from matrix isolation work (1,2). However, matrix techniques are very limited when kinetic information is desired. Additionally, the possibility of coordination of even relatively inert matrix gases, leading to a change in structure of the coordinated versus uncoordinated species exists.

It has long been recognized that it would be desirable to obtain real time kinetic information on coordinatively unsaturated organometallic compounds. Since coordination of hydrocarbon solvents by at least some coordinatively unsaturated organometallic species is a reality (14-16) and data is presented in this work that demonstrates that "solvents" as inert as Kr can be coordinated, the number of solvents that are truly non-coordinating may be quite limited. If a coordinatively unsaturated species has coordinated solvent, putative measurements of association rate constants can be effected by the finite rate of loss of the associated solvent molecules. As such it is desirable to be able to perform spectroscopic and kinetic measurements on "naked" coordinatively unsaturated species. This points to the gas phase as the phase of choice for such measurements. In this environment, these species can be studied in the absence of solvent molecules.

However, since coordinatively unsaturated species are very reactive, fast and sensitive spectroscopic techniques are needed for such studies. Since gas kinetic collisions occur on a microsecond timescale at a density of approximately 10^{15} molecules/cc, a system that is capable of detecting considerably fewer than 10^{15} molecules/cc on a submicrosecond timescale is desirable. Transient absorption systems based on a UV or visible probe that meet these criteria have been available for many years. However, many organometallic compounds and metal carbonyls, in particular, are extremely photolabile and their UV-visible spectra are not very structure sensitive (1-3). On the other hand, infrared spectra of the metal carbonyl absorption region is highly structure sensitive (1-3). Thus, time resolved IR is the spectroscopic probe of choice in these systems. Unfortunately infrared detectors are orders of magnitude less sensitive and slower than typical phototubes. Thus, time resolved infrared techniques have historically been plagued by a slow time response and/or a lack of sensitivity. Nevertheless, with advances in light sources and infrared detectors it has been possible to develop transient infrared spectrometers that are fast enough and sensitive enough to probe reactions of coordinatively unsaturated organometallic compounds that take place at the gas kinetic limit in a low pressure gas phase environment.

Much of the prior work in this area has been well summarized in a number of review articles (1,2). One of the interesting findings that comes out of this work is the uniformly large rate constants for the reaction of most coordinatively unsaturated metal carbonyls with the ligands studied to date. Most of these rate constants are within an order of magnitude of the gas kinetic limit and some are within a factor of two of this limit. An exception is the rate constant for reaction of $Fe(CO)_4$ with CO.

$$Fe(CO)_4 + CO \longrightarrow Fe(CO)_5$$

This reaction is almost 4 orders of magnitude slower than the gas kinetic limit. The source of this relatively small rate constant has been postulated to be the change in spin state in going from $Fe(CO)_4$, with a triplet ground state, to $Fe(CO)_5$, with a singlet ground state. Since a similar spin change has not been identified in other systems, the iron pentacarbonyl system currently offers the only available choice for additional study of the effect of this change in spin on rates and pathways for reaction. Thus, we felt it would be interesting to observe and quantify the effect of spin on reactions of coordinatively unsaturated metal carbonyls in the iron system with parent, CO and other small ligands. Much of this work has recently been reported on in detail in other publications (12,13).

We also report on a method that allows for the direct determination of bond energies based on measurements of the rate of dissociative loss of a ligand from the system under study. With this information and the rate constants for the association reaction of the relevant coordinatively unsaturated species with the ligand under study and with CO, the bond dissociation energy can be determined. This method has been applied to determine the bond energies for Group VI metal pentacarbonyls species complexed with heavy rare gases and allows for direct measurements of interactions between coordinatively unsaturated organometallic species and a wide variety of ligands (17). An area of particular interest to us is quantification of the interactions between coordinatively unsaturated metal carbonyls and typical solvent molecules.

We have also shown that measurements of bond energies and association rate constants allow us to determine whether a "poor" ligand does not form stable complexes because of kinetic or thermodynamic factors. We have applied this to both N_2O and CF_2Cl_2 to show that thermodynamic factors dominate in the making these "poor" ligands for binding to $W(CO)_5$ (18). For the N_2O system we have also applied time resolved Fourier transform infrared (TRFTIR) spectroscopy to the system which allows for extention of the convenient time range of our laser based systems to longer timescales.

Experimental

In the experiments under consideration we have used three different experimental set-ups. All the systems use either an excimer laser or tripled Nd:YAG laser to provide UV radiation to initiate the photolysis process leading to loss of one or more ligands. Since the wavelength of XeF excimer radiation at 351 nm and that for the frequency tripled output of the Nd:YAG laser at 355 nm are very similar it is not surprising that we have not observed any significant differences in product distributions with these two systems. The excimer laser has also been operated at 248 nm (KrF) and 193 nm (ArF). The Nd:YAG has been used in conjunction with a CO probe laser and the excimer has been used with either a CO or diode laser as the probe. For each of these systems the remainder of the experimental apparatus is similar and has been described in detail in prior publications (1,12,13,17-20).

Briefly, the output of either photolysis source makes a single pass through the sample cell. The photolysis laser pulses, at energies of typically less than 10 mj/cm^2, fill the entire cell volume. Various cell configurations were used with some cells having the capability for use of a purge gas to protect the cell windows from build up of involatile products. The specific advantages and difficulties associated with the use of such a configuration have been discussed in detail in previous publications (12,13). When flow cells were used the replenishment rate of the cell was typically 1–2 Hz with a photolysis rate of 1 Hz. When static cells were used the contents were manually replenished before a significant fraction of the contents of the cell were depleted. Static cells were typically employed when the reaction under study lead to regeneration of close to 100% of the starting material.

CO is always a product of photolysis of a metal carbonyl. This CO can be produced in vibrationally and rotationally excited states. Thus CO, along with the photogenerated coordinatively unsaturated metal carbonyls can produce a transient absorption signal. Since the CO laser by necessity operates on transitions that can be absorbed by internally excited CO it is possible for these CO absorptions to obscure the underlying absorptions of photogenerated coordinatively unsaturated metal carbonyls. This can be a significant problem when highly excited CO is produced that overlaps much of the spectral region in which the coordinatively unsaturated metal carbonyls also absorb. This problem was encountered in the $Fe(CO)_5$ system, particularly when a KrF or ArF laser was used as the photolysis source. This led us to use a diode laser for further study of this system (12,13). With the diode laser, which is continuously tunable, the frequency of operation can be chosen so that it does not overlap the discrete sets of frequencies that correspond to transitions for CO in various (v,J) states. The diode laser can also be tuned to higher frequencies than are convenient or possible with the CO laser. This facilitated the study of absorptions of a number of polynuclear metal carbonyls produced in reactions taking place in the $Fe(CO)_5$ system.

The TRFTIR apparatus is based on a Mattson rapid scan interferometer. In this system the entire interferrogram is recorded for a single laser pulse. We are currently able to acquire a spectrum in 27 msec at 8 cm^{-1} resolution and this acquisition procedure can be repeated at 60 msec intervals. There is a reciprocal trade-off between acquisition time and wavelength resolution. We view this system as complementary to our laser based TRIR systems. In the laser based systems long term noise places an effective limit on the timescale over which we can acquire a signal with suitable signal to noise levels. Though this timescale is a function of the wavelength and whether we use the diode laser or CO laser, signals with time constants longer than 10 msec are difficult to acquire without significant low frequency noise.

Experiments with the FTIR are implemented in an analogous manner to those using the laser based probes with the FTIR beam replacing the probe laser beam. The FTIR beam is incident on an external HgCdTe detector while the CO and diode laser beams are incident on InSb detectors with time response in the sub 100 nsec range.

In the TRIR experiments the probe device monitors the concentration of parent, reactant(s) and product(s) as a function of time. A transient spectrum is produced by having a computer connect points on transient signal taken at different wavelengths that are at a common time delay following the photolysis pulse. The time delay is then incremented to produce a nested set of spectra as shown in Figure 1. In all the figures in this paper, a positive going signal corresponds to less light falling on the detector and thus to production of a new species. A negative going feature corresponds to more light on the detector and thus less absorption by a species present in the cell.

For the TRFTIR experiments, manufacturer provided software was used to produce spectra very similar to those described above. However, for this system the interferrogram provided all the data points at a given time delay and subsequent interferrograms were the source of spectra at subsequent times. Thus, the basic difference between the two acquisition modes is that the TRIR system produces signals dense in time and sparse in wavelength (though with the diode laser the individual transients can be acquired at any arbitrary wavelength resolution) and the TRFTIR produces signals that are dense in wavelength and sparse in time. Other manufacturer provided software allows for the determination of decay rates at a given wavelength thus facilitating kinetic measurements.

When temperature dependent measurements were carried out, a jacketed cell was used with direct measurement of temperature using a number of thermocouples placed both outside and inside the cell. Temperature reproducibility is within 1°C and accuracy is estimated as ±1°C. $Cr(CO)_6$, $Mo(CO)_6$, and $W(CO)_6$ were obtained from Strem Chemicals, Alpha Products, and Aldrich Chemicals, respectively. $Fe(CO)_5$ was obtained from Alpha Products and all metal carbonyls were sublimed *in situ*. CO with a stated purity of 99.995% and N_2O and CF_2Cl_2 with the stated purities of 99% were obtained from Matheson and used without further purification.

Results

Reactions of Coordinatively Unsaturated Iron Carbonyls. We have monitored the reactions of coordinatively unsaturated iron carbonyls with parent and CO. Typical results are shown in Figure 1 for the reaction of $Fe(CO)_2$ with parent.

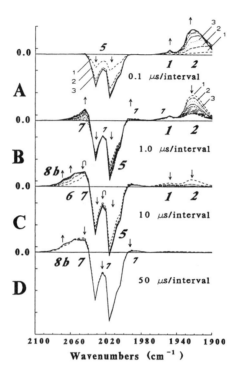

Figure 1. Time-resolved spectra generated upon 193-nm photolysis of Fe(CO)$_5$ for a sample containing 10 mTorr of Fe(CO)$_5$ and 10 Torr of argon buffer gas. Panels A-C are segmented into 10 equal time intervals with ranges of (A) 1.0 μs, (B) 10.0 μs, and (C) 100.0 μs. Panel D is segmented into 4 equal time intervals over a range of 200 μs. The ordinate is in arbitrary absorbance units with the zero of absorbance indicated. In panels A and B the first three traces are labelled. Arrows indicate the direction of change of various features during the indicated time intervals. See text for assignments. (Reprinted with permission from reference 12. Copyright 1991.)

The absorptions of the various coordinatively unsaturated iron carbonyls have been identified as a result of previous work (12,13). Products of reactions of coordinatively unsaturated iron carbonyls with parent can be identified since their rate of formation is the same as the rate of loss of parent and reactant. This procedure is illustrated in Figure 2 which displays plots of the rate of loss of $Fe(CO)_2$ and $Fe(CO)_5$ and the rate of formation of the reaction product, identified as $Fe_2(CO)_7$. These plots demonstrate that the reaction of $Fe(CO)_2 + Fe(CO)_5$ to form $Fe_2(CO)_7$ takes place. The effect of this reaction can also be seen in Figure 1 where the species labeled as 2 ($Fe(CO)_2$) reacts with parent (5) to form the species labeled as 7 ($Fe_2(CO)_7$). Assignments of polynuclear species can be further confirmed by observing the reaction of unsaturated polynuclear species with CO to form another polynuclear species which has previously been identified (1). The effect of this type of reaction can also been seen in Figure 1 where feature 7 decays at the same rate that feature 8 appears. Feature 8 has been shown to be due to the unbridged isomer of $Fe_2(CO)_8$. As expected, Features 7 and 8 also form as a consequence of the reaction of $Fe(CO)_2$ produced following 248 nm photolysis of $Fe(CO)_5$.

Another example of the latter identification procedure is shown in Figure 3 where the unbridged isomer of $Fe_2(CO)_8$, which forms as a product of the reaction of $Fe(CO)_3$ with $Fe(CO)_5$ following 351 nm photolysis of $Fe(CO)_5$, is observed to react with CO to form the known stable species, $Fe_2(CO)_9$. The unbridged isomer of $Fe_2(CO)_8$ has been identified based on its kinetic behavior and as a result of matrix isolation studies of isomers of $Fe_2(CO)_8$ (12). Note the isobestic point in Figure 3 which is characteristic of a direct A → B reaction.

Using these techniques we have been able to measure the rate constants for reaction of various coordinatively unsaturated iron carbonyls. A summary of these rate constants is shown in Table 1.

Measurement of Bond Energies. We have recently implemented a procedure to directly measure the bond dissociation energies of ligands bound to coordinatively unsaturated organometallic species (17,18). The procedure is based on the following kinetic scheme where an appropriate parent molecule is photolyzed in the presence of the ligand under study and CO.

$$W(CO)_6 \xrightarrow{\;h\nu\;} W(CO)_5 + CO \qquad (1)$$

$$W(CO)_5 + CO \xrightarrow{\;k_{CO}\;} W(CO)_6 \qquad (2)$$

$$W(CO)_5 + L \xrightarrow{\;k_L\;} W(CO)_5L \qquad (3)$$

$$W(CO)_5L \xrightarrow{\;k_d\;} W(CO)_5 + L \qquad (4)$$

In equations 1-4 $W(CO)_6$ is used as a prototypical parent molecule. This scheme involves a competition between CO and L for reaction with $W(CO)_5$. CO can react with $W(CO)_5$ to regenerated $W(CO)_6$. When L reacts with $W(CO)_5$ it forms

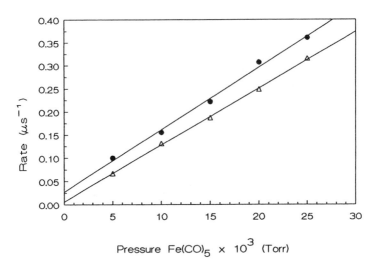

Pressure Fe(CO)$_5$ × 10^3 (Torr)

Figure 2. A plot of the pseudo-first-order rate constant for the reaction of Fe(CO)$_2$ + Fe(CO)$_5$. The rate of decay of Fe(CO)$_2$ at 1912 cm^{-1} (•) and the rate of formation of Fe$_2$(CO)$_7$ at 2044 cm^{-1} (Δ) are plotted against the pressure of Fe(CO)$_5$. The average slope of the linear least-squares fits gives a rate constant of 12 ± 1.8 μs^{-1}Torr^{-1} or (3.7 ±0.6) × 10^{-10} cm^3 molecule^{-1}s^{-1}. (Reprinted with permission from reference 12. Copyright 1991.)

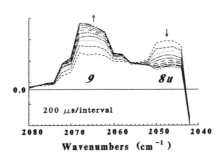

Figure 3. Time-resolved spectrum generated upon 351-nm photolysis of Fe(CO)$_5$. The photolysis flow cell contained 0.5 Torr of Fe(CO)$_5$ with 5 Torr of CO and 1.25 Torr of argon buffer gas. The data are shown in 10 equal intervals over a range of 2.0 ms. the reaction of unbridged Fe$_2$(CO)$_8$ + CO → Fe$_2$(CO)$_9$ leads to decay of feature 8u, assigned to unbridged Fe$_2$(CO)$_8$ in the text, and the rise of feature 9, assigned to ground-state Fe$_2$(CO)$_9$. Arrows indicate the direction of change of the absorption features. (Reprinted with permission from reference 12. Copyright 1991.)

$W(CO)_5L$. If L is weakly bound to $W(CO)_5$, $W(CO)_5L$ may dissociate and allow another opportunity for reaction of $W(CO)_5$ with either L or CO. Since $W(CO)_6$ is stable on the timescale of these experiments the reaction of $W(CO)_5$ + CO to form $W(CO)_6$ is considered to be irreversible (17). Reactant pressures and temperatures are typically chosen so that the rate of dissociation of $W(CO)_5L$ is rate limiting in the reaction to regenerate $W(CO)_6$. Under these circumstances, the kinetic scheme consisting of reactions (1-4) can be solved for the rate of reformation of $W(CO)_6$ using the steady state approximation. This leads to an expression for the rate of regeneration of $W(CO)_6$ which we designate as k_{obs} where

$$k_{obs} = \frac{k_d k_{CO}[CO]}{k_{CO}[CO] + k_{Xe}[Xe]} \tag{5}$$

This expression can be rearranged to give an expression for k_d, the rate constant for reaction (4).

$$k_d = k_{obs}\left[1 + \frac{k_{Xe}[Xe]}{k_{CO}[CO]}\right] \tag{6}$$

A measurement of the temperature dependence of k_d along with a direct measurement of the temperature dependence of the rate constant for the reverse reaction (3) can be used to obtain the temperature dependence of the equilibrium constant for the formation of $W(CO)_5L$. This of course can be used to obtain ΔH for the $W(CO)_5L$ bond. In practice, the reaction of coordinatively unsaturated species with either CO or L is typically unactivated indicating a reaction occurring without a barrier. Under these circumstances, the temperature dependence of k_d can be used to obtain the activation energy for reaction 4 which can be appropriately converted to a bond enthalpy (18).

A typical signal for k_{obs} is shown in Figure 4 where the reaction of $W(CO)_5$ with Xe is occurring and regeneration of $W(CO)_6$ is being monitored. From the temperature dependence of k_d and the rate constants for reaction of $W(CO)_5$ with Xe and CO it was possible to determine that the bond energy for the $W(CO)_5Xe$ complex and the other group VI $M(CO)_5Xe$ complexes is 8.5 ± 0.5 kcal/mole (17).

We have also employed this method in an effort to measure the bond dissociation energy for the $W(CO)_5(N_2O)$ complex. N_2O reacts rapidly with $W(CO)_5$, but at or near room temperature the rate of dissociative loss of N_2O was too slow to be conveniently followed using a CO laser (18). However, the kinetics of dissociative loss of N_2O from $W(CO)_5(N_2O)$ can be monitored using TRFTIR. A time resolved spectrum showing the disappearance of the absorption due to the $W(CO)_5(N_2O)$ complex is shown in Figure 5. The temperature dependence of this rate constant is shown in Figure 6. From this data, taken in conjunction with previous measurements of the association rate constants for CO and N_2O, a bond dissociation enthalpy of 14.6 ± 0.5 kcal/mole and a

Table I. Rate Constants for Reactions of Coordinatively Unsaturated
Iron Carbonyls

Species	Reaction Rates, cm^3 molecule$^{-1}s^{-1}$ +Fe(CO)$_5$	+CO	Spin Conserving
Fe(CO)	$(8.6\pm2.5) \times 10^{-11}$	a	yes
Fe(CO)$_2$	$(3.7\pm0.6) \times 10^{-10}$	$(3.0\pm0.5) \times 10^{-11}$	yes
Fe(CO)$_3$	$(2.9\pm0.4) \times 10^{-10}$	$(2.2\pm0.33)\times 10^{-11}$	yes
Fe(CO)$_4^b$	$(5.2\pm1.5) \times 10^{-13}$	$(5.2\pm1.2) \times 10^{-14}$	no
Fe(CO)$_4^*$	$(1.8\pm0.3) \times 10^{-10}$	a	yes
Fe$_2$(CO)$_7$	a	$(9 \pm 3) \times 10^{-11}$	yes
Fe$_2$(CO)$_8$ (unbridged)	a	$(7.8\pm2.3) \times 10^{-15}$	no

a. Not studied.
b. Rates of reactions of Fe(CO)$_4$ with H$_2$ and O$_2$ have also been measured.
*   = excited state

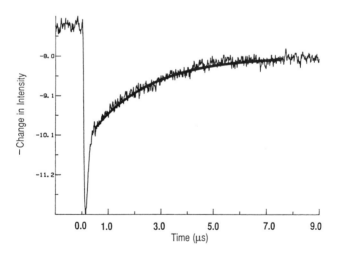

Figure 4. Transient signal obtained at 2000 cm^{-1} when 5-10 millitorr of W(CO)$_6$, 5.2 Torr Xe, 80 Torr He, and 1.2 Torr CO are photolyzed with 355-nm radiation. Decrease is due to loss of W(CO)$_6$ while increase in signal is return of W(CO)$_6$. Line drawn through slow part of W(CO)$_6$ regeneration is the computed fit, k$_{obs}$. (Reprinted with permission from ref. 17. Copyright 1992.)

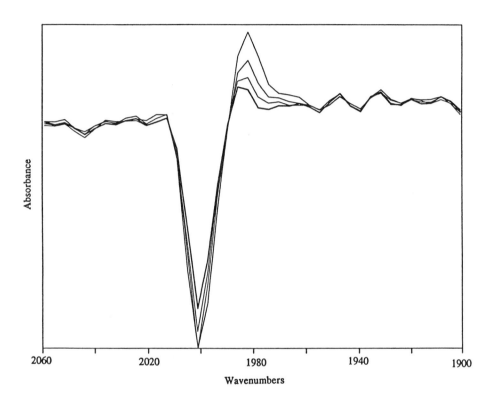

Figure 5. Time-resolved FTIR spectrum of the loss of a $W(CO)_5(N_2O)$ complex (1982 cm^{-1}) leading to regeneration of parent (2000 cm^{-1}). The photolysis cell contained 76 Torr N_2O and 2 Torr of CO in addition to 10 millitorr of $W(CO)_6$. The spectra which are shown at 90 msec time intervals are the result of an average of 200 shots from a frequency tripled YAG laser with an output at 355 nm. These spectra were acquired at 23 °C.

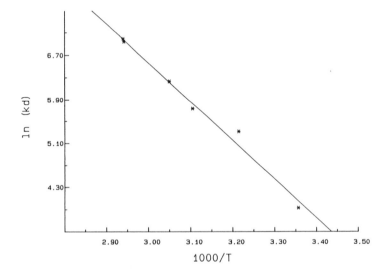

Figure 6. Temperature dependence of k_d (see equation 4) plotted as $\ln(k_d)$ vs. 1/T. A slope of 7.0 ± 0.4 is obtained from the data.

preexponential of 1×10^{12} sec^{-1} can be determined for the $W(CO)_5(N_2O)$ complex.

Discussion

Reactions That Produce Polynuclear Iron Carbonyls. Reactions of coordinatively unsaturated metal carbonyls with parent to form polynuclear carbonyls have been studied in a number of systems. In virtually all of these systems, except the iron system, all the measured rate constants are within an order of magnitude of the gas kinetic limit and many are within a factor of two of that limit (1,12). This is also true for coordinatively unsaturated species generated from other members of the Group VIII pentacarbonyls. Studies of reactions of coordinatively unsaturated osmium and ruthenium carbonyls with parent show they are very fast (19,20). The reaction of CO with the tetracarbonyls of the other group VIII metals, which have been calculated to have a C_{2v} geometry, as does $Fe(CO)_4$, are fast (19,20).

Thus, the iron carbonyl system appears to be unique. As previously mentioned, this is also the system that, to date, exhibits the smallest rate constant for addition of CO to a mononuclear coordinatively unsaturated metal carbonyl. These observations have been interpreted in terms of the effect of spin conservation on rate constants (1). The ground state of $Fe(CO)_4$ has been experimentally determined to be a triplet (21). The ground state of $Fe(CO)_3$ has been calculated to be a triplet (22). Thus addition of CO to $Fe(CO)_3$ to form $Fe(CO)_4$ will be spin allowed but the addition of CO to $Fe(CO)_4$ to form $Fe(CO)_5$ is spin disallowed. This could explain the source of the almost three orders of magnitude difference in the rate constants for these processes. The rate constant for addition of CO to $Fe(CO)_2$ is of the same magnitude as the rate constant for addition of CO to $Fe(CO)_3$. To explain this it has been postulated that $Fe(CO)_2$ also has a triplet ground state (1).

The rate constants we have measured for the reactions of coordinatively unsaturated metal carbonyls with parent and CO and for CO with polynuclear coordinatively unsaturated carbonyls, are consistent with this hypothesis. For example $Fe(CO)_3$ reacts with parent at a near gas kinetic rate to form the unbridged isomer of $Fe_2(CO)_8$. Within the context of our picture, this rapid rate of reaction implies spin conservation which in turn implies this isomer of $Fe_2(CO)_8$, first identified in matrix isolation studies, has a triplet ground state (23). Consistent with a triplet ground state, this isomer of $Fe_2(CO)_8$ has a relatively small rate constant for reaction with CO to give $Fe_2(CO)_9$, which has a singlet ground state, making this reaction formally spin disallowed. As expected the reaction of $Fe(CO)_4$ with $Fe(CO)_5$ to form $Fe_2(CO)_9$ is also relatively slow since this reaction is also formally spin disallowed.

Excited states of both $Fe(CO)_3$ and $Fe(CO)_4$ have been identified in prior studies of the iron pentacarbonyl system (12). The first excited states of these species have been predicted to be singlets (24). Though these species are difficult to study because they are rapidly collisionally relaxed to their ground states, what has been discerned about their reactive behavior indicates that rate constants for reaction are significantly influenced by the spin states of the reactants and products (15). For example, the rate constant for reaction of the excited state of $Fe(CO)_4$ with parent to form an isomer of $Fe_2(CO)_9$ is near gas kinetic and almost three orders of magnitude faster than rate constant for reaction of the ground state of $Fe(CO)_4$ with parent.

As a further test of the influence of spin states of reactants and products on rate constants for reactions, the reaction of H_2 and O_2 with $Fe(CO)_4$ have been measured (12). H_2 was chosen as another small diatomic molecule with a singlet ground state that would provide a complementary system to the $Fe(CO)_4 + CO$ system. O_2 was chosen because it has a $(X^3\Sigma_g^-)$ ground state. Thus a reaction of O_2 with $Fe(CO)_4$ might be expected to have a larger rate constant than a singlet reactant. In support of this hypothesis, triplet organic diradicals are typically observed to react rapidly with ground state O_2 while analogous singlet diradicals show greatly reduced reactivity (25).

The rate of reaction of O_2 with $Fe(CO)_4$ is more than 50 times larger than the rate of reaction of CO with $Fe(CO)_4$ under the same conditions. Since these studies were intended to compare these reaction rates under similar conditions, they did not involve a complete study of the pressure dependence of the measured rate constant for the reaction with O_2. As the authors point out, since the total pressure in the experiments involving O_2 was only 10 Torr it is possible that this reaction was not measured in the high pressure limit (12). If this were the case, the limiting (high pressure) rate constant for the reaction of $Fe(CO)_4$ with O_2 would be even larger and thus would approach the rate for the addition reaction of CO involving spin allowed processes.

The rate constant for reaction of H_2 to $Fe(CO)_4$ to form $H_2Fe(CO)_4$ is the same, within experimental error, as that measured for addition of CO to $Fe(CO)_4$. This provides evidence that it is not the specific reaction of CO with $Fe(CO)_4$ that is an anomaly but rather that the class of spin disallowed reactions of $Fe(CO)_4$ with the small ligands have anomalously small rate constants.

The results discussed above are compatible with spin conservation exerting a significant effect on the magnitude of the rate constant for reactions of coordinatively unsaturated metal carbonyls. However, it is clear that this is not the only factor that influences rate constants. There is an order of magnitude difference between rate constants for the formally spin disallowed addition reaction of CO to $Fe(CO)_4$ and to the unbridged isomer of $Fe_2(CO)_8$ (12). Differences in rate constants of similar magnitude have been observed for rapid reactions as well. The reaction of $Fe(CO)_3$ with ethylene is an order of magnitude faster than the corresponding reaction with CO (26). Reactions of $Fe(CO)_3$ and $Cr(CO)_4$ with other olefins are also very fast (27). The detailed explanations for these differences remain to be elucidated.

Bond Dissociation Energies. We have implemented a technique, described in the Results section, that involves a general method for determination of bond dissociation energies of organometallic species. Since laser based probes of reaction kinetics (TRIR) can be used to monitor fast association and/or dissociation reactions the technique can be used to determine bond dissociation energies of even very weakly bound ligands. The use of time resolved fourier transform spectroscopy (TRFTIR) has allowed us to extend our ability to monitor dissociative reactions to long time scales leading to a widely applicable technique for determining bond dissociation energies.

We have found that even the most weakly interacting of the ligands for which we have been able to measure reaction rates for ligand association reactions have rate constants that are of a similar order of magnitude as rate constants for association reactions involving strongly interacting ligands. It seems likely that it is generally the case that these species are "poor" ligands because they form weak bonds with the coordinatively unsaturated species. However, a bond of over 14 kcal/mole, as has been measured for the bonding of N_2O to $W(CO)_5$, is certainly not an insignificant bond energy. In fact this species has a halflife of ~.1 sec at 0°C. This is not far from the effective time limit for detection by normal infrared techniques.

This bond energy is somewhat less than that estimated in reference 18. It is worthwhile considering why original estimates for the bond dissociation energy of the $W(CO)_5(N_2O)$ complex were higher than the currently reported number. In the original set of experiments, it was not possible to directly measure the rate for dissociative loss of N_2O since it took place on a longer timescale than our laser based system was able to measure but on a shorter time scale than observable with conventional IR spectroscopy. As such our estimate of the rate of regeneration of $W(CO)_6$, due to dissociative loss of N_2O, was bounded by these time regimes. Based on these time regimes and the assumption of the same preexponential for dissociative loss of N_2O as we had measured for dissociative loss of another "poor" ligand, CF_2Cl_2, we were able to obtain estimates for the activation energy for the loss process and for the corresponding bond enthalpy. Our current measurements directly measure the rate of regeneration of $W(CO)_6$ and directly determine both the preexponential and the bond enthalpy for the loss of N_2O. Interestingly we find that the preexponential for loss of N_2O from $W(CO)_5$ is significantly smaller than the preexponential for loss of CF_2Cl_2 from the same parent. Though this change in preexponential is likely to reflect information about bonding geometries and force constants, more work is necessary before firm conclusions can be drawn regarding the significance of the change in magnitude of the preexponentials.

Data on the binding of Xe and Kr to the Group VI metal pentacarbonyls illustrates the extreme reactivity of coordinatively unsaturated metal carbonyls (17). This leads to the expectation that coordinatively unsaturated organometallic compounds will interact, in some cases strongly, with many solvents. A concern under these circumstances is that the species being studied in solution are not the "naked" coordinatively unsaturated species but a coordinated molecule. Attempts to study association reactions under these circumstances could lead to kinetic data that was influenced by dissociative or associative displacement of the solvent molecule(s). It is interesting to note that even though it may not have been generally recognized that coordinatively unsaturated metal carbonyls are sufficiently reactive to coordinate common solvents, data to support this point of view has been available for over 10 years. In the early '80s Bonneau and Kelly observed a 3 order of magnitude decrease in the rate of addition of CO to $Cr(CO)_6$ in cyclohexane versus perfluorocycloheptane (16). In 1982 Vaida and Peters reported that $Cr(CO)_5$ associated with simple solvents on a psec timescale (14).

Part of our future work in this area will involve measurements of the bond dissociation energies of common solvents coordinated to Group VI pentacarbonyls in an effort to quantify how significantly these interactions can and have effected solution phase kinetic studies of these species. We are also interested in further elucidating the bonding mechanism for such species.

Summary

Rate constants for reactions of coordinatively unsaturated iron carbonyls with $Fe(CO)_5$ are consistent with conservation of spin playing a role in determining the magnitude of these rate constants.

Rates of reaction of H_2 and O_2 with $Fe(CO)_4$ are also consistent with spin conservation being a factor in determining rate constants for reactions of metal carbonyls.

Variations in rate constants for spin allowed reactions of metal carbonyls indicate the factors other than spin conservation can significantly effect rate constants for association reactions.

A technique to measure bond enthalpies, which involves a real time monitor of the rate of dissociation of a organometallic species, is described.

Using this technique measurements have been made of the bond enthalpies for Xe bound to the Group VI pentacarbonyls. Bond enthalpies are 8.5 ± 0.5 kcal/mole for the three pentacarbonyl complexes.

Using this technique and time resolved fourier transform infrared spectroscopy, the bond dissociation enthalpy of a $W(CO)_5(N_2O)$ complex has been determined to be 14.1 ± 0.5 kcal/mole.

Acknowledgments. We acknowledge support of this work by the National Science Foundation under grants, CHE-88-06020 and 90-24509. We also would like to thank the donors of the Petroleum Research Fund administered by the American Chemical Society for partial support of this work under grant 18303-AC6,3. We thank Professor Martyn Poliakoff for a number of useful discussions and suggestions which were facilitated by NATO support under a collaborative grant. We also acknowledge Dr. Paula Bogdan for her role in implementing the technique for determining bond dissociation enthalpies.

References
1.   Weitz, E. *J. Phys. Chem.* **1987**, *91*,   3945.
2.   Poliakoff, M.; Weitz, E. *Adv. Organometallic Chem.* **1986**, *25*, 277; Poliakoff, M.; Weitz, E. *Acc. Chem. Res.* **1987**, *20*, 408.
3.   Geoffroy, G. L.; Wrighton, M. S. *Organometallic Photochemistry*; Academic: NY, 1979.
4.   Wrighton, M. S.; Ginley, D. S.; Schroeder, M. A.; Morse, D. L. *Pure Appl. Chem.* **1975**, *41*, 671; Wrighton, M. S.; Ginley, D. S.; Schroeder, M. A.; Morse, D. L. *Pure Appl. Chem.* **1975**, *41*, 671; Whetten, R. L.; Fu, K. J.; Grant, E. R. *J. Am. Chem. Soc.* **1982**, *104*, 4270; Weiller, B. H.; Grant, E. R. *J. Am. Chem. Soc.* **1987**, *109*, 1051.

5. Whetten, R. L.; Fu, K. J.; Grant, E. R. *J. Chem. Phys.* **1982**, *77*, 3769;Wrighton, M. S.; Ginley, D. S.; Schroeder, M. A.; Morse, D. L. *Pure Appl. Chem.* **1975**, *41*, 671; Schroeder, M. A.; Wrighton, M. S. *J. Am. Chem. Soc.* **1976**, *98*, 551; Schroeder, M. A.; Wrighton, M. S. *J. Am. Chem. Soc.* **1976**, *98*, 551.
6. Mitchener, J. C.; Wrighton, M. S. *J. Am. Chem. Soc.* **1981**, *103*, 975; Trusheim, M. R.; Jackson, R. L. *J. Phys. Chem.* **1983**, *87*, 1910; Jackson, R. L.; Trusheim, M. R. *J. Am. Chem. Soc.* **1982**, *104*, 6590; Darsillo, M. S.; Gafney, H. D.; Paquette, M. S. *J. Am. Chem. Soc.* **1987**, *109*, 3275.
7. Ray, U.; Brandow, S. L.; Bandukwalla, G.; Venkataraman, B. K.; Zhang, Z.; Vernon, M. *J. Chem. Phys.* **1988**, *89*, 4092; Venkataraman, B. K.; Bandukwalla, G.; Zhang, Z.; Vernon, M. *J. Chem. Phys.* **1989**, *90*, 5510.
8. Waller, I. M.; Hepburn, J. W. *J. Chem. Phys.* **1988**, *88*, 6658; Waller, I. M.; Davis, H. F.; Hepburn, J. W. *J. Phys Chem.* **1987**, *91*, 506.
9. Fletcher, T. R.; Rosenfeld, R. N. *J. Am. Chem. Soc.* **1988**, *110*, 2097; Rosenfeld, R. N.; Ganske, J. A. *J. Phys. Chem.* **1989**, *93*, 1959.
10. Yardley, J. T.; Gitlin, B.; Nathanson, G.; Rosan, A. M. *J. Chem. Phys.* **1981**, *74*, 370.
11. Rayner, D. M.; Ishikawa, Y.; Brown, C. E.; Hackett, P. A. *J. Chem. Phys.* **1991**, *94*, 5471; Ishikawa, Y.; Brown, C. E.; Hackett, P. A.; Rayner, D. M. *J. Phys. Chem.* **1990**, *94*, 2404.
12. Ryther, R. J.; Weitz, E. *J. Phys. Chem.* **1991**, *95*, 9841.
13. Ryther, R. J.; Weitz, E. *J. Phys. Chem.*, **1992**, *96*, 2561.
14. Welch, J. A.; Peters, K. S.; Vaida, V. *J. Phys. Chem.* **1982**, *86*, 1941.
15. Wang, L.; Zhu, X.; Spears, K. G. *J. Am. Chem. Soc.* **1988**, *110*, 8695; Simon, J. D.; Xie, X. *J. Phys. Chem.* **1986**, *90*, 6751; Lee, M.; Harris, C. B. *J. Am. Chem. Soc.* **1989**, *111*, 8963; Yu, S.; Xu, X.; Lingle, R. Jr.; Hopkins, J. B. *J. Am. Chem. Soc.* **1990**, *112*, 3668.
16. Bonneau, R.; Kelley, J. M. *J. Am. Chem. Soc.* **1980**, *102*, 1220; Kelly, J. M.; Long, C.; Bonneau, R. *J. Phys. Chem.* **1983**, *87*, 3344.
17. Wells, J. R.; Weitz, E. *J. Am. Chem. Soc.*, **1992**, *114*, 2783.
18. Bogdan, P. L.; Wells, J. R.; Weitz, E., *J. Am. Chem. Soc.* **1991**, *113*, 1294.
19. Bogdan; P. L.; Wells, J. R.; Weitz, E. *J. Am. Chem. Soc.* **1990**, *112*, 639.
20. Bogdan; P. L.; Wells, J. R.; Weitz, E. *J. Am. Chem. Soc.* **1990**, *111*, 3163.
21. Barton, T. J.; Grinter, R.; Thomson, A. J.; Davies, B.; Poliakoff, M. *J. Chem. Soc. Chem. Commun.* **1977**, 841.
22. Daniel, C.; Bénard, M.; Dedieu, A.; Wiest, R.; Veillard, A *J. Phys. Chem.* **1984**, *88*, 4805.
23. Poliakoff, M.; Turner, J. J. *J. Chem. Soc. A.* **1971**, 2403; Fletcher, S. C.; Poliakoff, M.; Turner, J. J. *Inorg. Chem.* **1986**, *25*, 3597.
24. Daniel, C.; Bénard, M.; Dedieu, A.; Wiest, R.; Veillard, A *J. Phys. Chem.* **1984**, *88*, 4805.
25. Berson, J. A. In *Diradicals*; Borden, T. W., Ed.; Wiley-Interscience: New York, **1982**; Chapter 4.
26. Hayes, D.; Weitz, E. *J. Phys. Chem.* **1991**, *95*, 2723.
27. Gravelle, S. J.; Weitz, E. *J. Am. Chem. Soc.* **1990**, *112*, 7839.

RECEIVED March 30, 1993

Chapter 13

Time-Resolved IR Spectroscopy of Transient Organometallic Complexes in Liquid Rare-Gas Solvents

Bruce H. Weiller

Mechanics and Materials Technology Center, The Aerospace Corporation, P.O. Box 92957, Los Angeles, CA 90009

Photolysis of $M(CO)_6$ (M = Cr and W) in low-temperature, rare-gas solutions leads to the formation of transient organometallic complexes between $M(CO)_5$ and weak ligands, including CO_2, N_2O, Xe and Kr. Time-resolved IR spectroscopy was used to capture IR spectra of the complexes over the temperature range of 150 to 200 K. A detailed kinetic investigation of the reaction of $W(CO)_5Xe$ with CO in liquid Xe is presented. From the temperature dependence of the kinetics, we obtain the binding enthalpy of Xe to $W(CO)_5$: $\Delta H = 8.6 \pm 0.4$ kcal/mol, in good agreement with independent gas-phase results. This establishes the utility of the technique for determining binding energies of weak ligands to electron-deficient metal centers.

Organometallic chemical vapor deposition (OMCVD) is a widely used technique for the deposition of thin-film materials for electronic, optical, and protective coatings. Central to this process are chemical reactions that occur between the organometallic precursor compounds, added reagents, and the heated surface. Reactions in the gas-phase and on the surface can have a profound influence on the properties of the resulting materials. In order to control and modify the properties of the materials produced by OMCVD, it is critical to elucidate the reaction mechanisms involved in the conversion of organometallic precursors to thin-film materials. This includes the structures of the reactive intermediates involved and the rates of their elementary reactions.

One reaction common to many OMCVD systems is complex formation between added reagents and electron-deficient metal centers. For example, in the deposition of III-V and II-VI semiconductors such as GaAs and ZnS, the reactants are often Lewis acid-base pairs such as $Ga(CH_3)_3$ and AsH_3 or $Zn(CH_3)_2$ and H_2S. Clearly one of the first reactions to occur in these systems will be complex formation. Laser-assisted OMCVD is another area where complex formation is important. Lasers can be used to deposit thin films at low temperatures with spatial selectivity by photolyzing organometallic precursors. Often metal carbonyls are used alone or in mixtures with other reagents in the laser-assisted OMCVD of metals and

0097–6156/93/0530–0164$06.00/0

ceramics. For example, CrO_2, a magnetic material commonly used in magnetic storage media, can be formed at room temperature by photolysis of mixtures of $Cr(CO)_6$ and O_2 or other oxidants (1). Photodissociation of CO from metal carbonyls is facile, and complex formation between coordinatively unsaturated fragments and added reagents is likely. The stability of the complexes formed and the subsequent reaction chemistry will play a significant role in determining the properties of the resulting material.

In order to study the intermediates important in OMCVD processes and their chemical reactions, we use a novel approach, laser photolysis in liquid rare-gas solvents with time-resolved IR spectroscopy. This is a powerful combination of techniques that slows reaction rates and provides accurate kinetic parameters by the use of the low temperatures. Furthermore, rare-gas solvents are inert, and the IR spectra of dissolved organometallics are simple with relatively narrow bands and no solvent absorptions. The utility of liquid rare-gas solvents for organometallic reaction studies (2) and for the synthesis of novel organometallic compounds (3) has been demonstrated. The advantage of combining liquid rare-gas solvents with time-resolved IR spectroscopy was shown in recent studies that reported the first direct rate measurement of organometallic C-H bond activation (4).

In the studies presented here, we have used a commercial FTIR spectrometer and liquid rare-gas solvents to examine the photochemistry relevant to the formation of metal-oxide films from metal carbonyls and N_2O or CO_2. During laser photolysis of mixtures of $Cr(CO)_6$ or $W(CO)_6$ and N_2O in liquid rare-gas solutions, relatively stable complexes are formed that are implicated in the laser-assisted OMCVD of metal oxides. CO_2, which is isoelectronic with N_2O, forms a much shorter-lived complex. In addition, we have investigated the complexes formed with the rare gases Xe and Kr. We present the first spectral data for $Cr(CO)_5Kr$ and $W(CO)_5Kr$ in fluid solution and also a detailed kinetic study for the reaction of $W(CO)_5Xe$ with CO. From the temperature-dependent kinetics, we obtain a value for the binding energy of Xe to $W(CO)_5$: 8.6 ± 0.4 kcal/mol ($\pm \sigma$). This result is in excellent agreement with a recent independent gas-phase measurement (5) and demonstrates the utility of the technique for determining binding energies.

Experimental

In these experiments, we used an excimer laser (Questek) to photolyze metal carbonyls dissolved in liquid rare-gas solvents and a commercial FTIR spectrometer (Nicolet, Model 800) that can be operated in a rapid-scan mode. To liquefy the rare gases, we used a specially designed high-pressure, low-temperature cell, shown in Figure 1. The cell, which was mounted in the sample compartment of the FTIR spectrometer, consists of a copper block with two perpendicular optical axes 1.43 cm and 5.0 cm in length. The cell is enclosed in an evacuated dewar and is cooled with liquid nitrogen. The temperature is measured with silicon diodes (± 0.1 K) and controlled via heaters. Samples containing metal carbonyls, reactants (N_2O, CO_2, or CO) and rare gas were prepared on a stainless-steel vacuum line. For the rapid-scan FTIR spectra, the interferometer mirror was moved at high velocity, and the laser was synchronized with the take-data signal of the spectrometer. No signal averaging was used. The laser beam was aligned along the long axis of the cell, and the IR beam passed through the short axis. A shutter (Uniblitz) served to block the laser until the start command was given at the keyboard of the computer. A TTL pulse from the computer then opened the shutter to let a *single* laser pulse irradiate the sample. Interferograms were then collected in only one mirror direction at equal

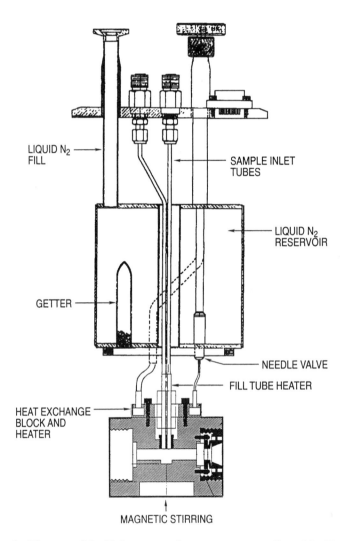

Figure 1. Diagram of the high-pressure, low-temperature cell used for IR spectroscopy of rare-gas solutions. Not shown is the external vacuum dewar.

time intervals after the laser pulse for the desired time period. For the experiments reported herein, spectra were collected at 8 cm^{-1} resolution and a time interval of 0.09 s using an MCT detector. Spectra were collected in the forward direction only, and the laser was synchronized to fire 2 ms prior to the start of data collection. After data collection, the interferograms were transformed, and converted to absorbance, and the difference was taken with reference spectra. For kinetic measurements, peak areas were used in order to avoid complications from potential distortions in peak widths. CO concentrations were determined from the integrated absorbance of the CO band using the gas-phase band strength (6) corrected for the index of refraction (7). Chemicals were obtained from the following suppliers: $Cr(CO)_6$ and $W(CO)_6$ (Alfa), Xe and Kr (Spectra Gases, research grade), CO_2 and CO (Matheson, research grade), and N_2O (Puritan-Bennett, medical grade). CO was purified with a liquid-N_2 cooled trap; the other chemicals were used without further purification.

Results and Discussion

$M(CO)_5(N_2O)$ **and** $M(CO)_5(CO_2)$. In order to understand the deposition of metal oxides by laser-assisted OMCVD, we have started an investigation into the complexes formed between metal carbonyl fragments and N_2O and CO_2. As discussed in the introduction, metal oxides can be formed at room temperature by photolysis of metal carbonyls and an oxidizer such as O_2, N_2O or CO_2. Relatively little is known about this process, but earlier studies in low-temperature matrices (8) have shown the existence of complexes between $M(CO)_5$ and N_2O and CO_2. Time-resolved experiments in the gas phase have also confirmed the existence of a complex with N_2O (9).

$M(CO)_5(N_2O)$. Figure 2 shows the IR spectra obtained when a solution of $W(CO)_5$ and N_2O dissolved in liquid Kr is photolyzed at 351 nm. The spectrum is presented as the difference between spectra before photolysis and at ~1-min. intervals after photolysis. A new species is formed as shown by the positive peaks at 2092, 1968, and 1954 cm^{-1} that slowly decays to reform $W(CO)_6$. Using the rapid-scan capability of the spectrometer, we have also observed N_2O complex formation in the gas phase. Very similar results were obtained for $Cr(CO)_6$ and N_2O in the gas phase and in solution (Table I). It should be noted that our observation of the high-frequency a_1 band is the first reported for these complexes and confirms their proposed C_{4v} symmetry. As shown in Table I, our results in the gas phase and in liquid Kr solution are in good agreement with independent gas-phase measurements. However, both sets of results diverge significantly from matrix isolation studies.

$M(CO)_5(CO_2)$. CO_2 is isoelectronic with N_2O and can also be used as an oxidant to form metal oxides. When solutions of $W(CO)_5$ and CO_2 in liquid Kr were photolyzed at 161 K, no stable complexes could be observed. However, using the rapid-scan mode of the spectrometer, it was possible to observe the formation of $W(CO)_5(CO_2)$, as shown in Figure 3. The time between spectra is 0.09 s, and the complex decays orders of magnitude faster than the analogous N_2O complex. A reasonable explanation for this observation is a weaker interaction between $W(CO)_5$ and CO_2. However, a definite explanation will have to wait for planned kinetic studies of the decay process. We can rule out polynulcear species since we see no CO_2 complex are in excellent agreement with the matrix isolation data, unlike the case for N_2O.

The apparent difference in the IR band positions of $M(CO)_5(N_2O)$ in the gas-phase and in liquid Kr versus those observed in low-temperature matrices is

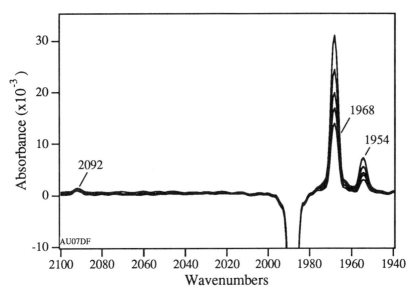

Figure 2. IR spectra of $W(CO)_5(N_2O)$ formed in liquid Kr at 180 K.

Table I. Observed IR Frequencies for CO_2 and N_2O Complexes

Complex	Frequencies (cm^{-1})			Conditions	Reference
$W(CO)_5(N_2O)$		1977	--	gas, 298 K	this work
		1980	1967	gas, 298 K	9
	2092	1968	1954	l. Kr, 180 K	this work
		1950	1925	s. Ar, 10 K	8
$Cr(CO)_5(N_2O)$		1979	--	gas, 298 K	this work
	2089	1972	1955	l. Kr, 180 K	this work
		1950	1925	s. Ar, 10 K	8
$W(CO)_5(CO_2)$		1957	1930	l. Kr, 180 K	this work
		1956	1926	s. Ar, 10 K	8
$Cr(CO)_5(CO_2)$		1952	1925	s. Ar, 10 K	8

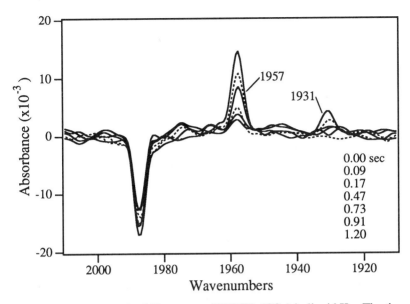

Figure 3. Time-resolved IR spectra of $W(CO)_5(CO_2)$ in liquid Kr. The times after the laser pulse are shown in the legend.

significant. The data for the CO_2 complex and for Xe complexes (see below) are in good agreement with matrix data and eliminates any doubt about the experimental evidence for bridging CO bands. Finally, we note that the observed bands for the technique. Therefore, we must look to other reasons to explain the large difference with the matrix isolation spectra ($\Delta v = 18$ cm^{-1}). One possibility is the formation of different isomers in the higher-temperature experiments (liquid Kr or the gas phase) versus the cryogenic ones (matrices). The nature of the bonding in the N_2O complexes is an open question, but calculations indicate that the N-bonded isomer is more stable (10). It may be that a less-stable isomer, such as an O-bonded species, is formed in the ultra-cold, rigid matrix that is not stable at higher temperatures (>150 K). Further investigations are underway to address this question and to quantify the bond energies in these complexes.

$M(CO)_5Kr$ and $M(CO)_5Xe$. The question of complex formation between $M(CO)_5$ and rare-gas atoms is not new. Early matrix isolation studies showed, from shifts in the lowest electronic transition, that $M(CO)_5$ interacts significantly with Ar, Kr, and Xe (11). The strength of the interaction was correlated with the polarizability of the rare-gas atoms. Direct evidence for a bonding interaction between $Mn(CO)_5$ and Kr was obtained from EPR spectra in Kr matrices (12). Finally, $Cr(CO)_5Xe$ was observed in liquid-Xe solutions with an FTIR spectrometer (13). These workers presented a rough estimate of the bond energy for Xe to $Cr(CO)_5$, but, until this symposium, there were no reliable values for the binding energy of Xe to metal centers. Now, from the gas-phase work (5) and this work in liquid Xe, we have two independent measurements of the value for $W(CO)_5Xe$.

$M(CO)_5Kr$. In order to determine the role of the solvent in liquid-Kr studies, we have examined the transients formed when $W(CO)_6$ and $Cr(CO)_6$ are photolyzed in liquid Kr with no added reagents. During the time-resolved experiments with

CO_2, we observed a short-lived transient prior to the formation of the CO_2 complex that was not observed in experiments with large concentrations of CO_2. When $W(CO)_6$ was photolyzed in liquid Kr in the absence of added reagents, we observed an identical transient, as shown in Figure 4. This new species, which we assign to $M(CO)_5Kr$, is short lived even at 154 K, with a lifetime of ~ 100 ms. Similar results were obtained with $Cr(CO)_6$ and with two samples of Kr from different suppliers. Unfortunately, due to the limited solubility of metal carbonyls in liquid Ar, it was not possible to use a more inert solvent than Kr to confirm the assignment. The observed frequencies for $Cr(CO)_5Kr$ are in agreement with data from matrix isolation experiments and fit the expected trend with polarizability of the rare-gas atom as shown in Table II. All of these data are consistent with an assignment of the transients to $M(CO)_5Kr$. To our knowledge, this is the first observation of $M(CO)_5Kr$ in fluid solution.

$M(CO)_5Xe$. In contrast to the short-lived complexes observed with Kr, complexes with significant stability were formed with Xe. When a solution of $Cr(CO)_6$ in liquid Xe was photolyzed, we observed the formation of a persistent complex that decayed on the time scale of *minutes* at 180 K. The observed frequencies, shown in Table II, are in excellent agreement with the results of Simpson et al. who demonstrated that the complex is $Cr(CO)_5Xe$ (13). Our observation of all three expected IR bands confirms their assignment. The same experiment with $W(CO)_6$ gave a complex with a lifetime of minutes at 198 K and three IR bands (Table II). Figure 5 shows that when the analogous experiment was performed using a mixture of 5% Xe in Kr, we observed very similar IR bands and lifetime. On this basis, we assign the complex to $W(CO)_5Xe$.

CO Substitution Kinetics in Liquid Xe. In order to determine the binding energy of Xe to $W(CO)_5$, we have examined the kinetics of the reaction of $W(CO)_5Xe$ with CO. Figure 6 shows the rapid-scan spectrum when $W(CO)_6$ is photolyzed in a mixture of 0.011 M CO in liquid Xe at 198.0 K. The decay of

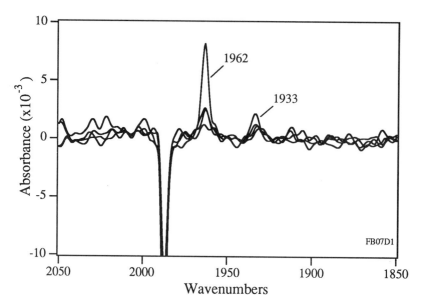

Figure 4. Time-resolved IR spectrum of $W(CO)_5Kr$ in liquid Kr at 154 K.

Table II. Observed IR Frequencies for $M(CO)_5(Q)$ Complexes

Complex	Frequencies (cm⁻¹)			Conditions	Reference
$W(CO)_5(Ar)$		1969	1935	s. Ar, 10 K	8
$W(CO)_5(Kr)$		1962	1933	1. Kr, 150 K	this work
$W(CO)_5(Xe)$	2090	1958	1930	1. Xe, 170 K	this work
		1963	1936	1. Kr, 183 K	this work
$Cr(CO)_5(Ar)$		1973	1936	s. Ar, 10 K	8
$Cr(CO)_5(Kr)$		1965	1938	1. Kr, 180 K	this work
		1961[a]	1933[a]	s. Ar, 10 K	11
$Cr(CO)_5(Xe)$	2087	1960	1934	1. Xe, 163 K	this work
		1960.3	1934	1. Xe, 175 K	13
		1956	1929	s. Xe, 20 K	11

[a]Estimated frequencies from ref. 11.

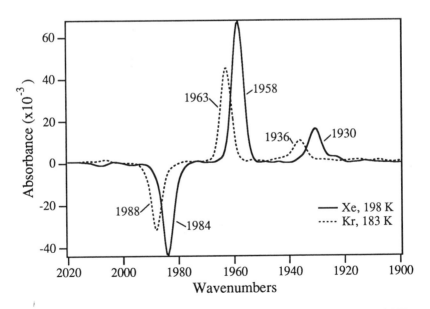

Figure 5. Comparison of $W(CO)_5Xe$ in liquid Xe and in 5% Xe in liquid Kr.

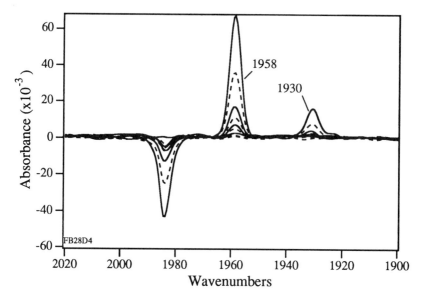

Figure 6. Time-resolved IR spectrum of $W(CO)_5Xe$ in liquid Xe with added CO. The time between spectra is 0.09 s.

$W(CO)_5Xe$ is well represented by an exponential with a decay constant $k_{obs} = 7.4 \pm 0.3$ s^{-1}, and $W(CO)_6$ recovers at the same rate within error limits. Figures 7 and 8 show the CO concentration dependence of the observed decay constants as a function of temperature from 173.0 K to 198.0 K. The decay constants are linear over the concentration range studied (0.003 to 0.03 M) and display roughly a factor of 2 increase every 5 K. At 8 cm^{-1} resolution, the time resolution of the spectrometer is ~0.1 s, and the fastest rate we can measure is 10 s^{-1}. The outlying data point in the 198.0 K data shows this limit and is not included in the fit.

A likely mechanism for this reaction is dissociative substitution in which Xe dissociates from $W(CO)_5Xe$ to form $W(CO)_5$, followed by reaction with CO to give $W(CO)_6$:

$$W(CO)_5Xe \xrightarrow{k_1} W(CO)_5 + Xe \qquad (1)$$

$$W(CO)_5 + Xe \xrightarrow{k_{-1}} W(CO)_5Xe \qquad (2)$$

$$W(CO)_5 + CO \xrightarrow{k_2} W(CO)_6 \qquad (3)$$

Using the steady-state approximation on the intermediate, $W(CO)_5$, results in an expression for the observed decay constant (k_{obs}):

$$k_{obs} = \frac{k_1k_2[CO]}{k_{-1}[Xe] + k_2[CO]} \sim \frac{K_1k_2[CO]}{[Xe]} \qquad (4)$$

Here we have assumed that $k_{-1}[Xe] \gg k_2[CO]$ because the values for k_{-1} and k_2 are comparable (5b), (14), and $[Xe] \gg [CO]$.

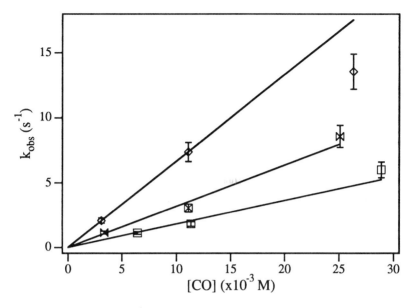

Figure 7. CO dependence of k_{obs} at 198 K (◇), 193 K (⋈), and 188 K (□). The highest [CO] point at 198 K is not included in the fit (see text).

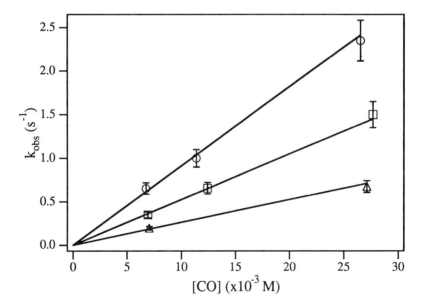

Figure 8. CO dependence of k_{obs} at 183 K (O), 178 K (□) and 173 K (△).

Another mechanism that needs to be considered is the associative pathway consisting of a bimolecular displacement of Xe by CO:

$$W(CO)_5Xe + CO \xrightarrow{\ k_a\ } W(CO)_6 + Xe, \qquad (5)$$

where k_a is the bimolecular rate constant. These two mechanisms can often be distinguished from the dependence of k_{obs} on CO concentration. The dissociative pathway should show limiting behavior in the concentration of CO. This can be seen from equation (3), which reduces to $k_{obs} \sim k_1$ when $k_2[CO] >> k_{-1}[Xe]$. For the associative mechanism, $k_{obs} = k_a[CO]$ and should show linear dependence on [CO]. Since the leaving ligand (Xe) is the solvent in this case, it was not possible with the current apparatus to obtain data under conditions where $k_2[CO] >> k_{-1}[Xe]$, due to the time-resolution limits of the spectrometer. Therefore, the two mechanisms are kinetically indistinguishable, and we must use other data to identify the mechanism.

The pre-exponential factor can be used to distinguish between these two mechanisms. If an associative mechanism is operative, then $k_{obs}/[CO]$ would be equal to the bimolecular rate constant. In this case, we would expect the pre-exponential factor derived from a simple Arrhenius plot to be in the range of 10^8 to 10^9 M^{-1}s^{-1}. Indeed, for Cp*Rh(CO)Xe, a compound expected to undergo associative substitution (15), the measured pre-exponential factor for CO substitution in liquid Xe is log(A) = 8.8 ± 0.3 (16). The value we derive in this work from a plot of $\ln(k_{obs}/[CO])$ vs $1/T$, log(A) = 12.2 ± 0.2, is not consistent with an associative mechanism.

The temperature dependence of the quantity K_1k_2 gives the binding energy of Xe to W(CO)$_5$. From the slopes of Figures 7 and 8 and the Xe density, we obtain values for K_1k_2 (17). Figure 9 shows that when the quantity $\ln\{K_1k_2\}$ is plotted

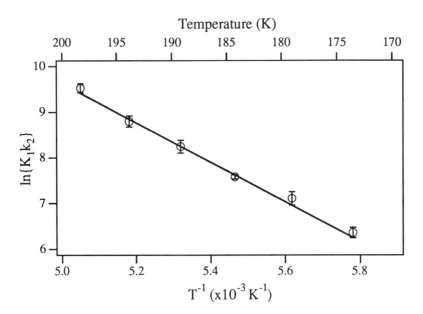

Figure 9. Arrhenius plot of $\ln\{K_1k_2\}$ versus $1/T$.

versus $1/T$, a straight line is observed. The significance of the slope and intercept of the fit can be seen from

$$\ln\{K_1 k_2\} = \frac{\Delta S_1}{R} + \ln(A_2) + \frac{-(\Delta H_1 + E_2)}{RT} \qquad (6)$$

The slope is equal to $-(\Delta H_1 + E_2)/R$, where ΔH_1 is the enthalpy for reaction (1), and E_2 is the activation energy for reaction (2). The intercept is equal to $\{\Delta S_1/R + \ln(A_2)\}$, where ΔS_1 is the entropy for reaction (1), and A_2 is the pre-exponential factor for the rate constant for reaction (2). The elementary rate constant for reaction (2) has been determined to be $k_2 = 3.0 \times 10^{10}$ $M^{-1}s^{-1}$ at 300 K in the gas phase (14). Since k_2 is within an order of magnitude of the gas kinetic value, it is reasonable to assume that the activation energy for this reaction is not significant ($E_2 \sim 0$). This allows us to determine the binding enthalpy of Xe to W(CO)$_5$; $\Delta H_1 = 8.6 \pm 0.4$ kcal/mol. In addition, we can use the value for k_2 as a measure of A_2 in order to determine ΔS_1. While there is some question concerning the relationship between A-factors in solution and in the gas-phase, a reasonable estimate for A_2 is the diffusion-controlled rate constant of 5×10^9 $M^{-1}s^{-1}$ (18). This leads to a value of $\Delta S_1 = + 18$ cal/molK that is slightly low for dissociation of a heavy atom like Xe. However, this value is not inconsistent with the small bond energy we derive for Xe-W(CO)$_5$. A weak Xe-W bond implies a low vibrational frequency and a significant amount of associated vibrational entropy that would be lost upon dissociation of Xe.

The pre-exponential factor derived from Figure 9 is in good agreement with expectations for the dissociative mechanism. If we use the approximation that k_2 and k_{-1} are roughly equal in magnitude, then $K_1 k_2 \sim k_1$. In this case, the intercept of Figure 9 should give a value consistent with a unimolecular A-factor for k_1. The value we find, $\log(A) = 13.6$, is in agreement with this expectation and is inconsistent with a bimolecular A-factor as discussed above. This further confirms the dissociative substitution mechanism.

Our value of $\Delta H_1 = 8.6 \pm 0.4$ kcal/mol is in excellent agreement with an independent determination in the gas phase of 8.5 ± 0.5 kcal/mol that was first presented at this symposium (5a). It should be noted that our approach directly gives a value for ΔH_1 and does not rely on any assumptions about the activation energy of the reverse reaction. However, we do require information about the activation energy for the CO recombination step. In all cases measured to date except one, Fe(CO)$_4$ + CO, these reactions are so fast that the activation energies are insignificant (19).

Since this is one of the first measurements of the Xe binding energy to an organometallic complex, it should be evaluated in light of the available related data. First, is a calculation of the binding energy for Mo(CO)$_5$Kr of 5.3 kcal/mol based on dispersion forces (20). Since Xe is more polarizable than Kr, we should expect a larger value. Second is thermal desorption data of Xe from metal surfaces; the binding energies of Xe to Pt(111) and W(111) have been determined to be 9.2 ± 0.9 (21) and 9.3 ± 1.3 (22) kcal/mol, respectively. Given this range of values, 8.6 ± 0.4 kcal/mol for W(CO)$_5$Xe appears to be quite reasonable.

Conclusions

We have shown that time-resolved IR spectroscopy in liquid rare-gas solvents is a useful technique for studying a range of transient complexes of M(CO)$_5$ with weak ligands, such as N$_2$O, CO$_2$, Xe and Kr. The N$_2$O and CO$_2$ complexes are implicated as intermediates in the formation of metal-oxide thin films by laser-assisted

OMCVD. We have presented the IR spectra of Cr(CO)$_5$Kr, W(CO)$_5$Kr, and W(CO)$_5$(CO$_2$) in fluid solution for the first time. In addition, we have carried out a detailed kinetic investigation of the reaction of W(CO)$_5$Xe with CO in liquid Xe. From the temperature dependence of the kinetics, we obtained the binding enthalpy of Xe to W(CO)$_5$: $\Delta H = 8.6 \pm 0.4$ kcal/mol. This value is in good agreement with an independent gas-phase determination and establishes the utility of the approach for determining binding energies. Future work will include bond energy measurements for N$_2$O and CO$_2$ complexes and investigations into the intermediates in the photochemical deposition of metal oxides from metal carbonyls.

Acknowledgments

This research was supported by the Aerospace Sponsored Research (ASR) Program.

Literature Cited

(1)   Perkins, F. K.; Hwang, C.; Onellion, M.; Kim, Y-G.; Dowben, P. A., *Thin Solid Films*, **1991**, *198*, 317.
(2)   see for example: Turner, J. J.; Simpson, M. B.; Poliakoff, M.; Maier, W. B., *J. Am. Chem. Soc.* **1983**, *105*, 3898.
(3)   Sponsler, M. B; Weiller, B. H.; Stoutland, P. O.; Bergman, R. G., *J. Am Chem. Soc.* **1989**, *111*, 6841.
(4)   Weiller, B. H.; Wasserman, E. P.; Bergman, R. G.; Moore, C. B.; Pimentel, G. C., *J. Am. Chem. Soc.* **1989**, *111*, 8288.
(5)   a) Weitz, E.; Wells, J. R.; Ryther, J. R.; House, P., this volume, b) Wells, J. R.; Weitz, E., *J. Am. Chem. Soc.* **1992**, *114*, 2783.
(6)   Pugh, L. A.; Rao, K. N., "Intensities from Infrared Spectra" *Molecular Spectroscopy: Modern Research*; Academic: New York, 1976, Vol. II.
(7)   Bulanin, M. O., *J. Mol. Struct.* **1986**, *141*, 315.
(8)   Almond, M. J.; Downs, A. J.; Perutz, R. N., *Inorg. Chem.* **1985**, *24*, 275.
(9)   Bogdan, P. L.; Wells, J. R.; Weitz, E., *J. Am. Chem. Soc.* **1991**, *113*, 1294.
(10) Tuan, D. F-T.; Hoffman, R., *J. Am. Chem. Soc.* **1985**, *24*, 871
(11) Perutz, R. N.; Turner, J. J., *J. Am. Chem. Soc.* **1975**, *97*, 4791
(12) Fairhurst, E. C.; Perutz, R. N., *Organometallics* **1984**, *3*, 1389.
(13) Simpson, M. B.; Poliakoff, M.; Turner, J. J.; Maier, W. B.; McLaughlin, J. G., *J. Chem. Soc. Chem. Comm.* **1983**, 1355.
(14) Ishikawa, Y.; Hackett, P. A.; Rayner, D. M., *J. Phys. Chem.* **1988**, *92*, 3863
(15) Rerek, M. E.; Basolo, F., *J. Am Chem. Soc.* **1984**, *106*, 5908.
(16) Weiller, B. H.; Wasserman, E. P.; Bergman, R. G.; Moore, C. B., submitted for publication. The measured activation energy is 2.8 kcal/mol. It should be noted that for the associative mechanism, the activation energy is only a lower bound on the binding energy.
(17) Temperature dependent values for the Xe density were taken from: Theeuwes, F.; Bearman, R. J., *J. Chem. Thermodynamics* **1970**, *2*, 501.
(18) Atkins, P. W., "Physical Chemistry," Oxford University Press, 1978.
(19) Ryther, R. J.; Weitz, E., *J. Am. Chem. Soc.* **1991**, *95*, 9841 and references contained therein.
(20) Rossi, A.; Kochanski, E.; Veillard, A., *Chem. Phys. Lett.* **1979**, *66*, 13.
(21) Rettner, C. T.; Bethune, D. S.; Scheizer, E. K., *J. Chem. Phys.* **1990**, *92*, 1442.
(22) Dresser, M. J.; Madey, T. E.; Yates, J. T., *Surface Sci.* **1974**, *42*, 533-551.

RECEIVED January 28, 1993

Chapter 14

Multiphoton Dissociation of Nickelocene for Kinetic Studies of Nickel Atoms

Carl E. Brown, M. A. Blitz, S. A. Decker, and S. A. Mitchell

Steacie Institute for Molecular Sciences, National Research Council of Canada, 100 Sussex Drive, Ottawa, Ontario K1A 0R6, Canada

The mechanism of nickel atom production by multiphoton dissociation of nickelocene at 650 nm is discussed with reference to the population distribution of nickel atoms among metastable excited states, and implications for kinetic studies of chemical reactions of ground state nickel atoms. A mechanism is proposed that involves direct dissociation of nickelocene on an excited state potential surface following absorption of three photons. Rate constants for collisional relaxation of metastable states are shown to correlate with electronic configuration of the excited state, in a manner consistent with qualitative considerations based on model potential curves for the complex NiCO.

Little is known about the gas-phase chemical properties of transition metal atoms. A useful method for producing metal atoms for studies of reaction kinetics near room temperature is laser induced multiphoton dissociation (MPD) of an organometallic precursor. We have investigated a variety of association reactions of transition metal atoms with simple molecules near room temperature, using MPD of volatile metal complexes for production of metal atoms in a static pressure reaction cell (1). Complexes of transition metal atoms with simple molecules are intriguing molecular species that are of interest as simple models for metal-ligand interactions in organometallic and surface chemistry.

For kinetic studies of metal atom reactions, details of the MPD process are unimportant, provided that production of metal atoms is effectively instantaneous, and thermalization of metastable excited-states of the atoms occurs rapidly compared with the ground-state chemical reaction of interest. In most cases that we have investigated these requirements are satisfied. However, certain reactions of nickel atoms produced by MPD of nickelocene at 650 nm are exceptional in this regard. This is due to the high reactivity of nickel atoms, coupled with the production of metastable excited-states which undergo relatively slow collisional relaxation processes. In this chapter we describe studies of MPD of nickelocene at 650 nm, which were undertaken to clarify

0097–6156/93/0530–0177$06.00/0
Published 1993 American Chemical Society

the role of metastable state relaxation in the reaction kinetics of nickel atoms with unsaturated molecules such as ethylene (2).

Production and Monitoring of Nickel Atoms

Production of nickel atoms was by pulsed visible laser photolysis of nickelocene at 650 nm, using a Lumonics model EPD-330 dye laser operating on DCM laser dye (Exciton), pumped by a Lumonics model 860-4 XeCl excimer laser. The temporal width of the photolysis pulse was ≈10 ns. The beam was roughly collimated (≈1.5 × 0.5 cm cross section), passed through a NRC model 935-5 variable attenuator and then focussed into the reaction cell using either a 25 cm or 50 cm focal length lens. Relative photolysis pulse energies were measured by directing a fraction of the beam onto a vacuum photodiode. Absolute pulse energies were measured using a Scientech energy meter.

Resonance fluorescence from nickel atoms was excited by a probe laser pulse from a second, independently triggered Lumonics excimer laser pumped dye laser, operating on PTP laser dye (Exciton). The photolysis and probe laser beams were collinear and counterpropagating. Laser induced fluorescence (LIF) was separated from scattered photolysis light by a 10 cm focal length monochromator (16 nm FWHM bandpass), and detected by a gated photomultiplier tube. Emission spectra produced by probe and/or photolysis laser excitation were recorded by scanning the monochromator. The pulsed output of the photomultiplier was amplified and then averaged using either a sample and hold circuit and digital voltmeter for computer storage, or by a boxcar integrator for display on a chart recorder. The time dependence of the atom population following the photolysis pulse was observed by scanning the time delay between the photolysis and probe laser pulses, and averaging the fluorescence signal over 10-60 laser shots at each increment of delay. Details of the experimental arrangement have been described elsewhere (1,3). For recording multiphoton ionization (MPI) spectra, a coaxial geometry ionization cell was used (4).

Nickelocene vapor above the crystal (Alfa Products) was admitted to the fluorescence or MPI cell using a metal vacuum line. The pressure of nickelocene in the cell was ≈5 mTorr. Nickelocene was found to be relatively stable as the crystal stored under vacuum, and in the vapor phase, and was convenient to use as a precursor for nickel atoms.

Population Distribution of Nickel Atoms. In Figure 1, an energy level diagram of atomic nickel is shown, that includes all states below 15,000 cm^{-1} excitation energy (5). On the left hand side of the diagram is shown a numbering scheme that is used in the following as a shorthand notation. In this scheme the first number represents a multiplet such as d^8s^2 ^3F, numbered consecutively from the ground state, and the second number specifies a particular J level within the multiplet. The ground state is thus **11**.

All of the states shown in Figure 1 have electronic transitions that are convenient for fluorescence excitation (6). The transitions that were used to monitor these states are listed in Table I. Strong signals were observed for all states up to and including **31**.

TABLE I: Relative Populations of Metastable States of Ni Following MPD of Nickelocene at 650 nm [a]

State	λ^{exc}/nm	LIF	F	Pop
11	339.10	1.00	2.0	1.00
21	339.30	0.73	2.6	0.94
22	344.63	0.42	2.0	0.42
12	341.35	0.30	2.6	0.39
23	342.37	0.15	2.3	0.17
13	338.09	0.10	2.6	0.13
31	338.06	0.09	2.9	0.12

(a) See Figure 1 for notation used for Ni states. λ^{exc} is the wavelength used to excite fluorescence, and LIF is the saturated fluorescence intensity, normalized to that for the ground state transition. Details of the transitions are given in ref. (6). F is a proportionality constant between relative fluorescence intensity and relative population of the lower state before the fluorescence excitation laser pulse. It includes electronic degeneracies of the lower and upper states of the transition, and the quantum yield for fluorescence in the bandpass of the detection monochromator (16). Pop is the relative population of the state following MPD of nickelocene at 650 nm (see the text).

However, no signals were observed for **41** or **51**, which indicates that these states were negligibly populated following photolysis at 650 nm. Estimates of relative populations were made from measurements of fluorescence intensities in the following way. For each transition investigated, the probe laser pulse energy was adjusted so that the transition was saturated with respect to the excitation radiant flux. Under these conditions, the upper and lower states of the transition are populated in the ratio of their electronic degeneracies, so a simple relationship exists between the excited state population at the end of the laser pulse and the total initial population. Knowledge of quantum yields for fluorescence within the bandpass of the detection monochromator thus allows estimates to be made of initial relative populations of the lower states. The required quantum yields are available from tabulated spontaneous transition probabilities (6). Our results are summarized in Table I. Uncertainties in the relative populations are estimated to be ± 30%, based on variations between measurements for the same state using different transitions. Systematic errors may have arisen from involvement of secondary laser excitation processes from excited states of the transitions investigated, or possibly from probe laser photolysis of a molecular fragment of nickelocene produced by the photolysis laser. The photolysis conditions used for these measurements were 0.75 mJ pulse energy at 650 nm, focused with a 25 cm focal length lens. Similar results were obtained when a 50 cm focal length was used. The total pressure in the cell was ≈5 mTorr, and the time delay between the photolysis and probe laser pulses was ≈80 ns. These conditions ensured that the population estimates were unaffected by collisions.

The relative populations in Table I are shown in the form of a Boltzmann plot in Figure 2. The electronic temperature is seen to be ≈3000 K. This is consistent with our observation of negligible population in the states **41** and **51**.

Emission spectra excited by the photolysis laser in the absence of the probe laser showed extremely weak features near 340 nm, most likely due to highly excited nickel atomic photofragments. The weakness of these emissions indicates that such excited nickel atoms were produced in very low abundance. Attempts to observe ionization signals produced by the photolysis laser alone were unsuccessful, although very large signals could be generated by tuning the tightly focused probe laser beam through nickel atomic resonance lines. This shows that multiphoton ionization occurred to a negligible extent at 650 nm, even under tightly focused excitation conditions.

MPD of nickelocene was found to occur readily at ≈445 nm and ≈555 nm. However, photolysis at these wavelength was not convenient for kinetic studies of nickel atoms because of the occurrence of photochemical depletion of nickelocene in the reaction cell. We attribute this to low photon order photofragmentation processes which deplete nickelocene in a large fraction of the laser beam volume. Photolysis at 650 nm induced MPD only in the small focal region of the photolysis beam.

Dependence on Photolysis Fluence. The dependence of the LIF signals for the **21** and **13** states of Ni on the photolysis laser pulse energy at 650 nm is shown in Figure 3. For these measurements the photolysis beam was focused with a 50 cm focal length lens, the total pressure in the cell was ≈5 mTorr and the delay between the photolysis and probe laser pulses was 80 ns. The **21** signal produced at the lowest photolysis

TABLE II: Rate Constants for Collisional Relaxation of Metastable States of Ni Atoms [a]

State (E/cm^{-1})	Config	$k/10^{-11}$ cm^3 s^{-1}			
		Ar	CO_2	CO	C_2H_4
21 (205)	d^9s^1 ^3D	7.6	8.3	>8	>8
22 (880)	d^9s^1	2.6	5.1	9.0	15
12 (1332)	d^8s^2 ^3F	0.03	1.6	0.88	8.7
23 (1713)	d^9s^1	6.2	4.2	13	18
13 (2217)	d^8s^2	0.01	2.8	0.13	6.7
31 (3410)	d^9s^1 ^1D	1.2	10	13	24

(a) See Figure 1 for notation used for Ni states. Excitation energies above the ground state are shown in parentheses. Config gives the electronic configuration, and, for the lowest energy component of each multiplet, the term designation of the multiplet. Rate constants were determined at 296 K.

and **12**, respectively, C represents ground state atoms and D represents the product of the Ni + NiCp$_2$ reaction. This scheme is appropriate because the **12** and **13** states are bottlenecks in the overall relaxation process for Ar buffer gas, as Table II shows. The parameters of the model function are the first-order rates for removal of **12**, **13** and ground state atoms, and the initial populations of the metastable states relative to the ground state. The parameters for the curve in Figure 5 are consistent with the rate coefficients and relative populations determined in this work, although it was necessary in order to achieve a good fit to use a slightly higher population for (**12** + **23**) than is indicated in Table I. Thus the kinetics of ground state atom removal can be modeled satisfactorily in this case. When collisional relaxation is induced by gases other than Ar, the bottlenecks in the relaxation process are less pronounced (see Table II), so the simplified kinetic model just described is not applicable.

The rate constants for collisional relaxation in Table II are correlated with the electronic configuration of the metastable state. The correlation is particularly evident for Ar and CO quenching gases. Thus, states that have the configuration d^{9}s^1 are more efficiently relaxed than those with the configuration d^{8}s^2. A simple rationalization of this effect may be given in terms of the properties of potential curves of the complex NiCO that correlate with the separated Ni + CO states listed in Table II. States associated with d^{9}s^1 ^{3}D$_J$ or d^{9}s^1 ^{1}D$_2$, for which relaxation is relatively efficient, are attractive in character, and correlate with bound molecular states, whereas those

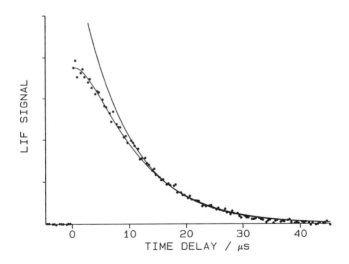

Figure 5. Kinetic trace showing removal of ground state Ni atoms by reaction with nickelocene (≈5 mTorr) in presence of 40 Torr Ar buffer gas at 296 K. Effects due to collisional relaxation of metastable excited state Ni atoms are shown by fitted curves (see the text).

associated with d^8s^2 ^3F_J are thought to correlate with states of NiCO that are repulsive at intermediate Ni-CO separation *(14)*. It is likely that these properties of the d^9s^1 and d^8s^2 configurations apply for other Ni atom - molecule interactions, since they arise from the ability of the d^9s^1 configuration to minimize σ-repulsion by polarization of s-electron density away from the molecule *(15)*. Reduced repulsive interactions allow closer collisions with enhanced probabilities for energy transfer by curve crossing processes. It is noteworthy that electron spin appears to be a much less important factor than electronic configuration. This is shown by the observation that the relaxation efficiency of the **31** state (d^9s^1 $^1D_2$) is in all cases relatively large, even though this state can only relax by a transition to a state of different spin multiplicity. This is consistent with the observation that the formally spin-forbidden association reactions of nickel atoms with CO and C_2H_4 occur readily at room temperature *(2)*.

Literature Cited

1   Mitchell, S. A. In *Gas-Phase Metal Reactions*; Fontijn, A., Ed.; Elsevier: Amsterdam, 1992, Chapter 12; pp 227-252.
2   Brown, C. E.; Mitchell, S. A.; Hackett, P. A. *Chem. Phys. Lett.* **1992**, *191*, 175.
3   Parnis, J. M.; Mitchell, S. A.; Hackett, P. A. *J. Phys. Chem.* **1990**, *94*, 8152.
4   Mitchell S. A.; Hackett, P. A. *J. Chem. Phys.* **1983**, *79*, 4815.
5   Corliss, C.; Sugar, J. *J. Phys. Chem. Ref. Data* **1981**, *10*, 197.
6   Fuhr, J. R.; Martin, G. A.; Wiese, W. L.; Younger, S. M. *J. Phys. Chem.Ref. Data* **1981**, *10*, 305.
7   Pilcher, G.; Skinner, H. A. In *The Chemistry of the Metal-Carbon Bond*; Hartley, F. R.; Patai, S., Eds.; Wiley: New York, 1982, Vol. 1; pp 43-90.
8   Hedaya, E. *Accounts Chem. Res.* **1969**, *2*, 367.
9   Leutwyler, S.; Even, U.; Jortner, J. *Chem. Phys.* **1981**, *58*, 409.
10 Gilbert, R. G.; Smith, S. C. *Theory of Unimolecular and Recombination Reactions*; Blackwell Scientific Publications, Oxford: 1990.
11 Aleksanyan, V. T. In *Vibrational Spectra and Structure*; Durig, J. R., Ed.; Elsevier: Amsterdam, 1982, Vol. 11, pp 107-167.
12 Lewis, K. E.; Smith, G. P. *J. Am. Chem. Soc.* **1984**, *106*, 4650.
13 Gordon, K. R.; Warren, K. D. *Inorg. Chem.* **1978**, *17*, 987.
14 Bauschlicher, C. W. Jr.; Barnes, L. A.; Langhoff, S. R. *Chem. Phys. Lett.* **1988**, *151*, 391.
15 Bauschlicher, C. W. Jr.; Bagus, P. S.; Nelin, C. J.; Roos, B. O. *J. Chem. Phys.* **1986**, *85*, 354.
16 Mitchell, S. A.; Hackett, P. A. *J. Chem. Phys.* **1990**, *93*, 7813.

RECEIVED March 31, 1993

Chapter 15

Ionization and Fluorescence Techniques for Detection and Characterization of Open-Shell Organometallic Species in the Gas Phase

William R. Peifer[1], Robert L. DeLeon[2], and James F. Garvey[1]

[1]Department of Chemistry, State University of New York at Buffalo,
Buffalo, NY 14214
[2]Physical and Chemical Sciences Department, Arvin/Calspan Corporation,
P.O. Box 400, Buffalo, NY 14225

Recent advances are discussed in the development of electronic spectroscopic probes for the study of excited-state structure and photodissociation dynamics of gas-phase organometallics. Because of the short timescale for intermolecular energy transfer within van der Waals clusters, the UV photodissociation dynamics of cluster-bound transition metal carbonyls differs considerably from the photodissociation dynamics of the naked species in the gas phase. It is therefore possible to employ multiphoton ionization to produce cluster-bound metal carbonyl photoions in high yield. *Resonant* photoionization (accomplished with tunable lasers) and mass-resolved detection allow one to probe the excited states of both closed-shell and open-shell neutral organometallics. Finally, a time-resolved two-laser technique employing fluorescent detection of atomic photoproducts is described. This technique allows one to study photodissociation dynamics of organometallics with a temporal resolution competitive with the fastest transient absorption techniques, and a level of sensitivity which is far superior.

Coordinatively unsaturated organotransition metal species play a central role in mechanistic organometallic chemistry. Such species are important both in stoichiometric reactions (*1,2*) and, perhaps more importantly, in a wide range of catalytic reactions: for example, synthetically important hydrogenation reactions (*3*); industrially important hydroformylation (*4*) and polymerization reactions (*5*); automotive engine combustion control and emission reduction (*6*); and biologically important processes such as nitrogen fixation (*7*) and oxygen transport (*8*). It has been suggested that the sin-

0097–6156/93/0530–0188$06.00/0

gle most important property of a homogeneous catalyst is a vacant coordination site(9). Clearly, the production of model coordinatively unsaturated species in the laboratory, and the development of fast, sensitive probes for detection and characterization of these species, impacts upon a broad scope of disciplines in the natural sciences.

A convenient route to the synthesis of coordinatively unsaturated molecules in the laboratory is the photolysis of appropriate saturated precursors by pulsed visible or UV laser irradiation. While single-quantum photodissociation in condensed phases is characterized exclusively by the breaking of a single bond and the production of a single site of coordinative unsaturation (10), photodissociation of a gas-phase molecule may lead to the breaking of several bonds (depending on the energy of the photon absorbed) and the production of multiple vacancies in the coordination sphere (11). These coordinatively unsaturated molecules are highly reactive: subject to the constraints of electronic spin conservation, they will undergo ligand recombination reactions in the gas phase at or near the gas-kinetic limiting rate (12), and will undergo coordination with solvent (or host) molecules in liquids and frozen matrices at diffusion-limited rates (13-19). In order to detect and characterize such highly reactive species, one must employ techniques which are sufficiently *sensitive* to allow detection at low number densities, and sufficiently *fast* to avoid interference from competing solvation and/or ligand recombination reactions.

Available Spectroscopic Probes for Organometallics

Vibrational Probes. One powerful method by which coordinatively unsaturated species have been identified and studied is time-resolved IR absorption spectroscopy, or TRIS (20). Such experiments are typically performed by pulsed UV laser photodissociation of an appropriate saturated precursor in a gas cell. Transient changes in the attenuation of a continuous infrared beam passed down the length of the cell, due to disappearance of the saturated precursor and appearance of the various photoproducts following UV photolysis, are monitored by a fast IR detector. Primary photoproduct identities are usually assumed on the basis of results from gas-phase chemical trapping studies (11) as well as known transition frequencies from matrix isolation studies (21), while photoproduct geometries can often be inferred from group theoretical analysis of the number and relative intensities of IR absorption features (22).

While the application of TRIS has led to advances in our understanding of the structure and reactivity of coordinatively unsaturated organometallic species (primarily those in the electronic ground state), the technique is limited by a few significant constraints. First, because this is an *absorption* technique, the product of the number density and extinction coefficient of a target molecule must be large in order to allow for the molecule's detection. The large extinction coefficients associated with C-O stretching modes of metal carbonyls allow for detection of these species at low number densities, but weaker absorbers must be present at higher number densities. For sufficiently weak absorbers, required number densities are large enough that reactive bimolecular collisions may take place during the course of product detection, thus complicating spectral interpretation. Second, because one employs infrared detectors, the temporal resolution of transient experiments is limited. Typical detector/preamp combinations for this type of experiment display simple exponential risetime constants in the range from 100 to a few hundred nanoseconds (*23*). The microscopic details of processes which take place on a much shorter timescale cannot be directly probed. Finally, since TRIS is a purely *vibrational* spectroscopic probe, one is unable to distinguish directly between vibrational transitions taking place in the *ground* electronic manifold and those taking place in an *excited* electronic manifold of the target molecule. The ability to make such a distinction is desirable, since one would expect that: (a) the excited states of coordinatively unsaturated organometallic species would represent a significant fraction of total photoproducts in some photophysical schemes, and (b) the structure and reactivity of electronically excited states of these species would differ significantly from those of the ground state.

Electronic Probes. Clearly, a spectroscopic technique which probes the *electronic* structure of coordinatively unsaturated organometallic species would allow the experimentalist to acquire information complementary to that provided by vibrational probes such as TRIS. However, interrogation of the electronic structure of these species is thwarted by the very property which makes them easy to produce and interesting to study in the first place; that is, the photolability of coordination compounds, both saturated and unsaturated. Electronic surface crossings among excited states of these molecules occur at rates which are often faster than those of optical transitions. Electronic relaxation generally occurs via non-radiative mechanisms and is often accompanied by the breaking of one or more metal-ligand bonds. Consequently, with a few notable exceptions (*24-26*), powerful techniques such as laser induced fluorescence (LIF) and multiphoton ionization (MPI) have not enjoyed general applicability to the study of electronic excited states of coordinatively unsaturated organometallic species.

The photodissociation behavior of organometallic molecules is clearly governed by the interplay between *intramolecular* energy partitioning, and *intermolecular* energy transfer. The photodissociation dynamics of these species evolves from behavior characteristic of the naked molecule (where energy disposal via bond cleavage predominates), to that characteristic of solvated molecules (where energy disposal via solvent heating predominates). By studying this behavioral evolution experimentally, one can in principle derive detailed insights about the electronic structure of these transient excited species. Weakly bound van der Waals clusters represent a convenient, *finite* experimental model of the complex, virtually infinite environment of condensed matter. Such clusters can be easily formed by expanding a buffer gas such as helium, seeded with a small proportion (perhaps a fraction of a percent) of a compound of interest, into a vacuum chamber through an orifice of small diameter. The cluster beam thus produced is amenable to interrogation by a broad array of spectroscopic techniques.

An area of ongoing research in our laboratory at SUNY, Buffalo, is the photodissociation dynamics of organometallics bound within van der Waals clusters. The ultimate goal of this research is to provide a better understanding of excited electronic states of organometallic molecules, particularly those which are coordinatively unsaturated, and the roles of these excited states in mediating energy transfer. It is hoped that a better understanding of these phenomena will lead to the development of novel approaches for fast, sensitive detection and characterization of this important class of molecules.

Photodissociation Dynamics of Homogeneous Metal Carbonyl van der Waals Clusters

$Fe(CO)_5$ **van der Waals Clusters.** Perhaps the earliest indications of the unusual photodissociation dynamics of cluster-bound organometallics came out of an investigation of the photoionization behavior of jet-cooled $Fe(CO)_5$ by Smalley and co-workers (27). Jet-cooled monomer from a seeded helium expansion was observed to undergo MPI following irradiation at 157 nm by an F_2 excimer laser, giving rise to the expected photoions (i.e., a very strong signal due to atomic metal ions, and very weak signals due to the iron mono- and dicarbonyl ions). At this wavelength, neutral $Fe(CO)_5$ clusters larger than the trimer were observed to undergo one-photon ionization, giving rise to a distribution of cluster ions of the type, $[Fe(CO)_5]_n^+$.

Drastically different results were obtained following irradiation at 193 nm from an ArF excimer laser. At this wavelength, two sequences of cluster ion signals were observed by time-of-flight mass spectrometry: a major sequence, attributed by Smalley and co-workers to bare metal cluster ions, Fe_x^+; and a minor sequence, attributed to $Fe_x(CO)_y^+$. Because ^{56}Fe (natural abundance 91.8%) has an atomic mass indistinguishable from twice the molecular mass of $^{12}C^{16}O$ (natural abundance 98.7%), it was not possible for Smalley and co-workers to assign a unique stoichiometry to any given cluster ion. They reasoned that since cluster ions containing *odd* numbers of CO ligands (and therefore falling between successive Fe_x^+ clusters) constitute only a small fraction of the total photoion yield, cluster ions containing *even* numbers of CO ligands probably constitute only a minor degree of isobaric interference. Photoion yields were found to be highly wavelength dependent; for example, MPI of the cluster beam at 266 nm failed to produce any significant cluster photoion signal. The photodissociation and ionization behavior of these van der Waals clusters is rather remarkable, since the photoproduct ion distribution implies extensive rearrangement of both strong (covalent) and weak (dispersive) bonds within the clusters following photoexcitation, and the pronounced wavelength dependence is observed over a narrow spectral region where absorption is diffuse.

$Mo(CO)_6$ van der Waals Clusters. Investigations of metal carbonyl van der Waals cluster photophysics in the SUNY laboratory began in late 1988 (*28*). $Mo(CO)_6$ was chosen for initial experiments for two reasons. First, much was already known about the one-photon photodissociation dynamics of the Group VIB hexacarbonyls in both gas and condensed phases, offering a body of existing knowledge upon which to build. Second, although the photodissociation dynamics of clusters of $Fe(CO)_5$ had received a certain amount of attention, clusters of homoleptic metal carbonyls from other triads had not yet been examined. Given the ready availability of the hexacarbonyls of chromium, molybdenum, and tungsten, it might then be possible to study the systematics of intracluster photochemistry in a series of analogous metal carbonyls. Although the quadrupole mass analyzer originally employed for photoproduct ion detection in the pulsed molecular beam apparatus was limited in terms of sensitivity and mass range, it enabled the collection of both electron-impact and MPI mass spectra of the $Mo(CO)_6$ cluster beam.

Mass Spectroscopic Analysis of the $Mo(CO)_6$ Cluster Beam. Mass spectra of the $Mo(CO)_6$ cluster beam collected following either electron impact or multiphoton ionization are shown in Figure 1. The electron impact mass spectra for the cluster beam and for an effusively introduced sample of $Mo(CO)_6$ were indistinguishable over the m/z range illustrated. The signals observed correspond to the daughter ions expected following prompt statistical decay of nascent $Mo(CO)_6^+$ parent ions. The MPI mass spectrum of naked $Mo(CO)_6$ is dominated by atomic metal ions and, as expected, displays no signals due to molecular ionization. The MPI mass spectrum of the $Mo(CO)_6$ cluster beam, however, displays totally unexpected features, which were attributed to the formation of metal oxide ions.

Possible Origins of the Metal Oxide Photoions. What is the mechanism by which these ions arise, and what inferences may be derived regarding cluster photophysics? From the dependence of the metal oxide ion signal upon the relative timing of the photoionizing laser and the molecular beam pulse, and from the absence of metal oxide ions in the MPI spectrum of naked $Mo(CO)_6$, we deduced that these unexpected photoions were arising from species present in the molecular beam rather than from oxygen-containing impurities in the vacuum chamber. However, no oxygen-containing species other than the carbonyl ligands themselves could be detected in the molecular beam gas mixture. This implied that some unusual bimolecular chemical reaction was taking place among metal carbonyl species in the molecular beam. Reactions occurring fast enough to reach completion within the ca. 20-nanosecond duration of the laser pulse would be most likely if they were occurring between

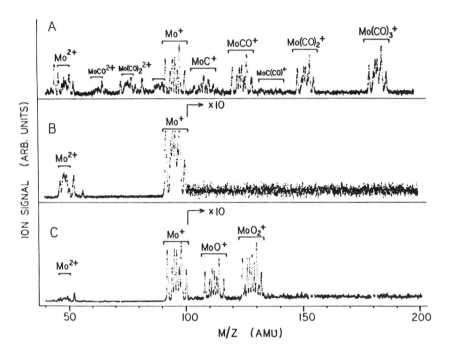

Figure 1. Mass spectra of Mo(CO)$_6$, 40-200 amu. Ion signals for the three spectra are in arbitrary units and are not normalized with respect to each other. Signals at 52 and 56 amu are due to trace impurities of Cr and Fe. (A) Electron impact ionization mass spectrum of [Mo(CO)$_6$]$_n$ van der Waals cluster beam. (B) MPI mass spectrum of Mo(CO)$_6$ monomer in a gas cell. (C) MPI mass spectrum of [Mo(CO)$_6$]$_n$ van der Waals cluster beam. (Reproduced from ref. 28.)

molecules *within* individual van der Waals clusters. We concluded that these intra-cluster reactions probably involved two neutral partners, rather than an ion and a neutral, since evidence of this bimolecular reactivity (i.e., appearance of metal oxide fragment ions) was noticeably absent following electron impact excitation. (Electron impact excitation of $Mo(CO)_6$ gives rise to a series of coordinatively unsaturated *ionic* species, while photodissociation gives rise to *neutral* molecular species.)

We speculated that irradiation of van der Waals clusters containing $Mo(CO)_6$ leads to production of cluster-bound, coordinatively unsaturated molybdenum carbonyl species. Such a photoproduct may utilize vacancies in its coordination sphere to accommodate one or more oxygen atoms from the carbonyl ligands of an adjacent $Mo(CO)_6$ molecule within the cluster. Although this mode of CO coordination involving symmetric bridging and four-electron donation is unusual, it is not without precedent in synthetic organometallic chemistry (*29-31*). Coordinatively unsaturated molybdenum carbonyls are known to undergo analogous bimolecular reactions with $Mo(CO)_6$ in the gas phase, although the structures of these dinuclear adducts are unknown (*32*). It was inferred that such a dinuclear adduct, when formed within the confines of a van der Waals cluster, can be subsequently photoionized and dissociated, giving rise to the observed metal oxide photoions. A cyclic adduct containing two bridging carbonyls might account for the appearance of both mono- and dioxomolybdenum photoions. It is surprising that complete ligand stripping of the inferred adduct does not take place, and that photoinduced activation of the C-O bond does. The strength of the C-O bond in such a symmetrically bridging carbonyl ligand should be considerably weaker than that of free CO. In fact, in the chemisorption of CO to Cr and Mo crystals in which the CO internuclear axis lies parallel to the metal surface, C-O stretching frequencies between 1000 and 1300 cm^{-1} have been observed (*33,34*). This mode of surface adsorption probably involves metal-ligand bonding somewhat analogous to that suggested for the cluster-bound adduct.

$Cr(CO)_6$ and $W(CO)_6$ van der Waals Clusters. The assumption that the dinuclear adduct formed following photoexcitation of an $Mo(CO)_6$ cluster is a six-membered metallacycle, containing two Mo atoms symmetrically bridged by two CO ligands, led us to predictions about intracluster photochemical behavior for clusters of other Group VIB hexacarbonyls (*35*). Given that the bridging carbonyls must be kept far enough apart to prevent overlap of filled CO p-orbitals within the plane of and protruding into the interior of the ring, there exists a limited range of metal-carbon and metal-oxygen bond lengths which can be accommodated within the constraints of the proposed metallacycle structure. While known metal-oxygen bond lengths for both tungsten- and molybdenum-containing organometallic species fall within this range, typical chromium-oxygen single-bond lengths are about 0.1 Ångstrom too short. We predicted that photoexcitation of $W(CO)_6$ clusters might lead to intracluster production of a dinuclear metal carbonyl having the cyclic structure suggested for the

$Mo(CO)_6$ system, but that photoexcitation of $Cr(CO)_6$ clusters would not (although a dinuclear chromium-containing adduct with some alternative structure certainly might be produced). It is interesting to note the resulting photoion yields following 248-nm MPI of cluster beams of $W(CO)_6$ and $Cr(CO)_6$. We observed the appearance of the anomalous metal oxide photoions in the former case, but not in the latter case (*35*). These observations lend credence to our structural model and inferences regarding metal carbonyl intracluster photochemistry.

Evidence from studies of metal carbonyl van der Waals cluster MPI suggests that the photophysical behavior of cluster-bound metal carbonyls is significantly different from that of naked organometallic molecules. It is quite remarkable that irradiation of these clusters does not result in evaporative dissociation of the cluster into its individual monomeric subunits, and that complete ligand stripping does not appear to precede photoionization. These inferences were formed on the basis of some very indirect observations. We were able to detect only those photofragment ions of fairly low m/z, and sensitivity was limited. Detection of primary intracluster photoproducts was complicated by the possibility of intracluster bimolecular reactions. In order to probe more directly the fundamental photophysical processes occurring in clusters containing metal carbonyls, we turned our attention to heterogeneous van der Waals clusters. Such clusters can form in the free jet expansions of gas mixtures containing a buffer gas (helium, for example) seeded with a small proportion of the metal carbonyl as well as a much larger proportion of some "solvent" molecule. Depending on the relative proportions of organometallic and solvent molecules in the beam gas mixture, the likelihood of forming clusters with more than one or two organometallic molecules per cluster is diminishingly small. The probability of intracluster bimolecular reactions between organometallic species is reduced, and complications from such reactions are thereby mitigated. We employed a time-of-flight mass spectrometer with single-ion detection capability to monitor cluster photoions, thus allowing greatly improved mass range and sensitivity. These experiments are described below.

Photodissociation Dynamics of Heterogeneous Metal Carbonyl van der Waals Clusters

Studies of the multiphoton ionization and fragmentation dynamics of metal carbonyl-containing heteroclusters were carried out using a differentially-pumped molecular beam time-of-flight mass spectrometer (R. M. Jordan Company). We chose to examine heteroclusters composed of $Cr(CO)_6$ and methanol (36,37). Methanol was chosen as a solvent molecule for two important reasons. First, methanol by itself was found to cluster quite easily, forming homogeneous clusters containing as many as a few dozen molecules. Second, nonresonant MPI of methanol clusters had already been well-characterized (*38*). The photoionization mass spectrum for the pure methanol clusters is fairly uncomplicated, consisting of ion signals attributed to the intracluster ion-molecule reaction products, $(CH_3OH)_nH^+$ and $(CH_3OH)_n(H_2O)H^+$. $Cr(CO)_6$ was preferred as the metal carbonyl component of the heteroclusters since the distribution of naturally-occurring chromium isotopes (^{52}Cr, 83.79%; three other isotopes, each less than 10%) simplified the interpretation of the mass spectra.

One of the questions we wished to address in studying the MPI of $CH_3OH/Cr(CO)_6$ heteroclusters involves the competition between photodissociation of metal-ligand bonds and photoionization of molecular species. Does multiphoton excitation of a cluster-bound metal carbonyl lead to the cleavage of all metal-ligand bonds and the eventual appearance of solvated atomic ions, or does it lead alternatively to the production of metal carbonyl photoions? If cluster-bound $Cr(CO)_6$ were to behave like naked $Cr(CO)_6$ in the gas phase, one would expect that ligand loss would be extensive, and that the heterocluster MPI mass spectrum would be dominated by signals due to solvated chromium ions. If, on the other hand, cluster-bound $Cr(CO)_6$ were to behave like $Cr(CO)_6$ in condensed (solution or matrix) phases, ligand loss from neutral $Cr(CO)_6$ would be less extensive, and production of metal carbonyl photoions would be likely. As discussed below, the dynamics of multiphoton dissociation and ionization was observed to be intermediate between these two extremes.

Cluster MPI Following 248-nm Irradiation at Moderate Fluence. We first examined the photoion yields following low-fluence 248-nm MPI of $CH_3OH/Cr(CO)_6$ heteroclusters (*36*). A portion of the MPI mass spectrum collected following irradiation of the cluster beam at 2×10^7 W/cm^2 is shown in Figure 2. It is immediately apparent that the observed photoions do not correspond simply to solvated chromium ions, but represent instead a series of solvated, coordinatively unsaturated metal carbonyl ions with the empirical formula, $S_nCr(CO)_x^+$, where S represents a methanol molecule and x = 0,1,2,5,6. We suggested that the mechanism for formation of these cluster ions does *not* involve initial photoionization of methanol, since neat methanol clusters could not be efficiently ionized by irradiating at the mild intensities employed here. Rather, these cluster ions appear due to photoionization of some cluster-bound chromium carbonyl species. The solvated penta- and hexacarbonyl ions appear for n+x≥7, while the remaining cluster ions appear for n+x≥6. Assignments of cluster ions were confirmed by examining photoion yields from MPI of a $CD_3OD/Cr(CO)_6$ heterocluster beam. Almost all of the chromium-containing cluster ions observed could be attributed to solvated *mononuclear* chromium carbonyl species, although some of the weaker ion signals at higher m/z were tentatively identified as *dinuclear* species.

One might imagine two alternative routes to the production of the metal carbonyl-containing cluster photoions. These two mechanisms can be experimentally distinguished, since the yield of photoions arising via the first mechanism should display a *different* dependence on the photoionization laser wavelength than the yield of photoions arising via the second mechanism. In the first mechanism, solvated

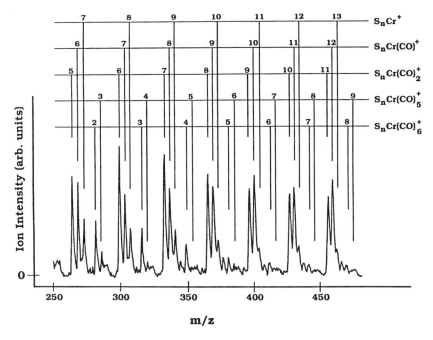

Figure 2. Low-fluence 248-nm MPI mass spectrum of $(CH_3OH)/Cr(CO)_6$ heterocluster beam, 250-475 amu. Integers above the peaks correspond to values of n, the number of CH_3OH molecules in an individual cluster ion. (Reproduced with permission from ref. 36. Copyright 1991 American Institute of Physics.)

$Cr(CO)_6$ is directly photoionized, and the distribution of coordinatively unsaturated metal carbonyl ions observed in the mass spectrum arise from fragmentation of this hexacarbonyl parent ion. The observation of solvated hexacarbonyl ions in the mass spectrum in Figure 2 indicates that at least *some* of the photoions must be produced by this mechanism. Under conditions of extremely high photoionizing laser fluence (e.g., 10^{13} W/cm^2), secondary photofragmentation of these solvated hexacarbonyl parent ions should be extensive, and the distribution of daughter photoion clusters should be insensitive to the laser's wavelength. Alternatively, these ions may form via a second mechanism. Some *neutral*, coordinatively unsaturated chromium carbonyl species, bound within a cluster of methanol molecules, may form following photodissociation of the solvated $Cr(CO)_6$. This neutral photoproduct may then be subsequently photoionized, eventually fragmenting into a series of solvated daughter ions. If the identity of the cluster-bound, *neutral* photoproduct depends on laser wavelength (as is the case for dissociation of naked, gas-phase $Cr(CO)_6$), then one might expect ion yields following MPI of the $CH_3OH/Cr(CO)_6$ heterocluster beam to display a dependence on photoionizing wavelength. The observation of such a wavelength dependence would indicate that the second alternative mechanism considered here is important in the MPI of the neutral heteroclusters.

Wavelength Dependence of Photoion Yields Following Heterocluster MPI.   Photoion yields following 248-nm MPI and 350-nm MPI of the $CH_3OH/Cr(CO)_6$ heterocluster beam were compared. At each wavelength, laser fluence was about 10^{13} W/cm^2. (Strong focusing was necessary at 350 nm in order to produce reasonable photoion signal, and was employed at 248 nm to facilitate comparison of the two spectra.) The 248-nm mass spectrum displays a sequence of peaks attributed to solvated chromium ions, and a second sequence attributed to cluster ions containing chromium and water. Water is known to be created in intracluster ion-molecule reactions in neat methanol cluster ions (*vide supra*). By contrast, the 350-nm mass spectrum displays a sequence of peaks attributed to solvated $Cr(CO)_5^+$, a second attributed to solvated $Cr(CO)_4^+$, and a third attributed to cluster ions containing both $Cr(CO)_5^+$ and water.

If MPI at 248 nm gave rise to the same *primary* photoion as MPI at 350 nm, and if subsequent fragmentation of this parent photoion were rigorously described by a statistical model, then one would expect the observed distribution of daughter photofragment ions to depend on the *total photon energy*, rather than the wavelength of the photons, absorbed by the nascent parent ion. Under conditions of strong focusing and extremely high laser fluence, one might therefore expect the two mass spectra to display nearly identical fragmentation patterns. The fact that these two spectra are *not* similar suggests that the primary photoion produced following 248-nm MPI is not the same as that produced following 350-nm MPI. Differences in the mass spectra may be accounted for by analogy with known single-photon dissociation behavior of naked $Cr(CO)_6$ in the gas phase. It is known that vibrationally excited $Cr(CO)_4$ is the major primary photoproduct of the 248-nm, single-photon dissociation of $Cr(CO)_6$ in the gas phase, while $Cr(CO)_5$ is the major primary photoproduct at 351 nm (*39*). It was suggested that single-photon dissociation of $Cr(CO)_6$ within methanol clusters might display the same wavelength dependence and give rise to the same (cluster-bound) coordinatively unsaturated photoproducts as are observed in simple gas-phase photodissociation (*37*). MPI of the $CH_3OH/Cr(CO)_6$ heterocluster beam was suggested to proceed *not* by initial ionization of $Cr(CO)_6$, but rather by initial formation of a *neutral* single-photon dissociation product within the cluster, followed by MPI of this solvated, coordinatively unsaturated species. Fragmentation of the nascent parent ion would then depend upon its internal energy. It was estimated that photoionization of hot $Cr(CO)_4$ (the photoproduct of 248-nm dissociation of solvated $Cr(CO)_6$) would require two additional 248-nm photons and would lead to a solvated $Cr(CO)_4^+$ parent ion with as much as 103 kcal/mol of internal energy; while photoionization of hot, solvated $Cr(CO)_5$ (the 350-nm single-photon photoproduct of solvated $Cr(CO)_6$) would require two additional 350-nm photons, but would lead to a solvated $Cr(CO)_5^+$ parent ion with only 21-36 kcal/mol of internal energy (*37*). One would then expect subsequent fragmentation of the nascent parent ion to be extensive following 248-nm MPI, but much less extensive (limited perhaps to loss of a single additional CO ligand) following 350-nm MPI. This predicted behavior is consistent with that observed in the mass spectra.

REMPI Spectroscopy of Cluster-Bound Metal Carbonyls. It was possible in the experiments described above to use a single pulse from the excimer-pumped dye laser (tuning range 346-371 nm) both to photolytically prepare solvated $Cr(CO)_5$ and to resonantly photoionize this pentacarbonyl (*37*). Photoion signal from a $Cr(CO)_5^+$-containing cluster ion was monitored while slowly stepping the laser wavelength through the tuning range of the dye. The resulting REMPI spectrum corresponds to an optical absorption spectrum of the neutral pentacarbonyl, displaying a resonant absorption feature which grows in at wavelengths shorter than about 353 nm and reaches a maximum somewhere outside the blue limit of the laser dye. Neat methanol clusters, by contrast, display a maximum in their REMPI spectra at about 362 nm. Since the $Cr(CO)_5^+$-containing cluster ions display no such resonant feature at 362 nm, they must arise not by initial photoionization of a methanol molecule within the cluster, but rather by ionization of a chromium carbonyl species.

Participation of Solvent Molecules in Energy Disposal. It is not possible in the cluster MPI experiments described here (*36,37*) to assess *directly* the importance of intracluster collisional relaxation or evaporation of solvent molecules as energy disposal mechanisms. It appears, however, that in the initial single-photon dissociation of solvated $Cr(CO)_6$, transfer of excess energy from excited photoproducts to

the solvent bath is not as efficient as in the case of condensed-phase photodissociation. If intracluster energy transfer to the solvent bath were highly efficient, one
might expect to see typical condensed-phase behavior; i.e., single-photon absorption
always leading to single-ligand loss. Solvent evaporation may be important in the
disposal of excess energy from cluster *ions*. We noted that in general, the daughter
ions produced following MPI of $CD_3OD/Cr(CO)_6$ heteroclusters display a lesser degree of fragmentation than those produced following MPI of $CH_3OH/Cr(CO)_6$ heteroclusters *(40)*. For example, both solvated $Cr(CO)_5^+$ and solvated $Cr(CO)_4^+$ photoions appear following 350-nm MPI of $CH_3OH/Cr(CO)_6$ heteroclusters; but only
the solvated pentacarbonyl photoions appear following 350-nm MPI of perdeuterated
$CD_3OD/Cr(CO)_6$ heteroclusters. This phenomenon was observed following MPI at
other wavelengths, and over a broad range of laser fluences. Apparently, CD_3OD is
more efficient than CH_3OH in collisionally relaxing excited chromium carbonyl photoions produced within these heteroclusters. It was suggested that intracluster energy
transfer from an excited metal carbonyl photoion to a surrounding solvent molecule
can occur through a highly non-statistical mechanism, provided that: (a) the energy
defect between the appropriate internal modes of the excited photoion and the adjacent
solvent molecule is very small; and (b) the transfer of energy from the donor mode to
the acceptor mode is not symmetry-forbidden. It is remarkable that intracluster energy transfer takes place in such an apparently non-statistical manner. These heteroclusters may constitute an interesting experimental model in which to study the dynamics
of intracluster energy partitioning and disposal.

**Future Directions -- Probing the Electronic Structure of Naked
Organometallics in the Gas Phase**

All of the experiments described above involve detection and characterization of
organometallic photoproducts trapped within van der Waals clusters. While much information on the electronic structure and dynamics of open-shell organometallics can
be learned from such studies, it would be desirable to develop probes for the interrogation of naked, open-shell organometallic species in the gas phase. The availability
of such probes would likely contribute to a better understanding of mechanistic
organometallic chemistry, refinements in metal-organic chemical vapor deposition
(MOCVD) technology, and further insights into the multiphoton dissociation dynamics of and metal atom production from gas-phase organometallics. Toward this end,
we have recently initiated a program of research on new time-resolved, two-laser
techniques for organometallic photoproduct generation and characterization. A given
coordinatively unsaturated organometallic is first produced by pulsed UV laser photolysis of an appropriate precursor, and a second laser pulse is then employed to induce multiphoton dissociation (MPD) of this photoproduct, yielding metal atoms in a
distribution of electronic states. The state distribution of these atoms can be deter-

mined by atomic emission and LIF. The electronic temperature of the *atomic* photoproducts is directly related to the internal energy of the *molecular* photoproduct created by the first laser pulse. For cases where MPD of the precursor species and MPD of the coordinatively unsaturated molecular photoproduct lead to different metal atom state distributions, it then becomes possible to detect the presence of the photoproduct with a level of sensitivity not attainable with conventional IR absorption techniques such as TRIS.

In the remainder of this section, studies of the dynamics of the MDP of organochromium compounds, which provided the conceptual basis for the two-laser technique described above, will be briefly reviewed. Preliminary results from a study of the photodissociation dynamics of jet-cooled $Cr(CO)_6$ using the two-laser technique will then be presented.

Multiphoton Dissociation Dynamics of $Cr(CO)_6$. Tyndall and Jackson have studied the 248-nm MPD dynamics of $Cr(CO)_6$ using fluorescence detection of the atomic photoproducts (*41*). On the basis of fluorescence quenching results, they indirectly determined that two mechanisms exist for production of chromium atoms following 248-nm irradiation of $Cr(CO)_6$. These two mechanisms can be described as direct and sequential, and are represented in Reactions (1) and (2), respectively.

$$Cr(CO)_6 \; + \; \geq 2\,h\nu \quad \rightarrow \; \{Cr^*\} \; + \; 6\,CO \qquad (1)$$

$$Cr(CO)_6 \; + \; h\nu \quad \rightarrow \; Cr(CO)_4^\dagger \; + \; 2\,CO \qquad (2a)$$

$$Cr(CO)_4^\dagger + \; h\nu \quad \rightarrow \; \{Cr^*\} \; + \; 4\,CO \qquad (2b)$$

MPD via the direct mechanism, represented in Reaction (1), gives rise to a statistical distribution of chromium electronic states characterized by a high temperature. Fluorescence from high-lying electronic states is found to be insensitive to the pressure of added buffer gas in the photolysis cell. However, MPD via the sequential mechanism, depicted in Reactions (2a) and (2b), proceeds through the intermediacy of a coordinatively unsaturated species (in this case, vibrationally excited $Cr(CO)_4$) and gives rise to low-lying states of chromium atoms characterized by a much lower electronic temperature. LIF signals pumped from non-emitting, low-lying states can be quenched by addition of buffer gas to the photolysis cell. Furthermore, it has been shown that in the *sequential* (but not the *direct*) MPD of organochromium precursors containing a variety of polyatomic ligands, the electronic temperature of the re●lting atomic photoproducts is inversely related to the vibrational state density of the molecular photoproduct's ligands (*42,43*). The fact that the $Cr(CO)_4$ intermediate's internal energy is reflected in the electronic temperature of the chromium atoms produced from the subsequent photodissociation of $Cr(CO)_4$ forms the conceptual basis of a two-laser technique to probe molecular photoproduct internal energy, electronic structure, and relaxation dynamics.

Effect of Precursor Temperature on 248-nm $Cr(CO)_6$ MPD Dynamics. In recent experiments at SUNY, we have examined the MPD dynamics of jet-cooled $Cr(CO)_6$, and compared the resulting chromium state distribution with that obtained following MPD of room-temperature $Cr(CO)_6$ in a gas cell experiment (44). Portions of the dispersed fluorescence spectra collected following MPD of both room-temperature and jet-cooled $Cr(CO)_6$ are shown in Figure 3. Emission signals in these spectra assigned to the septet system originate from excited chromium atoms prepared by direct MPD of $Cr(CO)_6$, while signals assigned to the quintet system originate from 248-nm LIF excitation of the two lowest quintet terms of chromium. These low-lying quintets are prepared by *sequential* MPD of $Cr(CO)_6$, through the intermediacy of an excited $Cr(CO)_4$ primary photoproduct. It is important to note that initial photodissociation of $Cr(CO)_6$, subsequent photodissociation of $Cr(CO)_4$, and final LIF excitation of the low-lying quintet terms of the chromium photoproduct, are all accomplished within the duration of a single 248-nm excimer laser pulse (20-nsec FWHM gaussian). While room-temperature $Cr(CO)_6$ appears to photodissociate through both the direct and sequential pathways, the fluorescence spectrum acquired following MPD of jet-cooled $Cr(CO)_6$ shows none of the quintet features indicative of sequential MPD.

One might expect the quintet signals to be quenched if the primary $Cr(CO)_4$ photoproduct were collisionally relaxed. However, number densities in the probed portion of the molecular beam were calculated to be too low for collisional quenching to be likely (45), and no evidence of radiation trapping, which might be expected under conditions of high number density, was revealed in measurements of excited state lifetimes. Given that excited $Cr(CO)_4$ is the primary 248-nm photoproduct of jet-cooled $Cr(CO)_6$ (46,47), the failure to observe evidence of sequential 248-nm MPD of jet-cooled $Cr(CO)_6$ seems most likely due to a slower rate of appearance of the excited $Cr(CO)_4$ photoproduct in the molecular beam. If $Cr(CO)_4$ does not appear within about 20 nanoseconds following initial photodissociation of the jet-cooled hexacarbonyl precursor (after which time the pulse of laser photons has passed), it cannot be detected by the atomic fluorescence technique.

Time-resolved Technique for Molecular Photoproduct Detection. If indeed $Cr(CO)_4$ is produced from 248-nm photolysis of jet-cooled $Cr(CO)_6$ at a slower rate than from photolysis of the room-temperature hexacarbonyl, then one should be able to detect it by irradiating the molecular beam with a *second* laser pulse which is delayed with respect to the first pulse. Slowly evolving $Cr(CO)_4$ created following initial photolysis of the jet-cooled hexacarbonyl should then be dissociated by the second pulse into the characteristic state distribution of chromium atoms, and emission lines in the quintet system associated with sequential MPD should once again appear in the fluorescence spectrum. The results of such a two-laser, time-resolved experiment are illustrated in the fluorescence spectrum in Figure 4. This spectrum was acquired by initial single-photon dissociation of jet-cooled $Cr(CO)_6$ at 222 nm, producing roughly equal amounts of $Cr(CO)_3$ and $Cr(CO)_4$ (*23*). These molecular photoproducts were then photodissociated at 248 nm by the pulse from a second excimer laser which was delayed 300 nanoseconds with respect to the first pulse. The quintet emission features characteristic of sequential MPD of the jet-cooled hexacarbonyl are once again visible. It is important to note that these features appear only for delay times longer than about 60 nanoseconds, reaching maximum intensity for delays greater than about 200 nanoseconds. The temporal profile for the growing in of these quintet signals can be fitted by a simple rising exponential with a 40-nanosecond time constant, convoluted with the integral of the first laser's temporal profile. Eventual decay of these quintet signals for delays in excess of one or two microseconds can be modeled simply in terms of the escape of primary molecular photoproducts from the photolysis probe volume at the nominal molecular beam velocity.

Implications for Photodissociation Dynamics. It is not presently clear why the rate of production of $Cr(CO)_4$ from photolysis of a jet-cooled precursor should be slower than from photolysis of a room-temperature precursor. Vernon and coworkers have suggested that the mechanism for production of $Cr(CO)_4$ following 248-nm photolysis of jet-cooled $Cr(CO)_6$ proceeds in three steps: non-statistical ejection of the first CO ligand; internal conversion of the nascent, electronically excited pentacarbonyl to its ground electronic surface; and finally, statistical loss of the second CO ligand (*45*). According to the calculations of Hay (*48*), the ground and first excited states of the pentacarbonyl are vibronically coupled through low-energy OC-M-CO bending modes. If these modes are not populated in the nascent pentacarbonyl, then the internal conversion step should occur at a slower rate. It is tempting to speculate that these modes cannot be directly populated by the one-photon absorption (and dissociation) process used to prepare the pentacarbonyl. In such a case, the only way these modes will be populated in the nascent pentacarbonyl is if the correlated bending modes are already populated in the hexacarbonyl precursor. One would expect that in a 300K molecule of $Cr(CO)_6$, such correlated modes *would* be thermally populated, whereas in a jet-cooled molecule of $Cr(CO)_6$, even these low-energy modes would not be significantly populated.

Concluding Remarks

The UV and near-UV laser techniques described in this chapter represent promising new approaches for the study of the electronic structure of open-shell organometallics, and the influence of excited electronic states on the dynamics of sin-

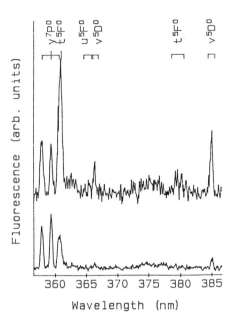

Figure 3. Atomic fluorescence spectra collected following 248-nm MPD of Cr(CO)$_6$. Term symbols for the emitting states of Cr are indicated across the top of the figure. Spectral features assigned to the septet system are due to emission from excited states of Cr, which appear following concerted MPD of Cr(CO)$_6$. Spectral features assigned to the quintet system are LIF transitions pumped out of the a^{5}S and a^{5}D states of Cr, which arise from sequential MPD of Cr(CO)$_6$. Upper spectrum: MPD of Cr(CO)$_6$ at 300K in gas cell. Lower spectrum: MPD of jet-cooled Cr(CO)$_6$.

gle- and multiphoton dissociation. The encapsulation of organometallic molecules within van der Waals clusters not only allows one to employ powerful laser-based probes for *spectroscopic* studies, but may provide the experimentalist with convenient models for the study fast *dynamical* processes such as intermolecular energy transfer within the quasi-condensed state.

Studies on the time-resolved, two-laser sequential MPD of transition metal carbonyls are continuing in our laboratory. This technique is characterized by sensitivity far better than that obtainable with transient absorption techniques, is not limited in terms of the identity of the attached ligands, and has the potential for superior temporal resolution (due to the use of visible rather than IR-sensitive detectors). Of particular interest is the influence of the electronic state of the intermediate molecular photoproduct on the state distribution of the subsequent atomic photoproduct. The ability to selectively prepare and detect specific excited states of naked, open-shell metal carbonyls may eventually allow for great advances in the study of mechanistic organometallic chemistry.

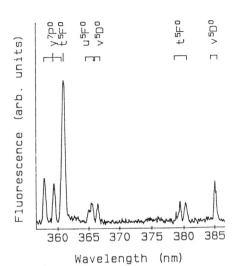

Figure 4. Atomic fluorescence spectrum collected following two-laser MPD of jet-cooled Cr(CO)$_6$. The hexacarbonyl was first photodissociated at 222 nm, and the coordinatively unsaturated photoproducts were probed 300 nanseconds later by further photodissociating at 248 nm to a distribution of atomic states of Cr. Intense LIF emission in the quintet system, indicative of sequential MPD of Cr(CO)$_6$, is observed for delays exceeding 60 nanoseconds.

Acknowledgments

We gratefully acknowledge the financial support of this work provided by the Office of Naval Research, the Petroleum Research Fund, administered by the American Chemical Society, and the Alfred P. Sloan Foundation. RLD wishes to acknowledge internal research and development funds provided by the Calspan-University at Buffalo Research Center (CUBRC).

Literature Cited

1. *Organic Synthesis via Metal Carbonyls*; Wender, I.; Pino, D., Eds.; Wiley-Interscience: New York, NY, 1968; Vol. 1.
2. Collman, J. P.; Hegedus, L. S. *Principles and Applications of Organotransition Metal Chemistry*; University Science Books: Mill Valley, CA, 1980.
3. James, B. R. *Adv. Organomet. Chem.* **1979**, *17*, 319.
4. Pino, P. *J. Organomet. Chem.* **1980**, *200*, 223.
5. Muetterties, E. L.; Stein, J. *Chem. Rev.* **1979**, *79*, 479.
6. Acres, G. J. K.; Cooper, B. J. *Platinum Metals Rev.* **1972**, *16*, 74.
7. Stewart, W. D. P. *Annu. Rev. Microbiol.* **1980**, *34*, 497.

8. Perutz, M. F.; Fermi, G.; Luisi, B. *Acc. Chem. Res.* **1987**, *20*, 309.

9. Collman, J. P. *Acc. Chem. Res.* **1968**, *1*, 136.

10. Geoffroy, G. L.; Wrighton, M. S. *Organometallic Photochemistry*; Academic Press: New York, NY, 1979.

11. (a) Nathanson, G.; Gitlin, B.; Rosan, A. M.; Yardley, J. T. *J. Chem. Phys.* **1981**, *74*, 361.

    (b) Yardley, J. T.; Gitlin, B.; Nathanson, G.; Rosan, A. M. *J. Chem. Phys.* **1981**, *74*, 370.

    (c) Tumas, W.; Gitlin, B.; Rosan, A. M.; Yardley, J. T. *J. Am. Chem. Soc.* **1982**, *104*, 55.

12. Weitz, E. *J. Phys. Chem.* **1987**, *91*, 3945, and references therein.

13. Welch, J. A.; Peters, K. S.; Vaida, V. *J. Phys. Chem.* **1982**, *86*, 1941.

14. Joly, A. G.; Nelson, K. A. *J. Phys. Chem.* **1989**, *93*, 2876.

15. Lee, M.; Harris, C. B. *J. Am. Chem. Soc.* **1989**, *111*, 8963.

16. (a) Simon, J. D.; Xie, X. *J. Phys. Chem.* **1986**, *90*, 6751.

    (b) Simon, J. D.; Xie, X. *J. Phys. Chem.* **1989**, *93*, 291.

    (c) Simon, J. D.; Xie, X. *J. Phys. Chem.* **1989**, *93*, 4401.

    (d) Xie, X.; Simon, J. D. *J. Am. Chem. Soc.* **1990**, *112*, 1130.

17. (a) Wang, L.; Zhu, X.; Spears, K. G. *J. Am. Chem. Soc.* **1988**, *110*, 8695.

    (b) Wang, L.; Zhu, X.; Spears, K. G. *J. Phys. Chem.* **1989**, *93*, 2.

18. Moore, J. N.; Hansen, P. A.; Hochstrasser, R. M. *J. Am. Chem. Soc.* **1989**, *111*, 4563.

19. Yu, S.-C.; Xu, X.; Lingle, R., Jr.; Hopkins, J. B. *J. Am. Chem. Soc.* **1990**, *112*, 3668.

20. Poliakoff, M.; Weitz, E. *Adv. Organomet. Chem.* **1986**, *25*, 277.

21. (a) Graham, M. A.; Perutz, R. N.; Poliakoff, M.; Turner, J. J. *J. Organomet. Chem.* **1972**, *34*, C34.

    (b) Perutz, R. N.; Turner, J. J. *Inorg. Chem.* **1975**, *14*, 262.

    (c) Perutz, R. N.; Turner, J. J. *J. Am. Chem. Soc.* **1975**, *97*, 4791.

    (d) Perutz, R. N.; Turner, J. J. *J. Am. Chem. Soc.* **1975**, *97*, 4800.

    (e) Burdett, J. K.; Graham, M. A.; Perutz, R. N.; Poliakoff, M.; Turner, J. J.; Turner, R. F. *J. Am. Chem. Soc.* **1975**, *97*, 4805.

22. Cotton, F. A. *Chemical Applications of Group Theory*; 2nd ed.; Wiley: New York, NY, 1971.

23. Ishikawa, Y.; Brown, C. E.; Hackett, P. A.; Rayner, D. M. *J. Phys. Chem.* **1990**, *94*, 2404.

24. Wrighton, M.; Hammond, G. S.; Gray, H. B. *J. Am. Chem. Soc.* **1971**, *93*, 4336.

25. Rosch, N.; Kotzian, M.; Jorg, H.; Schroder, H.; Rager, B.; Metev, S. *J. Am. Chem. Soc.* **1986**, *108*, 4238.

26. Preston, D. M.; Zink, J. I. *J. Phys. Chem.* **1987**, *91*, 5003.

27. Duncan, M. A.; Dietz, T. G.; Smalley, R. E. *J. Am. Chem. Soc.* **1981**, *103*, 5245.
28. Peifer, W. R.; Garvey, J. F. *J. Phys. Chem.* **1989**, *93*, 5906.
29. Shriver, D. F.; Alich, A., Sr. *Coord. Chem. Rev.* **1972**, *8*, 15.
30. Ulmer, S. W.; Skarstad, P. M.; Burlitch, J. M.; Hughes, R. E. *J. Am. Chem. Soc.* **1973**, *95*, 4469.
31. Blackmore, T.; Burlitch, J. M. *J. Chem. Soc., Chem. Commun.* **1973**, 405.
32. Ganske, J. A.; Rosenfeld, R. N. *J. Phys. Chem.* **1989**, *93*, 1959.
33. (a) Shinn, N. D.; Madey, T. E. *J. Chem. Phys.* **1985**, *83*, 5928.
    (b) Shinn, N. D.; Madey, T. E. *Phys. Rev. Lett.* **1984**, *53*, 2481.
34. Zaera, F.; Kollin, E.; Gland, J. L. *Chem. Phys. Lett.* **1985**, *121*, 464.
35. Peifer, W. R.; Garvey, J. F. *Int. J. Mass Spectrom. Ion Proc.* **1990**, *102*, 1.
36. Peifer, W. R.; Garvey, J. F. *J. Chem. Phys.* **1991**, *94*, 4821.
37. Peifer, W. R.; Garvey, J. F. *J. Phys. Chem.* **1991**, *95*, 1177.
38. Morgan, S.; Keesee, R. G.; Castleman, A. W., Jr. *J. Am. Chem. Soc.* **1989**, *111*, 3841.
39. Seder, T. A.; Church, S. P.; Weitz, E. *J. Am. Chem. Soc.* **1986**, *108*, 4721.
40. Peifer, W. R.; Garvey, J. F. In *Isotope Effects in Gas Phase Chemical Reaction and Photodissociation Processes*; Kaye, J., Ed.; ACS Books: Washington, D.C. 1992, ACS Symposiou Series 502; pp 335-355.
41. (a) Tyndall, G. W.; Jackson, R. L. *J. Am. Chem. Soc.* **1987**, *109*, 582.
    (b) Tyndall, G. W.; Jackson, R. L. *J. Chem. Phys.* **1988**, *89*, 1364.
42. Tyndall, G. W.; Larson, C. E.; Jackson, R. L. *J. Phys. Chem.* **1989**, *93*, 5508.
43. (a) Samoriski, B.; Hossenlopp, J. M.; Rooney, D.; Chaiken, J. *J. Chem. Phys.* **1986**, *85*, 3326.
    (b) Hossenlopp, J. M.; Samoriski, B.; Rooney, D.; Chaiken, J. *J. Chem. Phys.* **1986**, *85*, 3331.
44. Peifer, W. R.; Garvey, J. F.; DeLeon, R. L. *J. Chem. Phys.*, **1992**, *96*, 6523.
45. Miller, D. R. In *Atomic and Molecular Beam Methods*; Scoles, G., Ed.; Oxford University Press: New York, NY, 1988, Vol. 1.
46. Tyndall, G. W.; Jackson, R. L. *J. Chem. Phys.* **1989**, *91*, 2881.
47. Venkataraman, B.; Hou, H.; Zhang, Z.; Chen, S.; Bandukwalla, G.; Vernon, M. *J. Chem. Phys.* **1990**, *92*, 5338.
48. Hay, P. J. *J. Am. Chem. Soc.* **1978**, *100*, 2411.

RECEIVED January 14, 1993

Chapter 16

Reactions of Gas-Phase Transition-Metal Atoms with Small Hydrocarbons

James C. Weisshaar

Department of Chemistry, University of Wisconsin—Madison,
Madison, WI 53706

We compare the chemistry of gas phase transition metal atoms with alkanes and alkenes, surveying the different charge states M, M^+, and M^{2+}. Most monopositive cations are highly reactive, dehydrogenating or demethanating linear alkanes in exothermic reactions at room temperature. The dipositive cations abstract electrons or hydride anions from alkanes if the radii of long-range curve crossings permit. If not, M^{2+} can approach the hydrocarbon to close range, and the same kinds of H_2 and CH_4 elimination reactions occur as observed in M^+. The neutral atoms are inert to alkanes, but Sc, Ti, V, Ni, Zr, Nb, and Mo react with alkenes. The interplay of the energetics of low-lying metal states of different electron spin and electron configuration; the size of valence d and s orbitals; and the character of long-range forces determines chemical reactivity.

The chemistry of gas phase M^+ and M^{2+} is unusual (1,2). Certain ground state transition metal cations can break C-H bonds or even C-C bonds of linear alkanes at room temperature, leading to exothermic dehydrogenation or demethanation products. While solution phase chemists have discovered photochemically driven reactions in which metal centers insert in C-H bonds of alkanes (3) cracking of the C-C skeleton of linear alkanes at room temperature appears unique to the gas phase metal cations.

Comparison of the gas phase chemical reactivity of transition metal atoms of different charge, M, M^+, and M^{2+}, is becoming possible. This article focuses on reactions of such atoms with small hydrocarbons. The bare metal atoms themselves are electronically complex, but well understood (4). Our goal is to understand how metal atom electronic structure controls chemical reactivity. Here electronic structure refers to the *pattern* of ground and low-lying excited atomic energy levels, including configuration and electron spin, as well as the relative sizes of the valence nd and $(n+1)s$ orbitals.

To understand ground state reactivity, we must consider multidimensional intersections among diabatic potential energy surfaces emanating from both ground and low-lying excited states of reactants. The reason is that the ground

0097–6156/93/0530–0208$06.00/0

states themselves are often badly suited for forming chemical bonds due to high-spin coupling of the unpaired electrons or to unfavorable electron configuration (orbital occupancy). Often chemistry can occur only by mixing of excited-state character (hybridization) during the collision.

The gas phase itself is unique in allowing the study of isolated, bimolecular collisions between reactants. Both M^+ and M^{2+} share the enormous experimental advantage of easily tunable kinetic energy and straightforward identification of the mass of the charged product. In addition to traditional kinetics measurements, we can sometimes study collisions between hydrocarbon and a specific electronic state of the metal with simultaneous control of the kinetic energy as well.

Recent review articles *(1,2,5)* describe the chemistry of M^+ and of M^{2+} quite thoroughly. Here we seek common themes among reactions with different charge.

Overview of Electronic Structure

An atomic state is labeled by its electron configuration, total electron spin S, orbital angular momentum L, and total angular momentum J, as embodied in the symbol $^{2S+1}L_J$. This Russell-Saunders coupling scheme works best in the 3d series, where electron spin is a good quantum number. Spin is gradually degraded as a quantum number moving to the right within a series or downward in the periodic table from the 3d to the 4d to the 5d series.

The low-lying electronic states of bare metal atoms are well characterized experimentally for M, M^+, and M^{2+}, at least in the 3d and 4d series *(4)*. Both the sizes of atomic orbitals and their energies influence the strength of metal-ligand chemical bonding *(6)*. In the 3d series, the 4s orbital is much larger than 3d. In neutral M, most of the ground states (Sc, Ti, V, Mn, Fe, Co, and Ni) have $3d^{x-2}4s^2$ configurations; the lowest excited states are high-spin $3d^{x-1}4s^1$. In Cr and Cu, exchange interactions make $3d^{x-1}4s^1$ the ground state. Those ground states with $4s^2$ configurations have a closed-shell appearance at long range, suggesting they will be chemically inert. This is borne out by experiment. The M^+ ground states and low-lying excited states are either $3d^{x-1}4s^1$ or $3d^x$; the $3d^{x-2}4s^2$ states lie high in energy. Accordingly, the M^+ ground states are highly reactive. In M^{2+}, the ground states are $3d^x$, and again these are highly reactive species.

In the 4d series, the valence orbitals (5s and 4d) are larger than in the 3d series. Moreover, 5s and 4d are more similar in size and orbital energy than 4s and 3d. This enhances the of chemical bonding in the 4d series. The ground states are either $4d^{x-2}5s^2$ or high-spin $4d^{x-1}5s^1$. We might expect the 4d-series neutral atoms to be more chemically active than the 3d-series atoms, in agreement with the early experimental evidence. The 4d-series M^+ ground states are $4d^{x-1}5s^1$ or $4d^x$, and the M^{2+} ground states are $4d^x$.

In the 5d series, the lanthanide contraction stabilizes and contracts the 6s orbital relative to 5d. The energy level structure of the neutral atoms is superficially similar to the 3d series, with primarily $5d^{x-2}6s^2$ ground states for neutral M. The M^+ ground states are $5d^{x-1}6s^1$ on the left-hand side and $5d^x$ on the far right-hand side.

M^+ + Hydrocarbon Reactions

An impressive arsenal of complementary experimental techniques has been applied to the chemistry of M^+. These include guided ion beam measurements of cross sections (7); Fourier transform, ion-cyclotron resonance measurements of reaction rates and product identity (8); bulk kinetics measurements in fast flow reactors with He buffer gas (9); collision-induced dissociation (10); crossed ion-neutral beam measurements of total cross sections (11); drift cell measurements (12); and ion beam measurements of metastable kinetic energy release distributions on fragmentation (13).

Electron spin conservation during the rate-limiting step of M^+ insertion in C-H bonds (9) can qualitatively explain which ground-state, 3d-series M^+ cations react with alkanes in low energy collisions (Figure 1). The chemically active M^+ cations have either a ground state that can conserve spin during insertion (Co^+ and Ni^+) or a low-lying excited state that can do so (Sc^+, Ti^+, and the much less reactive V^+ and Fe^+). Here low-lying means smaller than ~ 1 eV, as shown in Fig. 1. The inert metal cations either have promotion energies to states of proper spin that are too large (in excess of 1.3 eV for Cr^+, Mn^+, and Cu^+) or they cannot form the 3d-4s hybrids necessary to make two σ-bonds (Zn^+).

Two particularly well-studied reactions are:

$$V^+ + C_3H_8 \rightarrow VC_3H_6^+ + H_2 \tag{1}$$

$$Fe^+ + C_3H_8 \rightarrow FeC_3H_6^+ + H_2 \tag{2a}$$
$$\rightarrow FeC_2H_4^+ + CH_4. \tag{2b}$$

State-specific cross sections as a function of kinetic energy have been measured for both reactions.

For the V^+ + C_3H_8 reaction, we (14) and others (15) have developed an appealing picture of how electronic factors control chemical reactivity (Figure 2). We assume that the rate-limiting step is M^+ insertion in a C-H bond of the alkane and consider the consequences of electron spin conservation and orbital correlation arguments along the path to the key H-M^+-C_3H_7 intermediate.

In V^+ the low-lying quintet terms $3d^34s(^5F)$ and $3d^4(^5D)$ are quite inert, while the triplet term $3d^34s(^3F)$ at 1.1 eV reacts efficiently to eliminate H_2 (16,17). The triplet's absolute reaction efficiency is 41% for the C_3H_8 reaction (absolute cross section of 37 ± 19 Å^2) at 0.2 eV collision energy. From these experimental results, we inferred the importance of electron spin conservation during V^+ insertion into a C-H bond of the hydrocarbon in determining the pattern of state-specific reactivity. A similar picture with the added feature of orbital specificity ($3d^n$ more reactive than $3d^{n-1}4s$) can explain the observed trends in M^+ + H_2 reactivity across the 3d series (7).

The Fe^+ + C_3H_8 reaction is puzzling when placed in the same conceptual framework that works well for V^+. The total reaction cross section is rather insensitive to changes in the initial Fe^+ electronic state (16,17), in contrast to the behavior of V^+ + C_3H_8. The initial states sampled vary from 0-1.1 eV in electronic energy, include both $3d^64s$ and $3d^7$ electron configurations, and include both sextet and quartet electron spins. In particular, the ^4F term has both the proper spin and orbital occupancy for C-H bond insertion. Nevertheless, the relative reactivity of ^6D, ^4F, and ^4D terms varies only a factor of four in Fe^+ + C_3H_8. Absolute cross sections show that the Fe^+ + C_3H_8 reaction is quite

(donor) overlaps the empty π^* on ethylene (acceptor) and doubly-occupied π on ethylene (donor) overlaps the empty 4s on the metal (acceptor).

TABLE I. Effective bimolecular rate constants (10^{-12} cm^3-molec^{-1}-sec^{-1}) for reactions of 3d-series, ground state neutral metal atoms at 0.80 torr He and 300 K. Rate constants precise to $\pm 10\%$ and accurate to $\pm 30\%$.

Reactant	Sc ^2D	Ti ^3F	V ^4F	Cr ^7S	Mn ^6S	Fe ^5D	Co ^4F	Ni ^3F	Cu ^2S
ethene	NR	NR	NR	NR	NR	NR	NR	0.5	NR
propene-h$_6$	9.5	6.2	9.6	NR	NR	NR	NR	11	NR
propene-d$_6$	8.3	5.0	9.6	--	--	--	--	21	--
C_3F_6	NR	NR	NR	NR	NR	NR	NR	0.3	NR
1-butene	14	7.1	14	NR	NR	NR	0.09	140	NR
t-2-butene	20	7.5	22	NR	NR	NR	NR	160	NR
c-2-butene	33	14	30	NR	NR	NR	NR	155	NR
isobutene	71	52	68	NR	NR	NR	NR	67	NR
1,3-butadiene	29	10	15	0.15	NR	NR	0.35	110	NR
n-butane	NR	NR	NR	NR	NR	NR	NR	NR	NR
cyclopropane	0.01	NR	0.02	NR	NR	NR	NR	10	NR
propane	NR	NR	NR	NR	NR	NR	NR	NR	NR

NR means no reaction observed; k less than 10^{-14} cm^3-s^{-1}.
Dash (--) means reaction not studied.

None of the neutral M atom ground states is properly prepared for a favorable interaction with C_2H_4. As M and alkene approach, the metal atom must hybridize to relieve repulsion between $4s^2$ or $4s^1$ and the doubly occupied π orbital on alkene. Two mechanisms are possible, sd hybridization and sp hybridization; in reality, the optimal combination of these will occur. In sd hybridization, $3d_{z^2}$ and 4s mix to form two hybrids, which we label sd_+ and sd_-. One hybrid (sd_+) concentrates probability along the z-axis; this becomes the acceptor orbital. The other (sd_-) concentrates probability in the xy-plane, thus relieving repulsion; this becomes the singly or doubly occupied non-bonding orbital. In sp hybridization, $4p_z$ and 4s mix to form two hybrids directed towards (sp_+) or away from (sp_-) the approaching C_2H_4 π-orbital. The sp_+ hybrid becomes the acceptor and the sp_- becomes the lone pair, polarized to the rear of the metal atom to relieve repulsion *(27,28)*.

Why are Sc, Ti, V, and Ni the only reactive 3d-series metal atoms? Compared with Fe and Co, Ni has a very small promotion energy to a state of low-spin, $3d^{x-1}4s$ character (10 kcal-mol^{-1} for Ni vs 34 for Fe and 21 for Co).

Only such a low-spin state can hybridize and accomodate two non-bonding electrons in the resulting sd hybrid. The avoided intersections of the repulsive surfaces from $3d^{n-2}4s^2$ ground state reactants and the attractive surfaces from low-spin $3d^{n-1}4s$ excited state reactants produce barriers to M-alkene adduct formation (Figure 4). The barrier height scales roughly as the energy difference between $3d^{n-2}4s^2$ and low-spin $3d^{n-1}4s$ atomic states. These ideas are familiar from the V^+ and Fe^+ + alkane examples. Ni overcomes the barrier at 300 K while Fe and Co cannot. The sd-hybridization scheme is preferred in Ni because the 4p orbitals lie very high in energy.

It is less obvious why Sc, Ti, and V react with alkenes at such similar rates. Electron spin permits both the sd- and sp-hybridization scheme to operate from the ground states of Sc, Ti, and V. Since the promotion energy to low-spin $3d^{n-1}4s$ states varies widely from Sc to Ti to V but the rate constants are quite similar, we suggest participation of 4p orbitals in the bonding.

The early data indicate that Zr ($4d^25s^2$ ground state) is much more reactive than its 3d-series congener Ti. The *energies* of corresponding types of terms are very similar in Ti and Zr. Consequently, we suggest that orbital sizes play an important role. In the 4d series, 4d and 5s orbitals have comparable size; in the 3d series, 3d is much smaller than 4s. Repulsive 4s-alkene interactions and bonding 4d-alkene interactions begin at more similar distances, attenuating the steepness of the repulsive surfaces from $4d^{x-2}5s^2$. In addition, the larger absolute size of 4d compared with 3d ultimately makes stronger chemical bonds due to better spatial overlap. Mo (high-spin, $4d^55s^1$ ground state) reacts with alkenes at 300 K, albeit slowly. In contrast, the 3d-series congener Cr ($3d^54s^1$) is inert. The same orbital size considerations come into play comparing Mo to Cr. In addition, electron spin is a poorer quantum number in the 4d series than in the 3d series. This means that crossings between nominal high-spin and low-spin surfaces will be more strongly avoided, making nominal spin-changing events more facile.

Summary and Prognosis

At a certain level, we are beginning to understand how the electronic structure of transition metal atoms determines their gas phase chemical reactivity. A recurring theme is the importance of intersections among different kinds of diabatic surfaces arising from the closely-spaced atomic asymptotes. It is really the overall *pattern* of low-lying states that determines reactivity, because the ground state itself is often ill-prepared to form chemical bonds.

Perhaps the presently perceived differences among M, M^+, and M^{2+} chemistry will someday seem superficial. The current data may show that chemical reactivity is governed by the ability of the metal atom to gain close access to the hydrocarbon, which is determined by the long-range part of the metal-hydrocarbon potential. That is, most of these systems may have low-energy pathways from M(hydrocarbon) at short range to elimination products, but only some of the systems, primarily those with +1 or +2 charge, see small enough barriers to access those pathways. We have only *begun* to probe the truly subtle chemical questions involving the geometries and stabilities of the intermediates between Fe^+ + C_3H_8 and $FeC_2H_4^+$ + CH_4, for example. Spectroscopic work could prove important here.

Finally, we can ask the question whether gas phase work is relevant to solution phase organometallic chemistry. There is an enormous literature of C-H bond activation in solution phase. Typically thermolysis or photolysis initiates

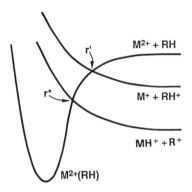

FIGURE 3. One-dimensional model of M^{2+} + alkane reactions, showing the radii of the electron transfer (r') and hydride transfer (r*) curve crossings.

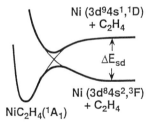

FIGURE 4. Schematic diabatic potential energy curves describing the approach of neutral Fe, Co, or Ni to an alkene. A barrier occurs on the lowest-energy adiabatic surface due to the avoided crossing of the curves.

the reaction by creating a coordinatively unsaturated metal center. Examples include complexes of Ru and Rh from the 4d series and of Ir and Pt from the 5d series. This is reminiscent of the list of M^+ cations that can dehydrogenate methane at low collision energy (Rxn. 3). The same metals would make a good list of heterogeneous catalysts as well. To some extent the highly active metals remember who they are, whether surrounded by vacuum, ligands and solvent, or other metal atoms!

Acknowledgments. I thank the National Science Foundation and the Donors of the Petroleum Research Foundation for continuing support of our research on the structure and reactivity of transition metal species. My former and current graduate students Dr. Lary Sanders, Dr. Russell Tonkyn, Dr. David Ritter, Dr. Scott Hanton, Dr. Andrew Sappey, Mr. Robert J. Noll, and Mr. John Carroll deserve most of the credit for the progress we have made.

Literature Cited

1. P.B. Armentrout in *Gas Phase Inorganic Chemistry*, edited by D.H. Russell (Plenum, New York, **1989**), pp. 1-42; S.W. Buckner and B.S. Freiser, *ibid.*, pp. 279-322; D.P. Ridge and W.K. Meckstroth, *ibid.*, pp. 93-113; R.R. Squires and K.R. Lane, *ibid.*, pp. 43-91; D.H. Russell, D.A. Fredeen, and R.E. Teckleburg, *ibid.*, pp. 115-135; M.F. Jarrold, *ibid.*, pp. 137-192; D.K. MacMillan and M.L. Gross, *ibid.*, pp. 369-401.

2. L.M. Roth and B.S. Freiser, Mass Spec. Rev., in press.

3. For example, see M. Hackett and G.M. Whitesides, J. Am. Chem. Soc. **110,** 1449 **(1988)** and references therein.

4. C.E. Moore, NBS Circ. No. 467 (U.S. Dept. of Commerce, Washington, D.C., **1949, 1952**); C. Corliss and J. Sugar, J. Phys. Chem. Ref. Data **14** (Supp. 2) 407 **(1985)**.

5. P.B. Armentrout and J.L. Beauchamp, Acc. Chem. Res. **22**, 315 **(1989)**; J. Allison in *Progress in Inorganic Chemistry*, edited by S.J. Lippard (Wiley-Interscience, New York, **1986**); P.B. Armentrout, Ann. Rev. Phys. Chem. **41,** 313 **(1990)**; P.B. Armentrout, Science **251**, 175 **(1991)**.

6. G. Ohanessian, M.J. Brusich, and W.A. Goddard III, J. Am. Soc. **112**, 7179 **(1990)**.

7. L.F. Halle, P.B. Armentrout, and J.L. Beauchamp, Organometallics 1, 963 **(1982)**; E.R. Fisher, R.H. Schultz, and P.B. Armentrout, J. Phys. Chem. **93,** 7382 **(1989)**; R. Georgiadis, E.R. Fisher, and P.B. Armentrout, J. Am. Chem. Soc. **111**, 4251 **(1989)**; S.K. Loh, E.R. Fisher, L. Lian, R.H. Schultz, and P.B. Armentrout, J. Phys. Chem. **93**, 3159 **(1989)**; R. Georgiadis and P.B. Armentrout, J. Phys. Chem. **92**, 7067 **(1988)**; R. Georgiadis and P.B. Armentrout, J. Phys. Chem. **92**, 7060 **(1988)**; L.S. Sunderlin and P.B. Armentrout, J. Phys. Chem. **92**, 1209 **(1988)**.

8. W.D. Reents, F. Strobel, R.G. Freas, J. Wronka, and D.P. Ridge, J. Phys. Chem. 89, 5666 **(1985)**; F. Strobel and D.P. Ridge, J. Phys. Chem. **93**, 3635 **(1989)**.

9. R. Tonkyn and J.C. Weisshaar J. Phys. Chem. **90**, 2305 **(1986)**; R. Tonkyn, M. Ronan, and J.C. Weisshaar, J. Phys. Chem. **92**, 92 **(1988)**.

10. D.B. Jacobson and B.S. Freiser, J. Am. Chem. Soc. **105**, 5197 **(1983)**; G.D. Byrd, R.C. Burnier, and B.S. Freiser, J. Am. Chem. Soc. **104**, 3565 **(1982)**.

11. J.C. Weisshaar in *Advances in Chemical Physics,* edited by C.Ng (Wiley-Interscience, New York, **1991**).

12. P.R. Kemper and M.T. Bowers, J. Am. Chem. Soc. **112**, 3231 (**1990**).

13. P.A.M. van Koppen, J. Brodbelt-Lustig, M.T. Bowers, D.V. Dearden, J.L. Beauchamp, E.R. Fisher, and P.B. Armentrout, J. Am. Chem. Soc. **113**, 2359 (**1991**).

14. L. Sanders, A.D. Sappey, and J.C. Weisshaar, J. Chem. Phys. **85**, 6952 (**1986**); L. Sanders, S.D. Hanton, and J.C. Weisshaar, J. Chem. Phys. **92**, 3485 (**1990**).

15. N. Aristov and P.B. Armentrout, J. Am. Chem. Soc. **108**, 1806 (**1986**); N. Aristov, Ph.D. Thesis, Univ. of California-Berkeley, Dept. of Chemistry (**1988**).

16. S.D. Hanton, R.J. Noll, and J.C. Weisshaar, J. Chem. Phys. **96**, 5165 (**1992**); J. Chem. Phys. **96**, 5176 (**1992**).

17. R.H. Schultz and P.B. Armentrout, J. Phys. Chem. **91**, 4433 (**1987**); R.H. Schultz, J.L. Elkind, and P.B. Armentrout, J. Am. Chem. Soc. **110**, 411 (**1988**); R.H. Schultz and P.B. Armentrout, in progress.

18. K.K. Irikura and J.L. Beauchamp, J. Am. Chem. Soc. **113**, 2769 (**1991**); T.J. McMahan, Y.A. Ranasinghe, and B.S. Freiser, J. Phys. Chem., submitted.

19. R. Tonkyn and J.C. Weisshaar, J. Am. Chem. Soc. **108**, 7128 (**1986**).

20. S.W. Buckner and B.S. Freiser, J. Am. Chem. Soc. **109**, 1247 (**1987**); S.W. Buckner, J.R. Gord, and B.S. Freiser, J. Am. Chem. Soc. **91**, 7530 (**1989**).

21. C.E. Brown, S.A. Mitchell, and P.A. Hackett, J. Phys. Chem. **95**, 1062 (**1991**) and references therein.

22. D. Ritter and J.C. Weisshaar, J. Am. Chem. Soc. **112**, 6425 (**1990**); D. Ritter and J.C. Weisshaar, J. Phys. Chem. **94**, 4907 (**1990**).

23. D. Ritter, Ph.D. Thesis, University of Wisconsin-Madison, **1990**.

24. C.E. Brown, S.A. Mitchell, and P.A. Hackett, Chem. Phys. Lett. **191**, 175 (**1992**).

25. P-O. Widmark, B.O. Roos, and P.E.M. Siegbahn, J. Am. Chem. Soc. **89**, 2180 (**1985**).

26. J. Carroll, K. Haug, and J.C. Weisshaar, work in progress.

27. C.W. Bauschlicher, Jr., Chapter in this book and references therein.

28. See, for example, M.R.A. Blomberg, P.E.M. Siegbahn, U. Nagashima, and J. Wennerberg, J. Am. Chem. Soc. **113**, 424 (**1991**).

RECEIVED January 21, 1993

Chapter 17

Gas-Phase Formation of Atoms, Clusters, and Ultrafine Particles in UV Laser Excitation of Metal Carbonyls

Yu. E. Belyaev, A. V. Dem'yanenko, and A. A. Puretzky

Institute of Spectroscopy, Russian Academy of Sciences, 142092 Troitsk, Moscow Region, Russia

Three main topics of laser-induced chemistry of organo-metallics are discussed - delayed luminescence of metal atoms, formation of electronically excited metal dimers and gas-phase production of ultrafine particles.

Gas-phase UV multiphoton excitation (MPE) and dissociation (MPD) of organometallic molecules has been the subject of intense studies in recent years. This process has a few specific features which make it attractive for both fundamental studies and various applications. One main goal of fundamental studies is describing the mechanism of bare metal atoms production. MPD of organometallic molecules leads to the formation of metal atoms in ground and excited states (1-5) and also gives rise to a set of highly reactive coordinatively unsaturated species (6-8) which can initiate and participate in secondary chemical reactions. That means that MPD of such molecules can be used for basic studies on Laser Chemistry of Organometallics. In this chapter we shall discuss three closely related and we believe central topics of laser chemistry of $Mo(CO)_6$.

The first topic is the formation of long lived highly excited Mo atoms. These energized atoms could start chemical reactions with the parent molecules and the coordinatively unsaturated compounds obtained in UV excitation of $Mo(CO)_6$. It was found that under relatively high laser fluences (300 to 500 mJ/cm^2) the so called "delayed" luminescence from high lying e^7D_j levels of Mo atoms was observed. The delayed luminescence pulse attains its maximum, about a few hundreds nanoseconds, after the end of the laser pulse and drops with a characteristic time of about a few microseconds, while spontaneous radiative lifetime of the observed transition z^7P°-e^7D is only a few nanoseconds. Detailed analysis has shown that this phenomenon can be explained by selective population of the e^7D_j levels through cascade transitions from the Rydberg states produced in UV MPE of the parent $Mo(CO)_6$ gas.

The second topic is related to the formation of electronically excited metal dimers. The attempts to produce ions of small metal clusters from stable organometallic compounds, such as $M_k(CO)_n$, (M is a metal atom, k=2-4, n=8,

0097–6156/93/0530–0220$08.50/0

10, 12) by eliminating ligands with UV laser radiation are known well (see Review (9)). When such organometallic compounds are photoionized by UV laser, the mass spectrum has, along with a metal ion, dimeric, trimeric and tetrameric peaks of M_2^+, M_3^+, M_4^+ (9). We have observed neutral electronically excited dimers Mo_2 upon excitation of $Mo(CO)_6$ gas with XeCl laser (10) The parent compounds for them are the intermediates $Mo_2(CO)_1$ formed from the $Mo(CO)_6$ by the first XeCl laser pulse. The second pulse of XeCl laser removed the ligands form $Mo(CO)_1$ and gave rise to Mo_2 dimers in the electronically excited state $A^1\Sigma_u^+$. These dimers were detected by luminescence. Electronic absorption spectra of Mo_2 were comprehensively studied in (11,12). Apart from such a laser-induced process, we have found out that electronically excited dimers, Mo_2, are formed in a collisional chemiluminescent chemical reaction. The possible mechanisms of formation of electronically excited dimers Mo_2 are discussed.

The third topic we are discussing in this chapter is the formation of ultrafine particles in UV MPD of $Mo(CO)_6$, $Cr(CO)_6$ and $W(CO)_6$ molecules in the gas phase. This process produces black soot which ICRFT mass spectrometer analysis has shown contains different molybdenum-carbon clusters. It has been shown that for $Mo(CO)_6$ reactant, bright emission is observed under excitation of ultrafine particles by the second XeCl-laser pulse. The main characteristics of this emission are given.

Experimental

The experimental setup comprised two XeCl lasers ($\lambda =308$ nm, pulse energies $E_1=30mJ$, $E_2=10mJ$, pulse widths ($\tau_1=15$ ns (FWHM), $\tau_2=10$ ns (FWHM)) and a stainless-steel chamber (350 mm x 350 mm x 120 mm) having quartz or NaCl windows ($\varnothing$ 60 mm). For two-pulse experiments, the laser beams were made to counter-propagate and focused into the chamber with quartz lenses ($f_1=20$ cm, $f_2=25$ cm, $\varnothing$ 60 mm). The cross-sectional areas of the both beams in the observation region were 2.5x8 mm^2 and 3.5x5 mm^2, respectively. The pulsed CO_2 laser used for studying delayed atomic luminescence produced 2 J pulses having a 100ns (FWHM) peak and 500 ns (FWHM) tail.

The chamber was evacuated to a residual pressure of about 10^{-5} Torr by means of an oil-diffusion pump fitted with a nitrogen trap. The container mounted directly on the chamber was filled with crystalline $Mo(CO)_6$. The $Mo(CO)_6$ pressure range used was 0.02 to 0.1 Torr. $Mo(CO)_6$ was commercial-grade and was sublimed and stored under vacuum before being used.

The luminescence spectra of dimers were analyzed by means of a monochromator (slit spectral width 20 Å/mm) and a gated optical multichannel analyzer (OMA) (the spectral response of this system was 350 to 850 nm, spectral resolution 2 Å, minimum gate width 150ns). The temporal evolution of luminescence at certain fixed wavelengths was studied using a photomultiplier (spectral response 200 to 800 nm).

To avoid the influence of the diffusive escape of particles from the observation zone in the time range 0-10 μs on the measured kinetics we had to choose a transverse observation geometry. The elongated exposed zone was projected (scale 1:1) by a lens (f=10 cm) not along the monochromator slit (height of the slit h=1.0 cm) but across it. A small region of this zone was registered. Our calculations have shown that with such a irradiation and detection geometry the expansion of the exposed zone even due to free particle flying apart does not affect considerably the measured time dependencies with $t \leq 10 \mu s$. Upon diffusive expansion of the exposed zone this time increased to about $50 \mu s$. To find the location of the formation region of ultrafine particles and roughly estimate their dimensions, light scattering measurements were made with a 2-mW, 1 mm diameter HeNe laser. The HeNe laser could be moved vertically to change the distance between the lasers. The scattered light was picked up with a photomultiplier through the upper window in the chamber.

Spectroscopic and Thermodynamic Parameters of $Mo(CO)_6$, $Mo(CO)_k$, Mo_2

$Mo(CO)_6$. The UV absorption spectrum of $Mo(CO)_6$ has been studied rather well (13). The XeCl laser frequency (32468 cm^{-1}) coincides with the long-wave wing of the UV absorption spectrum. The related section of the spectrum is formed by two comparatively weak absorption bands with their maxima at 31350 cm^{-1} (the dipole-forbidden transition $^1T_{1g} \leftarrow {}^1A_{1g}$) and 34880 cm^{-1} (the dipole-allowed transition $^1T_{1g} \leftarrow {}^1A_{1g}$). The absorption cross-section at the XeCl laser wavelength (λ =308 nm) is about 10^{-17} cm^2.

The removal of the first CO group requires the energy 1.76 eV (14), 1.73 eV (15). The energies needed to remove the subsequent CO groups are as follows. $Mo(CO)_5$ - 1.28 eV (8), 1.52 eV (16); $Mo(CO)_4$ - 1.34 eV (8), 1.30 eV (16). The removal of the remaining ligands, $Mo(CO)_{3-1}$, may be characterized by the average energy $E_{CO}=1.69$ eV (8). These energies were used to plot an energy diagram of removal of successive ligands by a laser photon with the energy $\hbar\omega_{las}=4.0$ eV (Figure 1).

$Mo(CO)_k$ (1< k< 6). The spectroscopy and thermodynamics of these fragments are still not clearly understood. There are just a few results on IR spectra of these radicals. They were used to measure the rate constants of reactions between $Mo(CO)_6$ molecules (k=3-5) (17). These results are given in Table I.

Table I. Rate Constants of Reactions of $Mo(CO)_k$ (k=3-5) with $Mo(CO)_6$[a]

Rate constant,	Torr$^{-1} \cdot$s^{-1}	Reaction
k_1'	$5 \cdot 10^6$	$Mo(CO)_5 + Mo(CO)_6 \xrightarrow{k'_1} Mo_2(CO)_{11}$
k_1''	$(2 \pm 1) \cdot 10^7$	$Mo(CO)_4 + Mo(CO)_6 \xrightarrow{k''_1} Mo_2(CO)_{10}$
k_1'''	$(2 \pm 1) \cdot 10^7$	$Mo(CO)_3 + Mo(CO)_6 \xrightarrow{k'''_1} Mo_2(CO)_9$

[a]Data is taken from (17).

Mo_2. The metal dimer Mo_2 has been studied rather well. Table II presents the lower electronic terms of Mo_2 and their related frequencies of nuclear vibrations (11,12). The dissociation energy of the ground state, $X^1\Sigma_g^+$, is D.=33000± 5000 cm^{-1} (11). The radiative lifetime of the state $\vartheta'=0$, $A^1\Sigma_u^+$ is 18 ns (12). Here we are going to consider the luminescence of Mo_2 at the transition $X^1\Sigma_g^+ \leftarrow A_1\Sigma_u^+$.

The specific feature of this transition is the zero shift of equilibrium position for nuclear vibrations under electronic excitation. As a result, the spectrum has a sequence with $\Delta\vartheta = 0$ only so that the frequency intervals between the sequential bands are determined by the vibrational frequency difference of the ground $X^1\Sigma_g^+$ and electronically excited $A^1\Sigma_u^+$ states, $\Delta\omega = 34$ cm^{-1}. The calculation of some lower electronic terms that explains the observed peculiarities of Mo$_2$ was done in (*18*).

Table II. Lower Singlet Electronic States of Mo$_2^a$

State	T_0,cm^{-1}	ω_e, cm^{-1}
$C\left(^1\Sigma_u^+\right)$	31400	408
$B\left(^1\pi_u\right)$	25651.65	412
$A\left(^1\Sigma_u^+\right)$	19303.70	443.4
$X\left(^1\Sigma_g^+\right)$	0	477.1

*Data is taken from (*11,12*).

Delayed Atomic Luminescence

Main experimental results. When Mo(CO)$_6$ gas is excited by an excimer laser with a fluence of about 0.5 J/cm^2, one can easily observe luminescence of metal atoms, i.e. a phenomenon which is well known for any organometallic compound. Figure 2 presents the typical section of luminescence spectrum consisting of atomic, Mo, and dimeric, Mo$_2$, lines. The mechanism for dimer formation and the luminescence spectrum of Mo$_2$ induced by excimer lasers will be discussed below. In this section we shall consider the peculiarities of the observed atomic luminescence.

The atomic luminescence spectrum shows two different types of lines. Ordinary ions appear rather fast, as a rule, during a laser pulse. The luminescence is quenched in accordance with the oscillator strength of the observed atomic transition. It should be mentioned that in some cases the atomic emission may be longer compared with the radiative lifetime due to radiation trapping. The phenomenon occurs with relatively high populations of the levels at which a particular transition is terminated (*4*).

The other group of lines (shaded lines in Figure 2) shows an interesting peculiarity. These lines appear when the laser pulse has already gone, i.e. the luminescence intensity reaches its maximum a few hundreds of nanoseconds after the maximum of the laser pulse. Table III gives an example of a few observed atomic lines which show pronounced delayed atomic luminescence.

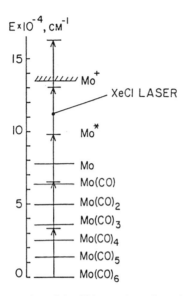

Figure 1.   Energetics of the 308 nm photodissociation of Mo(CO)$_6$

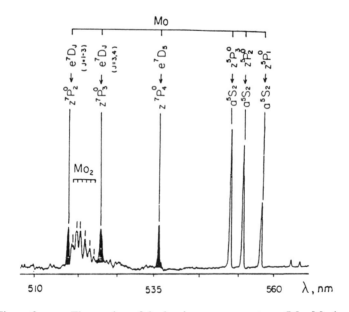

Figure 2.   The section of the luminescence spectrum (Mo, Mo$_2$)
              induced in Mo(CO)$_6$ gas by XeCl-laser. The spectrum is
              taken 150 ns after the laser pulse.

Table III. Delayed Mo luminescence lines[a]

Transition	$\lambda, \text{Å}$	Energy, cm^{-1}
$z^7P_2^{\circ} \leftarrow e^7D_1$	5174.18	25614 - 44936
$z^7P_2^{\circ} \leftarrow e^7D_3$	5172.94	25614 - 44941
$z^7P_2^{\circ} \leftarrow e^7D_3$	5171.08	25614 - 44947
$z^7P_3^{\circ} \leftarrow e^7D_3$	5240.88	25872 - 44947
$z^7P_3^{\circ} \leftarrow e^7D_4$	5238.20	25872-44957
$z^7P_4^{\circ} \leftarrow e^7D_5$	5360.56	26321- 44970

[a]Data is taken from (*19, 20*).

The lines assignment and the energies of the respective transitions are also given in the Table. Figure 3 shows the typical temporal profile of the $z^7P_4^{\circ} \leftarrow e^7D_5$ transition $(\lambda = 5360.56 \text{Å})$ following 308 nm excitation of $Mo(CO)_6$ with a pressure of 0.04 Torr. After the 15 ns (FWHM) laser pulse, the luminescence intensity starts to increase and reaches its maximum 600 ns after the laser pulse. The decay after the luminescence maximum was then fitted with an exponential function with the characteristic 1/e - decay time, τ_A. Further, to characterize the pulse of delayed atomic emission we will use also the time-integrated intensity, I_{int}. When the pressure of $Mo(CO)_6$ is varied from 0.02 to 0.1 Torr., τ_A changes slightly from 1 to 0.7 μs respectively. The integral intensity increases rapidly proportionally to $p^{1.8\pm0.3}$ when the pressure rises in the same pressure range.

Below we shall list the main experimental results explaining delayed atomic luminescence. They are the anomalous effect of neon in the set of noble buffer gases, the disappearance of delayed emission with addition of relatively small amounts of electronegative gases, SF_6 and $CC1_4$, and the burning out of the holes in an emission pulse by IR radiation of pulsed CO_2 laser.

Effect of buffer gases. The study of the influence of different buffer gases on the main parameters of the delayed luminescence pulse permits us to disclose a few interesting peculiarities. Figure 4a, b shows the I_{int} versus the pressure of the noble buffer gases: He, Ne, Kr, Ar and Xe. The behavior of these curves is typical of all the noble gases for which these measurements were conducted. The emission intensity starts to increase with the buffer gas pressure, then it reaches its maximum under the pressure which is specific of each buffer gas (see Table IV) and drops with the further increase in pressure.

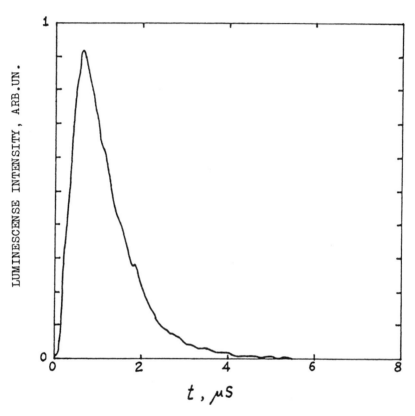

Figure 3.   Time profile of delayed atomic emission. The luminescence
              was picked up at 5360.56 Å.

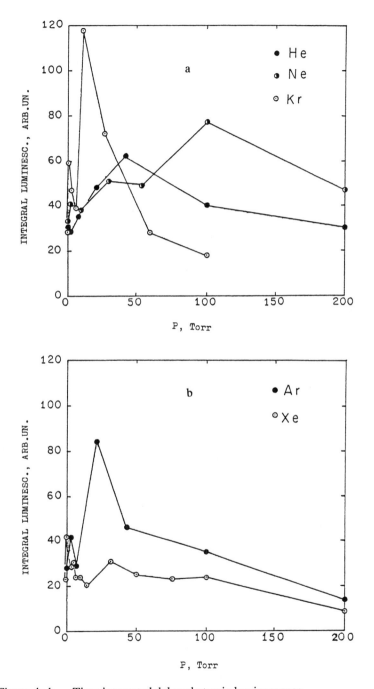

Figure 4a,b.   Time integrated delayed atomic luminescence
$\left(\lambda = 5360.56\text{Å}\right)$ vs the pressures of the noble gases - He, Ne,
Kr (a), Ar, Xe (b).

Table IV. Effect of Buffer Gases

Buffer gases	Pressure for max. I_{int}, Torr	Pressure for max. τ_A, Torr	Maximum $\tau_A, \mu s$
He	42	20	1.5
Ne	100	60	3.0
Ar	22	23	2.2
Kr	12	16	2.6
Xe	1	10	2.2
H_2	6	1	0.8
D_2	7	2	0.9
N_2	2	1	1.2
NH_3	0.5	0	0.7

The characteristic time, τ_A, shows similar pressure dependences. It is easy to see from Figure 4a, b and Table IV that Ne affects anomalously delayed emission compared with the other noble buffer gases. For all the buffer gases the delayed emission disappears when the buffer gas pressure is in excess of 500 Torr, with the exception of Ne.

The dependences $\tau_A(p_{buf})$ and $I_{int}(p_{buf})$ reach their maxima with 60 and 100 Torr respectively. Delayed luminescence is observed even with the Ne pressure exceeding 600 Torr.

Table IV also presents similar data for other buffer gases: H_2, D_2, N_2, NH_3. The effect of these gases on delayed emission is similar to that for the noble buffer gases. But a growth and a subsequent decrease of τ_A and I_{int} are observed with lower pressures especially for ammonia (see Table IV).

The electronegative gases SF_6 and CCl_4 which have relatively high electron affinities (1.48 and 2.12 eV) have an effect on delayed emission that differs qualitatively from the effect of other studies buffer gases. When relatively small amounts of these gases ($2 \cdot 10^{-3}$ Torr CCl_4 and 10^{-2} Torr SF_6) were added to the $4 \cdot 10^{-2}$ Torr of parent $Mo(CO)_6$ gas, the integral luminescence intensity and the decay time, τ_A, dropped very rapidly. Figure 5 shows the dependences of I_{int} on the pressures of CCl_4 and SF_6. The $Mo(CO)_6$ pressure was $4 \cdot 10^{-2}$ Torr. When the pressure of CCl_4 was about 0.5 Torr the delayed luminescence practically disappeared.

Two mechanisms of delayed atomic emission can be proposed - chemiluminescent and Rydberg. The chemiluminescent mechanism can be related to the formation of a highly excited intermediate complex decaying to form Mo atoms with their energies of about $45 \cdot 10^3$ cm^{-1}. But our kinetic calculations show that this mechanism does not enable us to describe the observed variations of the time profiles with the pressures of the buffer gases. We didn't find either a reasonable explanation of the observed anomalies with Ne and the effect of electronegative gases.

The second mechanism involves the formation of Rydberg Mo atoms following XeCl laser excitation of $Mo(CO)_6$. The decay of Rydberg atoms to the e^7D_5 state followed by the transition to the $z^7P_4^{\circ}$ states results in delayed atomic

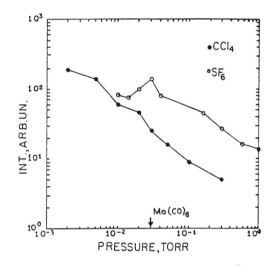

Figure 5.   Time integrated delayed atomic emission $\left(\lambda = 5360.56\text{Å}\right)$ vs the pressures of CCl_4 and SF_6.

emission. This mechanism can be checked by the subsequent ionization of Rydberg Mo atoms with pulsed CO_2-laser which must quench the emission by depopulation of the Rydberg states with energies $E \geq E_1 - \hbar\omega_{IR}$ (E_1 is the ionization energy of the Mo atom). IR laser excitation will result in hole burning in the delayed emission time profile which should be in coincidence with the onset of the CO_2-laser pulse.

Before going into details of the Rydberg mechanism we will discuss experimental results on UV-IR laser excitation of $Mo(CO)_6$.

Effect of the CO_2 - laser

To verify the Rydberg mechanism $Mo(CO)_6$ was excited by two delayed pulses of XeCl and CO_2 - lasers. The CO_2 - laser operated at 936.8 cm^{-1} (10P(28) line). Its pulse had a 100 ns (FWHM) peak followed by a 500 ns (FWHM) tail. The energy fluence in the observation region was 0.5 J/cm^2. It should be mentioned that $Mo(CO)_6$ does not have any intense absorption bands at 938.8 cm^{-1} (21). The excitation by a single CO_2 - laser didn't' give any luminescence.

Figure 6a presents two pulses of delayed atomic emission obtained with the excitation of $Mo(CO)_6$ by a XeCl-laser pulse (solid curve) and with both XeCl and CO_2 - laser pulses propagating with 200 ns delay. The onset of CO_2 - laser pulse is marked by an arrow. When the time delay was increased, the onset of the hole was also shifted. The difference between the time profiles of delayed atomic emission with and without CO_2 - laser pulse is shown in figure 6b. The CO_2 - laser pulse is shown for comparison.

Hole burning was observed when the onset of CO_2- laser pulse matched the maximum of the tail of the luminescence pulse. We didn't observe any hole burning in the leading edge of the emission pulse. Figures 6a, b show that the onsets of the hole and the CO_2 - laser pulse are in coincidence. Hole burning by CO_2 - laser pulse was observed only for relatively long pulses of delayed atomic luminescence. Therefore, the experiments on hole burning were carried out in conditions when some amounts of buffer gases were added to $Mo(CO)_6$.

Rydberg mechanism of delayed atomic luminescence

There are two processes which could lead to the formation of Rydberg atoms, Mo^{**}. In the first process Mo^{**} are created directly in UV laser excitation of $Mo(CO)_6$

$$Mo(CO)_6 + n\hbar\omega \rightarrow Mo^{**} + 6(CO) \qquad (1)$$

The second possibility is concerned with the recombination of slow electrons and ions, Mo^+, which results in the formation of neutral Mo atoms in different Rydberg states.

$$Mo(CO)_6 + n\hbar\omega \rightarrow Mo^+ + e + 6(CO)$$
$$Mo^+ + e + \left\{ \begin{array}{c} Mo(CO)_6 \\ Buff.gas \end{array} \right\} \rightarrow Mo^{**} + Mo(CO)_6 \qquad (2)$$

The subsequent radiative decay of Rydberg atoms populates the e^7D_{3-5} states with an energy of about $45 \cdot 10^3$cm and leads to the generation of delayed atomic

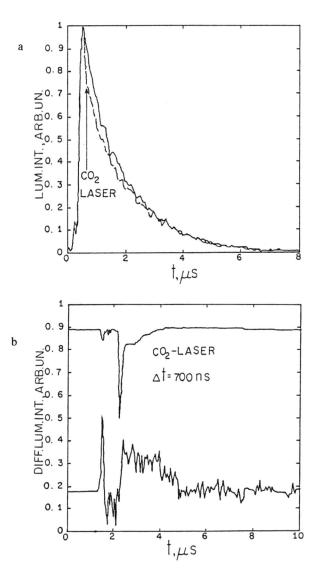

Figure 6a,b.   Time profiles of delayed atomic emission with (dashed
curve) and without (solid curve) CO_2-laser pulse. The arrow
shows the onset of the CO_2-laser pulse. Time delay between
XeCl- and CO_2- lasers was 200 ns (a). The difference
between time profiles without and with CO_2 - laser pulse,
following with the delay 700 ns after XeCl laser pulse (b).
The CO_2 - laser pulse is shown at the top.

emission at the transition $z^7P_4^\circ \leftarrow e^7D_{3-5}$. Figure 7 shows some of the transitions which populate the e^7D_J state. Three Rydberg series, $ns\ ^7S_3$, $nd\ ^7D_J$ and $np\ ^7P_J^\circ$ studied in (22) are shown in this Figure.

The appearance of a delayed maximum of the luminescence pulse in the process (1) could be related to collisional relaxation populating the states whose dipoles transitions to the e^7D_J states are allowed. The nature of the time delay in the process (2) is defined by the rate constant of collisional recombination. Suppose that under the recombination (2) Mo atoms are produced in the required Rydberg states. In this case we can describe the process (2) by the following kinetic equations

$$\left[\mathring{A}^+\right] = -\sum_i k_i \left[A^+\right]^2 [M] \tag{3a}$$

$$\left[\mathring{A}^{**}_i\right] = -k_{1i}\left[A^{**}_i\right] + k_i\left[A^+\right]^2 [M] \tag{3b}$$

$$\left[\mathring{A}\right] = \sum_i k_{1i}\left[A^{**}_i\right] - k_{rad}[A] \tag{3c}$$

Here $\left[A^+\right]$ is the Mo$^+$ concentration, $\left[A^+\right]_o = \alpha[M]$ is their initial concentration, $[M] >> \left[A^+\right]_o$. Rate constants, k_1, govern the population of i-th Rydberg state and k_{1i} stand for radiative decay of these states to the state from which the observed luminescence transition starts.

The process (1) could be described by the same equations (3) if we make the substitution of the term $k_i[A^+]^2[M]$ with $k_i\left[A^{**}_o\right][M]$. Here $\left[A^{**}_o\right]$ is the concentration of Rydberg atoms in an initial Rydberg state prepared by laser radiation.

The equations (3) were solved numerically and only five Rydberg levels were considered for simplicity. The radiative constants k_{1i} were taken inversely proportional to the third power of the main quantum number in accordance with the well known relation (23), i.e. $k_{1i}=1.5 \cdot 10^{-6}$ s^{-1}, $k_{15}=8.7 \cdot 10^{-6}$ s^{-1}, $k_{rad}=10^{-8}$ s^{-1}. These calculations are in good agreement with the dependences of the pulse shape on the pressures of the parent and buffer gases. The calculated time profile for the process (2) fits well the experimental pulse of delayed atomic emission. For example, the calculation gives that the luminescence intensity is proportional to $p^{1.74}$.

The effect of the CO_2 - laser pulse was taken into account by means of setting the population of one or a few Rydberg states to zero. A rectangular 150 ns CO_2 - laser pulse was used. Figure 8 shows the calculated time profile of the delayed luminescence pulse with the subsequent ionization of Rydberg atom by CO_2 - laser taken into account. The calculated time profile fits the experimental pulses. It should be mentioned that the process (2) does not take into account the collisional relaxation among different Rydberg states and does not enable us to describe the increase of τ_A when some buffer gas is added.

The Rydberg mechanism explains the observed anomalous behavior with

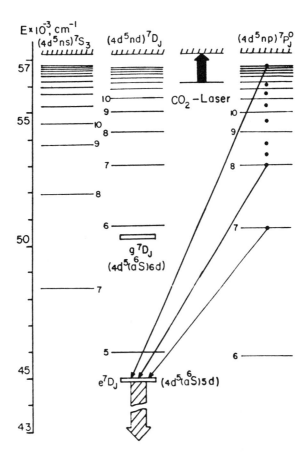

Figure 7:   Energy levels for the ns^7S_3 , nd^7P_J of Rydberg series Mo atoms. The energies were taken from (22).

Ne and the effect of the electronegative buffer gases. The relaxation process of Rydberg atoms with different buffer gases are well studied (24). The parameter with defines the scattering cross-section, σ, is the scattering length, L. Table V gives the values of scattering length for the noble gases.

Table V. Electron-scattering lengths L (a.u.) for noble gases[a]

Noble gas	He	Ne	Ar	Kr	Xe
L	1.1	0.2	-1.4	-3.1	-6.1

[a]Data is taken from Ref. (24)

For slow electrons σ and L are related by the simple equation

$$\sigma = \pi L^2 (v_e \to 0) \tag{4}$$

From Table V one can see that neon has an anomalously low cross-section.
   The electronegative gases SF_6 and CCl_4 can effectively trap slow electrons or ionize Rydberg atoms. The attachment rate constants are $2.7 \cdot 10^{-7}$ and $3.6 \cdot 10^{-7}$ cm^3/s for SF_6 and CCl_4 required for effective quenching of delayed atomic luminescence.
   Finally, we have found that the XeCl-laser generates delayed atomic emission. The mechanism which involves the Rydberg atoms formation under UV laser excitation of the parent molecule explains well the observed delayed atomic emission.

Formation of Dimers in UV Excitation of Mo(CO)$_6$
The idea of metal dimer generation under UV laser excitation of Mo(CO)$_6$ is as follows. The first excimer laser removes the ligands, CO, and results in the formation of cordinatively unsaturated metal carbonyl species, Mo(CO)$_k$ (k<6). The fragment distribution over k is laser wavelength and fluence dependent (8). These fragments react very rapidly with the parent molecules or fragments (Table I) which gives rise to the formation of Mo$_2$(CO)$_m$ producing metal dimers, Mo$_2$. This process is easy to realize experimentally. The peak of dimer luminescence was observed which matches the second excimer-laser pulse (10). Besides this laser-induced process, we were able to observe collisional chemiluminescent process of the formation of electronically excited dimers, Mo$_2$. The dimer formation and the main characteristics of dimer emission are given in detail in (10). Here is a short review of the main experimental results given together with the new data we have obtained recently.

Two-Pulse Experiments. The excitation of Mo(CO)$_6$ gas by two XeCl-laser pulses gives rise to a luminescence pulse whose onset coincides with the second laser pulse. This luminescence pulse is shown in Figure 9. It should be noted that the laser fluences were chosen so that neither of the lasers alone produced luminescence. The luminescence induced by the joint action of laser pulses has rather complex kinetics consisting of two parts: a pulse peak coincident with the second laser pulse and a tail being much longer than the peak. The luminescence

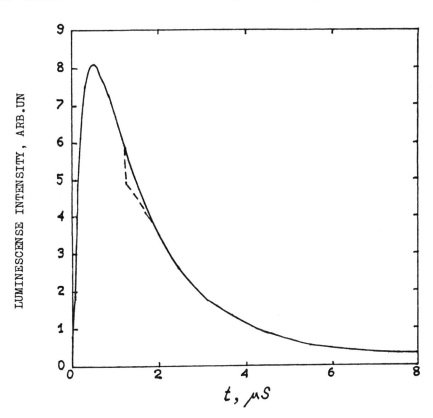

Figure 8.   The calculated time profile for delayed atomic emission without (solid curve) and with CO_2 - laser (dashed curve).

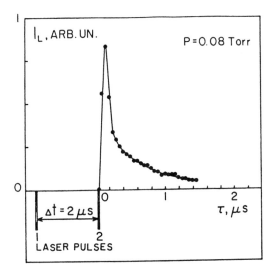

Figure 9.   The Mo_2 luminescence pulse induced by two-pulse excitation of $Mo(CO)_6$.

spectra of the peak and the tail show that it is the dimer Mo_2 that luminescences in the both cases (the luminescence spectra are given below). The complex shape of the luminescene time profile indicates that the mechanisms of formation of electronically excited Mo_2 in these cases are quite different.

Luminescence pulse peak. The luminescence spectrum of Mo_2 measured for the pulse peak is shown in Figure 10a. The luminescence spectrum consists of a band sequence with $\Delta \vartheta = 0$, and the intensity distribution of individual lines gives the population distribution of individual levels ϑ' of the electronically excited state $A^1\Sigma_u^+$. Therefore, it can be concluded from the spectrum that the electronically excited dimers Mo_2 which give rise to the peak are characterized by a highly nonequilibrium distribution over the vibrational states of the electronically excited term $A^1\Sigma_u^+$. There are at least two ϑ' levels whose populations exceed the populations of the adjacent vibrational states. These are the levels $\vartheta'=1, 3$ $A^1\Sigma_u^+$. Figure 11 shows the spectrum-integrated intensity of luminescence band $I_{int}^{(D)}$ as a function of the time delay, Δt, between the laser pulses when the $Mo(CO)_6$ gas pressure in the chamber is 0.042 Torr. The luminescence intensity attains its maximum value when $\Delta t=1.5 \mu s$. It rises drastically when Δt increases from zero to Δt_{max} and drops rather slowly when $\Delta t > \Delta t_{max}$. In the pressure range studied $I_{int}^{(D)}$ is proportional to $p^{1.8\pm0.3}$ for $\Delta t_\circ$ and to $p^{0.8\pm0.2}$ for Δt_{max}.

Let's consider the processes which may result in the observed dependences. The first laser removes ligands and gives rise to a set of fragments $F_k=Mo(CO)_k$ where $0\leq k \leq 6$, with distribution of fragments over k depending on laser fluence

$$M \xrightarrow{n\hbar\omega(1)} F_k, \qquad (5)$$

where $M=Mo(CO)_6$ denotes parent molecules. Then, the F_k fragments reacting either with the parent molecules M or with the slightly stripped fragments F_k, form molecular dimers $D_1=Mo_2(CO)_1$. These dimers reacting with M or F_k, with the rate constant k_2 produce a certain product P

$$F_k+M(F_k,)\xrightarrow{k_1} D_1+M(F_k)\xrightarrow{k_2} P$$
$$\downarrow n\hbar\omega(2) \qquad (6)$$

$$\overbrace{D_1, \quad D_o, \quad D_o^*, \quad D_o^{**},...}$$
$$\quad\quad \wr \quad \wr \quad \hbar\omega_L$$

The second laser pulse coming with a variable delay, Δt, removes the ligands from the molecular dimer D_1 thus producing either highly stripped molecular dimers D_1, or metal dimers $D_0=Mo_2$ distributed over electronically excited states. By monitoring the luminescence of Mo_2 from a definite electronically excited state we can follow the kinetics of formation of molecular dimer, D_1.

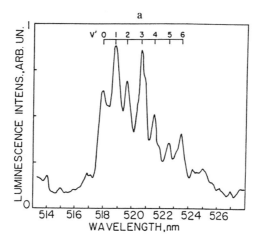

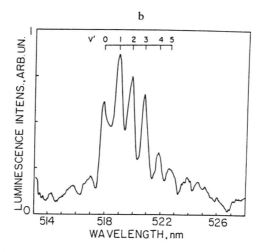

Figure 10a, b. Mo$_2$ luminescence spectra for the peak (a) and the tail (b) of
the luminescence pulse. (Adapted from ref. 10)

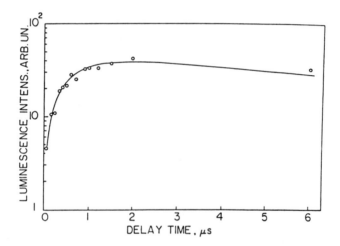

Figure 11.   Spectrum integrated Mo_2 luminescence intensity vs time
delay between XeCl-laser pulses. (Reproduced with the
permission from ref. 10. Copyright 1990.)

The formation and disappearance of molecular dimers D_1 is described by simple rate equations. As a result, we can derive from them an expression for the concentration, D_1

$$[D_1] = [F_k]_0 \cdot \frac{k_1}{k_2 - k_1} \cdot \{\exp(-k_1[M]t) - \exp(-k_2[M]t)\}, \tag{7}$$

when $[F_k]_0$ is the initial concentration of F_k fragments. The curve obtained from Eq. (7) is shown in Figure 11 by a solid line. From comparison of experiment and calculation for the constants k_1 is in a good agreement with the rate constants for the reactions of $Mo(CO)_4$ and $Mo(CO)_3$ with the parent molecules $Mo(CO)_6$ given in (17) (Table I).

With the laser fluences varying between 100 and 200 mJ/cm^2 the probability of induced single-photon absorption is $\sigma\Phi_1 \cong 3$. Therefore, the volume under irradiation contains a high concentration of relatively long-lived slightly stripped fragments F_k, with k'=5. It is these fragments that along with the parent molecules M are obviously the most probable partners for F_k in reaction (6).

The integral intensity of luminescence is proportional to $[D_1]$ which shows that with Δt_{max}, I_{int}^D is proportional to p. When Δt is small, it follows from (7) that

$$I_{int}^D \cdot [D_1]_o = k_1 [F_k]_o \cdot [M] \sim p^2 \tag{8}$$

The both cases are in a good agreement with the experimental dependences (10).

Luminescence pulse tail. The luminescence pulse tail of Mo_2 dimers has some specific features which are not described by the mechanism for the peak discussed above. First of all, this is relatively long luminescence. It should be noted that the radiative time of luminescence for the state $\vartheta' = 0, A^1\Sigma_u^+$ is just 18 ns (12). Figure 10b shows a luminescence spectrum of Mo_2 dimers measured for the luminescence pulse tail. The time parameters in this case were: the delay between the laser pulses $\Delta t = 1 \ \mu s$, the gate duration $\Delta g = 500 ns$, the delay between the onset of the second laser pulse and the gate $\Delta t = 700 ns$. This spectrum differs greatly from the similar spectrum measured for the peak of Mo_2 dimers by the intensity distribution of the related lines. It should be also noted that the population of the vibrational level $\vartheta' = 0, A^1\Sigma_u^+$ is lower compared to the vibrational level $\vartheta' = 1$. The characteristic feature of the luminescence spectrum is retained when the delay Δt is varied. It has been shown that $I^{(D)}$ is proportional to p$^{3\pm0.3}$, i.e. the pressure dependence is stronger for the pulse tail than for the peak. This explains the qualitative effect of disappearance of the pulse tail as the $Mo(CO)_6$ pressure decreases.

The experimental dependences of the luminescence intensity on different experimental parameters can be explained by the following mechanism. Some dimers, Mo_2, (below referred to as active dimers $D_o^{(a)}$) produced by the second laser pulse react with the parent gas $Mo(CO)_6$ or with the fragment F_k, forming highly excited unstable complexes, like $T_n^x = Mo_3(CO)_n^{**}$ which dissociate to form electronically excited dimers Mo_2 in the state $A_1 \Sigma_u^+$

$$D_o^{(a)} + M(F_{k'}) \xrightarrow{k_3} T_n^x \xrightarrow{k_4} D_o^* + A + nL \xrightarrow{k_5} D_o + \hbar\omega_L \tag{9}$$

Here, A is the Mo atoms, L is the CO ligands and $K_5 = 1/\tau_{rad}$. Under the assumption that $k_4 \gg k_3$, k_5 the sequence of the processes (9) is described by rate equations from which it follows that

$$I^D \sim k_5 \left[D_o^* \right] \approx \left[D_o^* \right] \cdot [M] k_3 \cdot \exp\left(-k_3[M]t\right) \tag{10}$$

Here, $t = \Delta\tau$. It is assumed in (10) that $k_5 \gg k_3[M]$. From (10) it follows that with small delays Δt between the laser pulses, when $D_o^{(a)} \sim p^2$ and $\Delta t = \Delta t_{max}$, when $D_o^{(a)} \sim p$,

$$I^{(D)} \sim p^2, p^3, \tag{11}$$

if Δt is small enough.

Single-pulse Excitation. In the case of MPE by one XeCl-laser pulse with fluences 300 -500 mJ/cm^2 one can observe a luminescence band corresponding to the transition $X^1\Sigma_g^+ \leftarrow A^1\Sigma_u^+$ of the Mo_2 dimer (Figure 2).

The luminescence induced by a single laser has two characteristic features. First, the luminescence spectrum of Mo_2 dimers measured under one pulse XeCl laser excitation is similar to that measured under two pulse excitation for the luminescence pulse tail. The comparison (see Figure 2 and 10b) reveals the same anomaly in the intensity of the band $\vartheta'' = 0, X^1\Sigma_g^+ \leftarrow \vartheta' = 0, A^1\Sigma_u^+$ and a similar intensity ratio for other bands $\vartheta'' \leftarrow \vartheta'$. Second, the time evolution of the intensity of dimer luminescence differs from that observed before because for the single-laser excitation there is a well defined maximum with certain delays.

The first specific feature indicates that the nature of dimer luminescence caused by single-pulse excitation seems to be the same as for the pulse tail in the case of two-pulse excitation. This means that with single-pulse excitation the chemiluminescent mechanism (9) takes place, too, and produces electronically excited dimers D_o^*. The only difference here is the nature of primary dimers for initiating the process (9). In the case of single-pulse excitation they, probably, appear as a result of chemical reaction of fragments $Mo(CO)_k$ formed by UV MPE of $Mo(CO)_6$ with the parent molecules or with $Mo(CO)_k$, (k < k') fragments. It is this additional stage that explains the above said second peculiarity - the presence of a maximum in the time profile of dimer luminescence.

The probable mechanism of dimer luminescence upon single-pulse UV excitation of $Mo(CO)_6$ is reduced to

$$F_k + M(F_{k'}) \xrightarrow{k_6} D_o^{(a)} + M(F_{k'}) \xrightarrow{k_7} T_n^x \xrightarrow{k_4} D_o^* + A + nL \xrightarrow{k_8} D_o + \hbar\omega_L \quad (12)$$

From the rate equations it is easy to deduce an expression for the dimer luminescence intensity

$$I^{(D)} \sim k_8 \cdot [D_o^*] \approx \frac{k_6 k_7 [M][F_k]_o}{(k_7 - k_6)} \cdot \left\{ \exp(-k_6 [M]t) - \exp(-k_7 [M]t) \right\} \quad (13)$$

For direct comparison of (13) with the experimental time evolution of dimer luminescence the knowledge of the specific values of the constants k_6 k_7 is required. These values can be taken from two-pulse experiment. Indeed, it can be seen that the first stage of (12) agrees with the first stage of (9) for the luminescence pulse tail. Thus, $k_6 \approx k_1, k_7 \approx k_3$. The observed kinetics of the single-pulse process (13), with its rate constants determined above describe well the experimental data.

Vibrational Distribution in the Electronically-Excited State, $A^1\Sigma_u^+$.
As mentioned above, the Mo_2 luminescence spectra (Figures 2 and 10a, b) show nonequilibrium vibrational distribution in the $A^1\Sigma_u^+$ state. But XeCl-laser excitation of $Mo(CO)_6$ results in the generation of short-lived nonequilibrium emission (about 100 ns) whose spectrum overlaps that for dimers (*10*). We have demonstrated that 248 nm KrF- laser and 351 nm XeF- laser also produce electronically excited Mo_2. But in these cases the spectra of short-lived luminescence differ considerably from those with XeCl-laser and we were able to study the time evolution of the Mo_2 spectrum immediately after the laser pulse.
    Figure 12 shows the time evolution of the Mo_2 spectrum formed by KrF and XeF- laser excitation of $Mo(CO)_6$. There is no dimer sequence when $\Delta t \leq 100 ns$. Later on, a dimer spectrum appears so that Mo_2 is formed at the vibrational states $\vartheta' = 1,2$ (XeF-laser) and $\vartheta' = 2$ (KrF-laser). When the time is increased one can see a trend to equilibrium with the population distributed among the vibrational levels of $A^1\Sigma_u^+$ term. But even at $\Delta T \approx 600$ ns the equilibrium distribution is not achieved.
    The Franck-Condon factors for $X^1\Sigma_g^+ \leftarrow A^1\Sigma_u^+$ transition of Mo_2 were calculated in accordance with (*27*). It was shown that for $\Delta\vartheta = 0$ and $\vartheta' = 0 - 5$ they did not differ from each other more than by 2%. That means the intensity distribution in the spectrum directly gives relative populations of particular vibrational states of $A^1\Sigma_u^+$. From the data presented in Figure 12, one can conclude that excimer laser production of electronically excited Mo_2 gives substantially nonequilibrium distribution over the vibrational stats of $^1\Sigma_u^*$ electronic term.

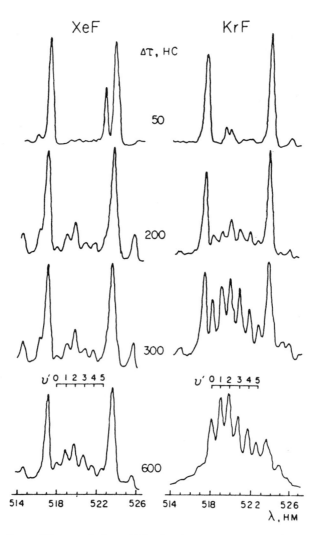

Figure 12.   Time evolution of Mo$_2$ luminescence spectrum in XeCl- and KrF-laser excitation of Mo(CO)$_6$.

Effect of Buffer Gases on Mo_2 Luminescence Spectrum. The buffer gases change the Mo_2 luminescence spectra in the spectral range 515-540 nm dramatically. Figure 13 presents the luminescence spectra of Mo_2 obtained when He was added to 0.06 Torr of $Mo(CO)_6$. $Mo(CO)_6$ was excited by two XeCl-laser pulses with the time delay $\Delta t = 1\mu s$. The luminescence was registered by gated optical multichannel analyzer (OMA) with the gate length, $\Delta g = 300ns$, so that the onset of the gate was delayed by $\Delta t = 300ns$ after the second laser pulse. Even small amounts of He (about 1 Torr) result in remarkable changes of vibrational distribution. The intensity of the first line of the luminescence spectrum drops but its general shape does not change strongly. With a higher He pressure (more than 20 Torr) the spectrum changes considerably. Its maximum is shifted by 15 cm^{-1} a the new structure appears which differs from the for Mo_2.

These changes are more pronounced in the spectra registered with a higher resolution (Figure 14a, b). The pressure of $Mo(CO)_6$ was 0.08 Torr and the other experimental parameters were $\Delta t = 1\mu s$, $\Delta t = 100ns$, $\Delta_g = 300ns$. The tail luminescence was picked up. The characteristic feature of neon in this case was that even 50-100 Torr of this gas do not change the initial vibrational distribution. The narrowing of the luminescence band occurs with a relatively high pressure of Ne (about 700 Torr). With He and other noble buffer gases the same narrowing occurs with a pressure of about 100 Torr. Our analysis of the luminescence spectrum with a buffer gas shows that these spectra can be convoluted from two sequences shifted by 15 and 30 cm^{-1} to the red relatively to the 0-0 transition for Mo_2 (the dashed and with lines in Figure 14a). The intervals between the adjacent lines for the dashed and withe sequences and also for dimer, Mo_2, are presented in Table VI.

Table VI. Intervals between adjacent lines in the observed Mo_2 sequences Δ

Lines	0-1	1-2	2-3	3-4	4-5	5-6	
	30.1	31.5	34.0	35.4	35.2	36.6	a
Δ, cm^{-1}	27.1	30.0	31.4	35.3	35.2	29.6	b
	28.6	32.5	32.4	37.8	36.6	-	c

[a]Without buffer gas.
[b]55 Torr Ne (dashed sequence, Figure 14a).
[c]55 Torr Ne (white sequence, Figure 14a).

The same values for the interval between the adjacent lines and the shape of the vibrational distribution similar to those for Mo_2 permit one to suppose that the buffer gas results in the formation of weakly bound complexes with Mo_2. In the simplest case the recombination of electronically excited Mo_2 and CO can lead to the formation of $Mo_2(CO)_k$. Two different types of electronically-excited compounds, for example, $Mo_2(CO)$ and $Mo_2(CO)_2$ can be result in the appearance of two sequences in the luminescence spectrum. It should be underlined that the masses of Ne and CO do not differ greatly. That means the collisional removal of translational energy from CO by Ne should be the most effective compared to other noble gases.

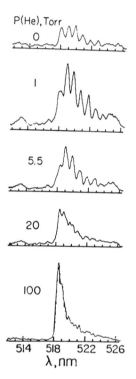

Figure 13.   Effect of the buffer gas, He, on the Mo_2 luminescence
spectrum. The pressure of $Mo(CO)_6$ and He was 0.065 and
100 Torr.

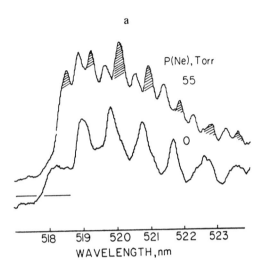

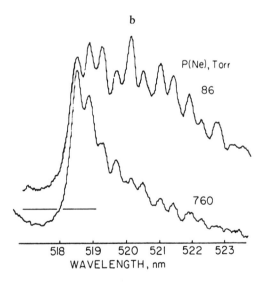

Figure 14.   Effect of neon on the Mo_2 luminescence spectrum. The neon pressure was 0 and 55 Torr (a), 86 and 760 Torr (b).

Besides, the band with its maximum at 518.5 nm with addition of a buffer gas there are two more bands with maxima at 527.5 nm (Fig. 15) and 537.5 nm. The band with its maximum at 527.5 nm shows a slightly pronounced structure. Note also that the study of the spectra of Mo_2 and Mo_3 in matrices at low temperatures show that the absorption peaks of these bands lie in the regions of 518 and 529nm, respectively (28,29).

It should be also noted that the main band at 518.5 nm sharply drops in the region of 525 nm which can be hardly explained by the existing vibrational distribution. It can be assumed that this drop of the luminescence band is related to predissociation, and the band at 527.5 nm is merely a long-wave tail of the main band at 518.5 nm.

Ultrafine particles formation.
An exciter laser photolysis of gas phase $Mo(CO)_6$ produces a black powder. The Mo content of this powder was determined to be about 61%. The rest of the powder consists of carbon. The powder analysis by ICRFT mass spectrometer confirmed the results of chemical analysis. Figure 16a, b shows the negative (a) and positive (b) ions, produced in the ICR cell of the mass spectrometer. We assigned the three main peaks in Figure 16a to $MoC_4^-, Mo_2C_8^-$ and $Mo_3C_{12}^-$. Heavier clusters are also presented in the spectra. The same three species formed positive ion mass spectrum, which is more complicated because of fragmentation.

The XeCl laser excitation of the ultrafine particles formed in $Mo(CO)_6$ gas gives rise to bright emission. One can observe the emission by excitation of the $Mo(CO)_6$ gas with two XeCl-laser pulses, propagating with a relatively large time delay, $\Delta t = 0.1 - 1s$. Despite the relatively large time delay, Δt, of the second laser pulse, the observed luminescence region coincides spatially with the radiation volume of the first laser pulse. That means that no particles with the masses compared with the parent molecule $Mo(CO)_6$ could create the observed emission. In the opposite case the expansion of the luminescent region with result in its dimensions, $R \cong (4D\Delta t)^{1/2} \cong 10cm$, where D=25 cm^2/s is the diffusion coefficient with the pressure of $Mo(CO)_6$ about 0.1 Torr, $\Delta t = 1s$. The absence of distinct diffusive expansion means that the observed emission is caused by heavy species (ultrafine particles). Figure 17 shows the dependences of the intensity of observed emission versus time delay, Δt. The emission starts to appear when $\Delta t = 5 \cdot 10^{-4}$ ns. The emission spectrum is wide and structureless with its maximum at around 850 nm.

Similar ultra fine particles are formed under 248 nm KrF-laser excitation. We were not able to observe this emission when the XeCl-laser pulse was followed by KrF-laser. We failed to generate similar emission under excimer laser excitation of other metal carbonyls - $Cr(CO)_6$ and $W(CO)_6$.

Ultrafine particles in the particular regions of the excimer laser beam can be easily visualized by HeNe-laser light scattering. Figure 17 (dashed line) shows the scattered light intensity vs time after the XeCl-laser pulse. The XeCl and HeNe-laser beams were made to counterpropagate. This experiment visualizes directly the ultrafine particles with an average size of about 1μ are formed. HeNe-laser probing showed also that the regions of ultrafine particle formation coincide with the observed zone of particle emission. When shifting the HeNe-

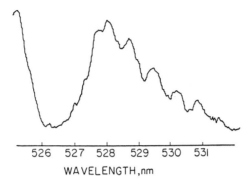

Figure 15.   The second band in the luminescence spectrum of Mo_2 appearing with the He buffer gas. The pressure of $Mo(CO)_6$ and He were 0.065 and 100 Torr.

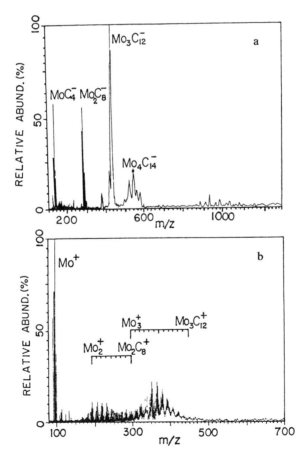

Figure 16.   ICR FT mass spectra for the negative (a) and positive (b) ions of the powder obtained under photolysis of $Mo(CO)_6$ with XeCl-laser.

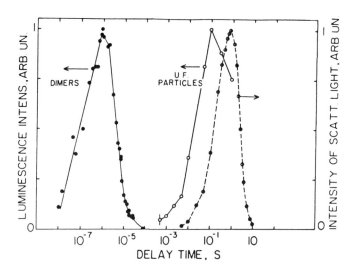

Figure 17.   Dimer and ultrafine particle luminescence intensity vs time
delay between laser pulses (solid curves). The time profile
of the scattered HeNe-laser light is shown by dashed line.

laser beam downwards form XeCl-laser beam we could observe a fall of ultrafine
particles and hence measure their velocities. As a result, a conclusion was made
that the average size of ultrafine particles is XeCl-laser fluence dependent (10).

Conclusions

XeCl-laser excitation of $Mo(CO)_6$ gas generates delayed emission of Mo atoms at

the transition $z^7 P_J^{\circ} \left(26 \cdot 10^3\, cm^{-1} \right) \leftarrow e^7 D_J \left(45 \cdot 10^3\, cm^{-1} \right)$. The population of the

initial $e^7 D_J \left(45 \cdot 10^3\, cm^{-1} \right)$ levels by the transition from Rydberg Mo states
explains the observed experimental results.

The electronically excited dimers, Mo_2, are produced by two different
ways - laser induced stripping of $Mo_2(CO)_6$ and chemiluminescent reaction of
highly energized fragments.

Excimer laser excitation of metal carbonyls produces ultrafine particles
consisting of metal and metal carbon clusters.

Acknowledgments

We would like to acknowledge the assistance of Dr. R.L. Hettich from Chemical
Physics Section of Oak Ridge National Laboratory in ICRFT Mass Spectrometer
analysis.

Literature Cited

1.   Carne, Z. ; Naaman, R. ; Zare, R.N. *Chem. Phys. Lett.* (1978), *59*, 33
2.   Gedanken, A.; Robin, M.B.; Kuebler, N.A. *J. Phys. Chem.* (1982), *86*,
       4006
3.   Samoriski, B.; Chaiken, J. *J. Chem. Phys.* (1989), *90*, 4079
4.   Tyndall, G.W.; Jackson, R.L. *J. Chem. Phys.* (1988), *89*, 1364

5. Tyndall, G.W.; Jackson, R.L. *J. Phys. Chem.* (**1991**), *95*, 687
6. Weitz, E. *J. Phys. Chem.* (**1987**), *91*, 3945
7. Ishikawa, Y.; Brown, C.E.; Hackett, P.A.; Rayner, D.M. *J. Phys. Chem.* (**1990**), *94*, 2404
8. Rayner, D.M.; Ishikawa, Y.; Brown, C.E.; Hackett, P.A. *J. Chem. Phys.* (**1991**), *94*, 5471 (and references therein)
9. Hollingsworth, W.E.; Vaida, V. *J. Phys. Chem.* (**1986**), *90*, 1235
10. Dem'yanenko, A.V.; Puretzky, A.A. *Spectrochimica Acta*, (**1990**), *46A*, 509
11. Efremov, Yu.M.; Samoilove, A.N.; Kozhukhovsky, V.B.; Gurvich, L.V. *J. Chem. Phys.* (**1983**), *78*, 1627
12. Hopkins, J.B.; Langridge-Smith, P.R.R.; Morse, M.D.; Smalley, R.E. *J. Chem. Phys.* (**1983**), *78*, 1627
13. Gray, H.B.; Beach, N.A.; *J. Am. Chem. Soc.* (**1963**), *85*, 2922
14. Lewis, K.E.; Golden, D.M.; Smith, G.P. *J. Am. Chem. Soc.* (**1984**), *106*, 3905
15. Poliakoff, M.; Weitz, E. *Acc. Chem. Res.* (**1987**), *20*, 408
16. Venkataraman, B.; Hou, H.; Zhang, Z.; Chen, S.; Bandukwalla, G.; Vernon, M. *J. Chem. Phys.* (**1990**), *92*, 5338
17. Ganske, J.A.; Rosenfeld, R.N. *J. Phys. Chem.* (**1989**), *93*, 1959
18. Goodgame, M.M.; Goodard III, W.A. *Phys. Rev. Lett.* (**1982**), *48*, 135
19. Corliss, C.H.; Bozmann, W.R. N.B.S. *Monograph* (U.S. GPO, Washington, D.C., **1962**), *53*
20. Moore, C.E. *Atomic Energy Levels*, (**1970**), NSRDS, *Vol. 34*, National Bureau of Standards; U.S. GPO: Washington, D.C.
21. Pilcher, G.; Ware, M.J.; Pittam, D.A. *J. of the Less-Common Metals* (**1975**), *42*, 223
22. Samoriski, B.; Chaiken, J. *Phys. Rev. A.* (**1988**), *38*, 3498
23. Stebbings, R.F.; Dunning, F.B. *Rydberg States of Atoms and Molecules*, Eds. Cambridge University Press, Cambridge-Sydney, **1983**
24. Heber, K.-D.; West, P.J.; Matthias, E. *Phys. Rev. A.* (**1988**), *37*, 1438
25. Christophorou, L.G.; McCorkle, D.L.; Carter, J.G. *J. Chem. Phys.* (**1971**), *54*, 253
26. Davis, F.J.; Compton, R.N.; Nelson, D.R. *J. Chem. Phys.* (**1973**), *59*, 2324
27. Chen, K.; Pei, C. *J. Mol. Spectr.* (**1990**), *140*, 401
28. Ozin, G.A.; Klotzbuecher, W. *J. Mol. Catalysis* (**1977/78**), *3*, 195
29. Pellin, M.J.; Foosnaes, T.; Gruen, D.M. *J. Chem. Phys.* (**1981**), *74*, 5547

RECEIVED February 8, 1993

Chapter 18

Reaction Kinetics and Simulation of Growth in Pyrolytic Laser-Induced Chemical Vapor Deposition

D. Bäuerle, N. Arnold, and R. Kullmer

Angewandte Physik, Johannes-Kepler-Universität Linz, A—4040 Linz, Austria

This overview summarizes model calculations on the gas-phase kinetics and the simulation of growth in pyrolytic laser-induced chemical vapor deposition. The theoretical results are compared with experimental data.

Laser-induced chemical vapor deposition (LCVD) permits the single-step production of microstructures and the fabrication of thin extended films. An overview on the various possibilities of this technique, on the different systems investigated, and on the experimental arrangements employed is given in (1).

The decomposition of precursor molecules employed in laser-CVD can be activated mainly thermally (photothermal or pyrolytic LCVD) or mainly non-thermally (photochemical or photolytic LCVD) or by a combination of both (hybrid LCVD). The type of process activation can be verified from the morphology of the deposit, from measurements of the deposition rate as a function of laser power, laser wavelength, substrate material etc. and, last not least, from the analysis of data on the basis of theoretical models.

In this chapter we concentrate on the analysis of the reaction kinetics and the simulation of growth in *pyrolytic* laser-CVD of microstructures.

Photochemical laser-CVD based on selective electronic excitations has also been applied for the fabrication of microstructures (1). Nevertheless, the main importance of this technique is found in the field of thin film fabrication.

This chapter is organized as follows: We start with a few remarks on precursor molecules (Sect. 1) and on the type of microstructures fabricated by pyrolytic LCVD (Sect. 2). In Sect. 3 we discuss different theoretical approaches for the description of the gas-phase kinetics in pyrolytic LCVD of microstructures. The simulation of growth on the basis of one-dimensional and two-dimensional models is treated in Sect. 4. While we concentrate on a few model systems for

0097–6156/93/0530–0250$06.75/0
© 1993 American Chemical Society

which the most complete data are available, the discussion of results
is very general in the sense that most of the trends, features and
results apply to all the corresponding systems that have been
studied. The investigations presented are of relevance to both the
application of laser-CVD in micro-patterning and the elucidation of
the fundamental mechanisms in laser micro-chemistry.

1. Precursor Molecules

The precursor molecules most frequently employed in laser–CVD are
halogen compounds, hydrides, alkyls, carbonyls, and various
organometallic coordination complexes.

For the application of laser-CVD in micro-patterning, the proper
selection of precursor molecules is of particular importance. Halogen
compounds and hydrides require, in general, higher temperatures for
thermal decomposition than organic molecules. In photolytic LCVD,
organic precursors permit a greater flexibility in the sense that it
is easier to find a precursor molecule which matches the requirements
for high yield photodecomposition at the available laser wavelength.
On the other hand, utilization of organic precursors is often linked
with the incorporation of large amounts of impurities into the
deposit, in particular of carbon. Such impurities deteriorate the
physical and chemical properties of the deposited material. Thus, the
various precursor molecules and the different activation mechanisms
for decomposition all have their specific advantages and
disadvantages (1).

2. Types of deposits

The fabrication of microstructures by pyrolytic laser-CVD has been
investigated for spots, rods and stripes (1). In such experiments, a
semi-infinite substrate is immersed in a reactive gaseous ambient and
perpendicularly irradiated by a laser beam. Henceforth, we assume
that the laser light is exclusively absorbed on the surface of the
substrate or on the already deposited material. The laser-induced
temperature rise on this surface heats also the adjacent gas phase.
Thus, laser radiation can photothermally activate a reaction of the
type

$$\zeta_{AB} \; AB + \zeta_C \; C + M \; \underset{k_2}{\overset{k_1,k_3}{\rightleftharpoons}} \; \zeta_A \; A(\downarrow) + \zeta_{BC} \; BC + M \qquad (2.1)$$

where ζ_j are stoichiometric coefficients which can depend on the
particular reaction path. The forward reaction between AB and C can
take place heterogeneously at the gas–solid interface and/or
homogeneously within the gas phase just above the irradiated
substrate surface. The surface and gas-phase forward reactions shall
be characterized by the rate constants k_1 and k_3, respectively. M is
an inert carrier gas. A and BC are products of the forward reaction.
A shall be the relevant species for surface processing. If A is
generated in the gas phase, it must first diffuse to reach the
substrate surface to be processed. Species A can just stick on the

surface and form a deposit or it can react further. In the backward reaction, condensed species A react with BC to form the original constituents. This backward reaction is characterized by k_2. For thermally activated reactions, the rate constants can be described by an Arrhenius law $k_j(T) = k_{oj} \exp\{-\Delta E_j/[k_B T]\}$.

We now consider some practical cases. Assume the reaction (2.1) describes the laser-induced chemical vapor deposition of tungsten according to

$$WF_6 + 3H_2 + M \underset{k_2}{\overset{k_1, k_3}{\rightleftarrows}} W(\downarrow) + 6HF + M \qquad (2.2)$$

This process has been studied in great detail. Figure 1 shows different types of W spots deposited according to (2.2) by means of Ar^+-laser radiation. The spot shown in Figure 1a is well *localized* and surrounded by a transparent ring and a very thin film (see Figure 1a'). Spots of the type shown in Figure 1c are very flat and have a *diffuse* shape. They are also surrounded by a thin film. Spots of the type shown in Figure 1b represent an intermediate situation. At fixed laser power, P, and laser-beam illumination time, τ_ℓ, the shape of spots depends on the ratio of partial pressures $p^+ = p(H_2)/p(WF_6)$. On the other hand, if p^+ and τ_ℓ are fixed, the shape of spots changes with increasing power from the 1a via the 1b to the 1c type.

Another practical example is the deposition and etching of Si according to

$$SiCl_4 \underset{k_2}{\overset{k_1, k_3}{\rightleftarrows}} Si(\downarrow) + 2Cl_2 \qquad (2.3)$$

In this reaction one can add H_2 and/or an inert gas as well. If the chemical equilibrium is shifted to the left side, (2.3) describes the etching of Si in Cl_2 atmosphere (1).

With the saturation of growth in spot diameter, the continuation of laser-beam irradiation can result in the growth of a rod along the laser-beam axis. This has been described in detail in (1).

If the laser beam is scanned with respect to the substrate, one can directly write lines, or complicated patterns. Examples for direct writing can be found in (1).

3. Gas-phase Kinetics, Heterogeneous and Homogeneous Reactions

The models employed for investigating the gas-phase kinetics in pyrolytic LCVD are shown in Figures 2 and 3. The reaction zone is either represented by a hemisphere or by a thin circular film placed on a semiinfinite substrate. The hemispherical model is particularly suited to investigate different types of gas-phase transport processes. The cylindrical model, on the other hand, permits to

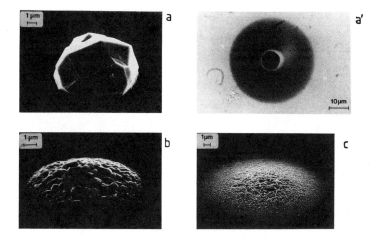

Figure 1. Scanning electron microscope (SEM) pictures (a-c) and optical transmission microscope picture (a') of W spots deposited from a mixture of WF_6 and H_2 on fused quartz (SiO_2) substrates by means of Ar^+-laser radiation (λ = 514.5 nm, $w_o(1/e) \approx 1.1$ μm). a,a') P = 120 mW, τ_ℓ = 0.2 s, p^+ = 2 [10/5 $\equiv$ 10 mbar H_2 + 5 mbar WF_6] b) P = 110 mW, τ_ℓ = 0.5 s, p^+ = 5 [25/5] c) P = 120 mW, τ_ℓ = 0.5 s, p^+ = 50 [250/5] (Reproduced with permission from reference 2. Copyright 1992 Springer-Verlag).

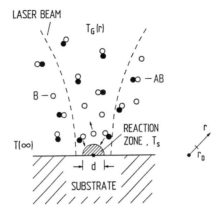

Figure 2. Schematic for laser-induced pyrolytic processing. The reaction zone is represented by a hemisphere of radius r_D = $d/2$ whose surface temperature, T_s , is uniform. The gas-phase temperature is $T_G(r)$. The origin of the radius vector r is in the center of the hemisphere. The laser radiation is exclusively absorbed on the surface r = r_D . Carrier gas molecules, being possibly present, are not indicated.

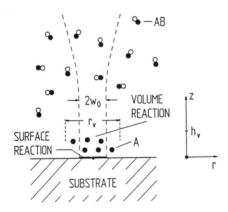

Figure 3. Schematic for pyrolytic laser-induced chemical processing. The origin of the coordinate system is on the substrate surface in the center of the laser beam. The laser light is exclusively absorbed on the surface z = 0. The temperature distributions on the surface z = 0 and in the gas-phase are $T_s(r)$ and $T(r,z)$, respectively (Reproduced with permission from reference 6. Copyright 1992 Elsevier Sequoia).

include, in a simple way, the effect of volume reactions and to describe qualitatively the shape of deposits.

In the model shown in Figure 2 we assume the temperature of the hemisphere (deposit) to be uniform and the reaction to be purely heterogeneous. The reaction rates can then be calculated analytically with respect to temperature- and concentration-dependent transport coefficients, and with respect to the effect of thermal diffusion and chemical convection. The results were published in (3-5). These results permit to explain qualitatively many features observed in pyrolytic laser-CVD. One example is the effect of thermal diffusion, that has been studied experimentally during steady-growth of carbon rods deposited from pure C_2H_2 and from admixtures with H_2, He, and Ar. For further details the reader is referred to (1,3). The calculations elucidate the relative importance of single contributions to the reaction rate, and also reveal new effects that originate from the coupling of fluxes (5).

With the model shown in Figure 3, the influence of heterogeneous *and* homogeneous contributions to the reaction flux has been calculated (6). This model certainly applies to "flat" structures only. Nevertheless, the solutions help to interpret processing profiles observed in laser-CVD and laser-chemical etching. Let us consider a reaction of the type (2.1) and discuss only the most general case which must in fact be considered with reactions (2.2) and (2.3). Figure 4 shows only a few of the many different shapes that can be obtained in such a situation. Here, we have plotted the normalized particle flux, $-J^*(r^*)$, which is proportional to the reaction rate, $W^*(r^*)$. At lower center temperatures than those shown in the figure, the maximum reaction flux appears in the center of the laser beam at $r^* = 0$ and thus the thickness of the deposit decreases monotoneously with distance r^*. Above a certain center temperature, however, the maximum reaction flux may occur at a distance $r^* \neq 0$ (Figure 4). At higher temperatures and distances $r^* \geq 1$ etching may even dominate (see curves for $T_c = 2700$ K and 3000 K). For even larger distances, the thickness of the deposit becomes finite again. Here, a thin film is formed via diffusion of atoms A out of the reaction zone. Further examples and details on the calculations can be found in (6).

Comparison with experiments. Figure 4 reveals that the transparent ring observed in Figure 1a' can be explained by etching, and the thin film by species generated within the gas phase. Additional calculations presented in (6) describe qualitatively the changes in spot shape mentioned in Sect.2. The hollow seen in Figure 4 near the center $r^* \approx 0$ is a consequence of the gas-phase transport limitation of species towards the reaction zone. Figure 5 may represent experimental evidence for such a behavior. The relative importance of deposition and etching reactions depends on the relative size of rate constants k_j. This, in turn, depends on the temperature, on the apparent activation energies ΔE_j, and on the

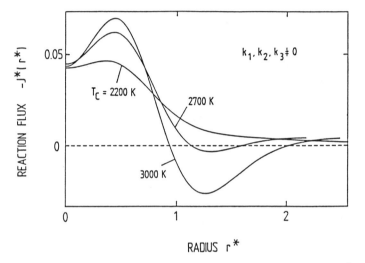

Figure 4. The normalized reaction flux as a function of r^* with $k_1 \neq 0$, $k_2 \neq 0$, $k_3 \neq 0$, $\Delta E_1/k_B$ = 15000 K, $\Delta E_2/k_B$ = 8000 K, $\Delta E_3/k_B$ = 10000 K. The different curves refer to T_c = 2200 K, 2700 K, 3000 K (Reproduced with permission from reference 6. Copyright 1992 Elsevier Sequoia).

partial pressures of species (these enter the preexponential factor k_{oj}).

The investigations summarized in this section permit the qualitative description of many features observed in pyrolytic laser-CVD. The main drawback in this type of approach is related to the assumption that the laser-induced temperature distribution remains unaffected during the deposition process. This can be considered, to some extent, as a reasonable approximation only as long as the thermal conductivities of the deposited material and the substrate are about equal (1).

4. Simulation of Growth

It has already been outlined in (1) that the laser-induced temperature distribution depends strongly on the geometry of the deposit. Therefore, any quantitative or semi-quantitative analysis of the deposition process requires a self-consistent treatment of the equation of growth and the laser-induced temperature distribution. In this section we present two models.

In paragraph 4.1 we employ a one-dimensional approach and we approximate the laser-induced temperature distribution by an analytic equation which is solved simultaneously with the equation of growth. This model can be applied to laser direct writing.

In paragraph 4.2 the equation of growth and the heat equation for the deposit are solved self-consistently for a two-dimensional model. Clearly, only numerical techniques can be employed in this case. Here, even with the assumption of a purely heterogeneous reaction and the omission of gas-phase transport, considerable computational efforts are required.

The type of approaches presented in this section ignore any gas-phase transport and assume a purely heterogeneous reaction. Nevertheless, they permit a semi-quantitative description of experimental data.

4.1 One-Dimensional Approach to Direct Writing. In this section we simulate pyrolytic direct writing in a one-dimensional self-consistent calculation.

The model employed in the calculations is depicted in Figure 6. It is applied to laser direct writing of a stripe (deposit) onto a semiinfinite substrate. The respective thermal conductivities and temperatures are denoted by κ_D, T_D and κ_S, T_S. Scanning of the laser beam is performed in x-direction with velocity v_S. We consider quasi-stationary conditions with the coordinate system fixed with the laser beam. Thus, in this system, the geometry of the stripe remains unchanged; its height behind the laser beam is h, and its width d = $2r_D$. The temperature distribution induced by the absorbed laser-light intensity, I_a, is indicated by the dashed curve. Within the center of the laser beam at x = 0 the temperature is denoted by T_c, while at the forward edge of the stripe at x = a it is denoted by T_e, and with x → − ∞ by T(∞). As outlined in (1), changes in laser-induced temperature distribution related to changes in the

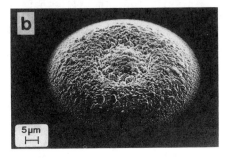

Figure 5. SEM pictures of W spots deposited from 0.49 mbar WCl_6 + 400 mbar H_2 by means of 680 nm dye-laser radiation a) P = 300 mW, τ_ℓ = 20 s b) P = 240 mW, τ_ℓ = 40 s (Reproduced with permission from reference 7. Copyright 1992 Elsevier Sequoia).

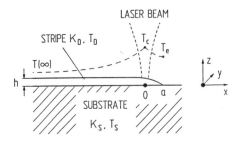

Figure 6. Schematic for direct writing. The coordinate system is fixed with the laser beam. The center of the laser beam with radius w is at the origin x = 0; the forward edge of the stripe is at x = a. T_c and T_e are the corresponding temperatures. The temperature profile is indicated by the dashed curve. The height of the stripe is h and its width is d = $2r_D$ (Reproduced with permission from reference 8. Copyright 1993 Elsevier Science Publishers B.V.).

geometry of the deposit are the more pronounced, the more the ratio of thermal conductivities differs from unity. For this reason we consider the case $\kappa^* = \kappa_D/\kappa_S \gg 1$. This applies, for example, to direct writing of metal lines onto glass substrates. For further details on these calculations the reader is referred to (8).

4.1.1 Temperature Distribution. The laser-induced temperature distribution can be calculated from the energy balance

$$\kappa_D F \frac{\partial^2 T_D}{\partial x^2} - \eta \kappa_S \left[T_D - T(\infty) \right] + P_a \delta(x) = 0 \qquad (4.1)$$

and the boundary conditions

$$\frac{\partial T_D}{\partial x} \bigg|_{x=a} = 0$$

$$T_D(x \longrightarrow -\infty) = T(\infty)$$

The first and the second term in (4.1) describe the transport of heat along the stripe and into the substrate, respectively. η is a dimensionless geometrical parameter. The source term is given by the absorbed laser power. The influence of scanning has been ignored. This is a good approximation as long as $v_s r_D^2/[D_S \ell] \ll 1$. D_S is the thermal diffusivity of the substrate, and ℓ some characteristic length (see below). All material parameters and the cross section of the stripe, $F \equiv F(x)$, have been assumed to be constants. Because $\kappa^* \gg 1$ we ignore within the stripe any temperature gradient in y- and z-direction, i.e. $T_D(x,y,z) = T_D(x,0,0) \equiv T_D$ and set $T_S(z=0) \approx T_D$ The radius of the laser beam, w, was assumed to be small compared to r_D so that the absorbed laser power per unit length can be replaced by $P_a \delta(x) = PA \delta(x)$. A is the absorptivity. Finally, we introduce the variable

$$\ell^2 = \frac{F \kappa^*}{\eta} \qquad (4.2)$$

ℓ characterizes the drop in laser-induced temperature in x-direction. With x < 0 the solution of (4.1) is

$$T_D(x) = T(\infty) + \Delta T_C \exp \left[\frac{x}{\ell} \right] \qquad (4.3)$$

where

$$\Delta T_C \equiv \Delta T(x=0) = \frac{P_a}{2 \eta \kappa_S \ell} \left\{ 1 + \exp \left[-\frac{2a}{\ell} \right] \right\}$$

and with $0 < x < a$

$$T_D(x) = T(\infty) + \Delta T_e \cosh \left[\frac{x - a}{\ell} \right] \tag{4.4}$$

where

$$\Delta T_e \equiv \Delta T(x = a) = \frac{P_a}{\eta \, \kappa_s \, \ell} \quad \exp \left[- \frac{a}{\ell} \right]$$

For the determination of the unknown quantities ΔT_c, ΔT_e, a, F, ℓ, h, and r_D we need four additional equations beside of (4.2)-(4.4). The cross section of the stripe is

$$F \approx \zeta \, h \, r_D \tag{4.5}$$

where ζ is a dimensionless geometrical coefficient ($\zeta \approx 2$ for a rectangular stripe, $\zeta \approx 4/3$ for a parabolic cross section, etc.). The width of the stripe is characterized by

$$r_D \approx \xi \, a \tag{4.6}$$

ξ is again dimensionless and of the order of unity. It determines the position of the laser beam. This Ansatz implies that the temperature distribution is approximately of axial symmetry near the tip of the stripe. This is confirmed by both experimental observations and more accurate numerical simulations of the growth process (see paragraph 4.2.2). We assume that the temperature at the tip of the stripe, T_e, is equal to the threshold temperature T_{th} at which deposition becomes significant (7), i.e. we set

$$T_{th} \approx \Delta T_e + T(\infty) \tag{4.7}$$

The fourth equation is given by (4.9).

4.1.2 Simulation of Growth. In a coordinate system that is fixed with the laser beam, the height of the stripe is given by

$$\frac{\partial \, h}{\partial \, t} = W(T_D) + v_s \frac{\partial h}{\partial x} \tag{4.8}$$

where $h = h(x,t)$. $W(T)$ is the growth rate as described by the Arrhenius law, $W(T) = k_o \exp [-\Delta E^*/T]$ where k_o [$\mu m/s$] is the preexponential factor and $\Delta E^* = \Delta E/k_B$ the apparent chemical activation energy. Clearly $\partial h/\partial x \neq 0$ only, if $W(T) \neq 0$, i.e. if $T > T_{th}$. Subsequently, we assume stationary conditions, $\partial h/\partial t = 0$. If we use for the position $x = 0$ the approximation $\partial h/\partial x \approx - h/[\gamma a]$ where h is now the (constant) height behind the laser beam, we obtain

$$W(T_c) - v_s \frac{h}{\gamma a} = 0 \tag{4.9}$$

γ is again dimensionless and of the order of unity. Equations (4.2), (4.3-7), and (4.9) permit to calculate all relevant dependences.

From (4.1) and (4.9) we find that for constant v_s the lengths r_D, h, a, and ℓ scale linearly with the absorbed laser power, P_a. In this case T_c is independent of P_a. The dependence of these quantities on the velocity v_s is more complex. From (4.3) and (4.4) together with (4.7) we obtain

$$\mu = \frac{a}{\ell} = \text{arc cosh} \left[\frac{\Delta T_c}{\Delta T_{th}} \right] \tag{4.10}$$

where $\Delta T_{th} = T_{th} - T(\infty)$. From (4.4) we find

$$\ell = \frac{P_a}{\eta \, \kappa_s \, \Delta T_{th}} \, \exp[-\mu] \tag{4.11}$$

With (4.10) this yields

$$r_D = \xi a = \xi \ell \, \mu = \frac{\xi \, P_a}{\eta \, \kappa_s \, \Delta T_{th}} \, \mu \, \exp[-\mu] \tag{4.12}$$

From (4.2), (4.5), and (4.6)

$$h = \frac{\eta}{\zeta \, \xi \, \kappa^*} \, \frac{a}{\mu^2} = \frac{P_a}{\zeta \, \xi \, \kappa_D \, \Delta T_{th}} \, \mu^{-1} \, \exp[-\mu] \tag{4.13}$$

From (4.9) we obtain

$$v_s = \frac{\gamma \, \zeta \, \xi}{\eta} \, \kappa^* \, \mu^2 \, W(T_c) \tag{4.14}$$

Equations (4.10) and (4.12-14) are a parametric representation for the determination of h and r_D as a function of v_s. Note, that both T_c and μ increase with v_s.

4.1.3 Comparison of Theoretical Results with Experimental Data. According to (4.12) and (4.13) the width and height of stripes produced by pyrolytic laser direct writing increases linearly with the laser power. This is in agreement with experimental data obtained with many different systems for which the model assumption $\kappa_D/\kappa_s \gg 1$ holds (1).

The dependence of r_D and h on scanning velocity is more complicated. Figures 7 show the normalized quantities $h \, \kappa_D T(\infty)/P_a$ and

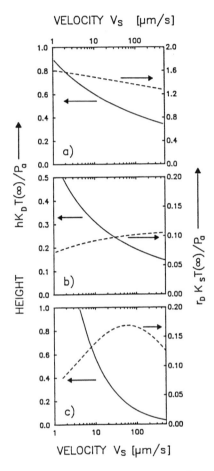

VELOCITY V_S [μm/s]

VELOCITY V_S [μm/s]

Figure 7. Normalized width and height of stripes as a function of scanning velocity. The respective parameter sets employed are in case a) k_o = 5.3 ·10^{12} μm/s, $\Delta E^*/T(\infty)$ = 45, $T_{th}/T(\infty)$ = 1.17, κ^* = 30; in case b) k_o = 6.6·10^9 μm/s, $\Delta E^*/T(\infty)$ = 90, $T_{th}/T(\infty)$ = 3.7, κ^* = 15; and in case c) k_o = 16.05 μm/s, $\Delta E^*/T(\infty)$ = 5.7, $T_{th}/T(\infty)$ = 2.7, κ^* = 17. The geometrical factors are the same in all cases: η = 1.6, ζ = 4/3, ξ = 1.25, and γ = 1.3 (Reproduced with permission from reference 8. Copyright 1993 Elsevier Science Publishers B.V.).

$r_D \kappa_s T(\infty)/P_a$ as a function of v_s, as calculated from (4.12-4.14). The respective parameter sets employed are listed in the figure caption. The figure shows that with all parameter sets, the height of stripes (full curves) decreases monotonously with increasing v_s. This is quite obvious, because the dwell time of the laser beam decreases with $1/v_s$. The width of stripes, $d = 2r_D$, shows a more complex behavior. In Figure 7a, r_D decreases monotonously with increasing scanning velocity, while in Figure 7b it increases monotonously. In Figure 7c, r_D first increases up to a maximum value r_D^{max}, which occurs at a velocity v_s^{max}, and then decreases for $v_s > v_s^{max}$. The different behavior can be understood from the fact that $r_D(v_s)$ shows a maximum at $\mu = 1$. For this point, we obtain

$$T_c^{max} = T(\infty) + \Delta T_{th} \cosh[1] \approx T(\infty) + 1.5 \, \Delta T_{th}$$

$$r_D^{max} = \xi \, a = \xi \, \ell = \frac{\xi \, P_a}{\eta \, \kappa_s \, \Delta T_{th}} \, e^{-1} \tag{4.12a}$$

$$h^{max} = \frac{P_a}{\zeta \, \xi \, \kappa_D \, \Delta T_{th}} \, e^{-1} \tag{4.13a}$$

$$v_s^{max} = \frac{\gamma \, \zeta \, \xi}{\eta} \, \kappa^* \, W(T_c^{max}) \tag{4.14a}$$

Thus, the different behavior shown in Figures 7a-c is related to the fact that the maximum in r_D can or cannot be observed within a reasonable range of scanning velocities, depending on the parameter set employed. For the material with the *low* deposition threshold (case *a*) the maximum velocity $v_s^{max} < 1 \; \mu m/s$ and r_D will thereby decrease monotonously with increasing v_s. For a material with a rather high deposition threshold, r_D may increase monotonously with v_s (case *b*). Case *c* represents an intermediate situation. Qualitatively, the occurence of a maximum in $r_D(v_s)$ is related to the increase in T_c with increasing v_s. This increase in T_c, in turn, is related to the diminished heat flux along the stripe.

Let us now compare the behavior of $h(v_s)$ and $r_D(v_s)$ as shown in Figure 7 with experimental data.

The parameter set employed in the case of Figure 7a refers, approximately, to pyrolytic laser-CVD of Ni from $Ni(CO)_4$. In fact, experimental investigations on the direct writing of Ni lines have shown that r_D decreases with increasing scanning velocity (1).

The parameters employed for calculating the curves presented in Figure 7b describe, approximately, the deposition of Si from SiH_4 and of C from C_2H_2. For these systems, however, no systematic investigations on the direct writing of lines onto thermally insulating substrates are known.

The parameters employed in Figure 7c are typical for the deposition of W from $WCl_6 + H_2$ (7). The behavior shown in the figure is, in fact, in qualitative agreement with the experimental results obtained on the direct writing of W lines onto insulating substrates.

Subsequently, we compare the theoretical results with the experimental data obtained for direct writing of W lines deposited from $WCl_6 + H_2$ by means of cw Ar^+-laser radiation ($\lambda = 514.5$ nm; w $\equiv$ $w_0(1/e) = 7.5$ μm) onto quartz (SiO_2) substrates. The substrates employed in the experiments were frequently coated with a $h_\ell = 700$ Å thick film of sputtered W. This film permits a well-defined initiation of the deposition process (1). Its influence on the temperature distribution can be ignored if $\kappa^* \cdot h_\ell / r_D << 1$. Other experimental examples and further details are outlined in (1,2,7).

Figure 8 shows experimental data for the height (squares) and width (triangles) of W stripes as a function of scanning velocity for two different WCl_6 pressures. The effective incident laser power was in both cases P = 645 mW. Full and dashed curves are calculated from (4.12-4.14) with the same geometrical factors as in Figure 7. The other parameters are noted in the figure caption. The height of stripes decreases with increasing velocity v_s. The width of stripes, however, first increases up to a maximum and then decreases with increasing v_s. As expected from (4.14a), the maximum value occurs at a higher velocity with the higher WCl_6 partial pressure. This is a consequence of the preexponential factor, k_0, in the reaction rate W(T). For a partial reaction order of unity with respect to the WCl_6 concentration, k_0 is proportional to $p(WCl_6)$. This describes, at least qualitatively, the trend in the experimental results in Figures 8 to 10. Figure 9 shows the velocity v_s^{max}, corresponding to the maximum width as a function of laser power for $p(WCl_6) = 0.49$ mbar (circles) and 1.1 mbar (squares). The figure shows that v_s^{max} increases by almost a factor of 2 to 3 when the WCl_6 pressure is increased from 0.49 mbar to 1.1 mbar. This is in qualitative agreement with (4.14a).

Equations (4.12a) and (4.13a) suggest that the maximum height, h^{max}, and width, d^{max}, depend only on threshold temperature and increase linearly with laser power. This is also confirmed by the experiments. Figure 10 shows the results for two different WCl_6 partial pressures. In fact, h^{max} and d^{max} show only a slight

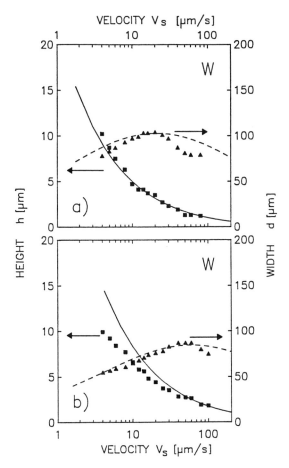

Figure 8. Width and height of W stripes as a function of scanning velocity for two mixtures of $WCl_6 + H_2$. The H_2 pressure $p(H_2)$ = 50 mbar. The full and dashed curves have been calculated as described in the text. The parameters were P = 645 mW, A = 0.55, w_o = 7.5 μm, T(∞) = 443 K, $\Delta E^*/T(\infty)$ = 5.7, κ^* = 17, κ_s = 0.032 $Wcm^{-1}K^{-1}$; η = 1.6, ζ = 4/3, ξ = 1.25, γ = 1.3 a) $p(WCl_6)$ = 0.49 mbar, k_o = 7.15 μm/s, $T_{th}/T(\infty)$ = 2.4 b) $p(WCl_6)$ = 1.1 mbar, k_o = 16.05 μm/s, $T_{th}/T(\infty)$ = 2.7 (Reproduced with permission from reference 8. Copyright 1993 Elsevier Science Publishers B.V.).

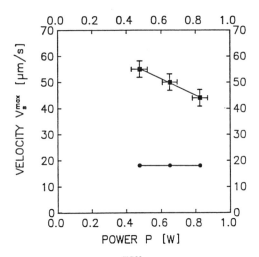

Figure 9. Scanning velocity v_s^{max} referring to the maximum width of stripes achieved in laser direct writing of W lines from WCl_6 + H_2 . The pressure was $p(H_2)$ = 50 mbar and ● $p(WCl_6)$ = 0.49 mbar, ■ $p(WCl_6)$ = 1.1 mbar (Reproduced with permission from reference 8. Copyright 1993 Elsevier Science Publishers B.V.).

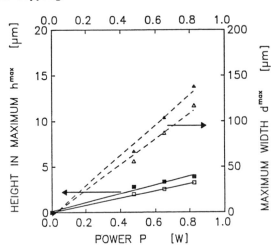

Figure 10. Maximum width and height of W stripes produced by laser direct writing from WCl_6 + H_2 . The H_2 pressure was $p(H_2)$ = 50 mbar. ■, ▲ $p(WCl_2)$ = 0.49 mbar ; □, △ $p(WCl_2)$ = 1.1 mbar (Reproduced with permission from reference 8. Copyright 1993 Elsevier Science Publishers B.V.).

dependence on pressure. This can be understood from the (small) dependence of the threshold temperature on $p(WCl_6)$ (7).

With the approximations made in this model, the agreement between the theoretical predictions and the experimental data must be considered to be quite reasonable. Besides of the simplyfiing assumptions made, important parameters such as k_o, ΔE^*, and T_{th} can only be estimated from previous experimental investigations. This is true also for reaction orders. The temperature dependences in κ_D, κ_s, and A can be taken into account in a similar approach. Even in this case a linear dependence of r_D, h, a, and ℓ on absorbed laser power is predicted; the maximum center temperature, T_c^{max}, will still depend only on T_{th} and the materials' properties. With realistic parameters, the changes in quantities are smaller than 30%.

4.2 Two-dimensional model. Laser-CVD has also been simulated in a two-dimensional self-consistent calculation. The deposit is again placed on a semiinfinite substrate. The origin of the coordinate system is fixed with the laser beam. The shape of the deposit is described by an arbitrary function h(x,y) with 0 < z < h(x,y). As derived in (9), the temperature distribution can be calculated from

$$\theta_s(\theta_D) = \theta_s^o + \frac{\kappa^*(T(\infty))}{2\pi} \left\{ \nabla_2 [h \nabla_2 \theta_D] * \frac{1}{|r|} \right\} \qquad (4.15)$$

where

$$\theta_s^o = \frac{1}{2\pi \kappa_s(T(\infty))} \left\{ I_a * \frac{1}{|r|} \right\}$$

is the (linearized) temperature distribution without the deposit. θ_s and θ_D are the linearized temperatures for the substrate and the deposit, respectively. r is a two-dimensional radius vector within the xy-plane, and $*$ denotes the convolution integral. With (4.15) the shape of the deposit can be calculated from (4.8). Further details on the model calculations and their applications can be found in (9).

In the following we will present only some results on the simulation of growth of W and Ni spots and on the direct writing of W lines. The substrate material considered is quartz (SiO_2).

For convenience, we introduce normalized quantities and indicate them by an asterisk. h, x, y, v_s, and k_o are normalized to the radius of a Gaussian laser beam, w_o. Correspondingly, all temperatures and activation energies are normalized to the temperature $T(\infty)$. The normalized maximum intensity is $I_o^* = I_o w_o / [T(\infty) \kappa_s(T(\infty))]$.

4.2.1 Growth of Spots. Figure 11 shows the (normalized) height of W spots calculated for various stages of growth as a function of the (normalized) distance from the center of the laser-beam. The kinetic data employed, $k_o^* = 2.14$, $\Delta E^* / T(\infty) = 5.68$, were taken from experimental investigations on the deposition of W from $WCl_6 + H_2$ (7). The other parameters used were $T_{th} / T(\infty) = 2.71$, $\delta T_{th} / T(\infty) = 0.01$, $\kappa^* (T(\infty)) = 50.56$, $AI_o^* = 10.3$. δT_{th} characterizes the sharpness of the threshold. The thermal conductivity of the deposited W is approximated by $\kappa_D(W) = c_1 + c_2/T - c_3/T^2$ where $c_1 = 42.65$ $Wm^{-1}K^{-1}$, $c_2 = 1.898 \cdot 10^4$ Wm^{-1} and $c_3 = 1.498 \cdot 10^6$ $Wm^{-1}K$. This corresponds to one half of the heat conductivity reported in (10)]. For the SiO_2 substrate we choose $\kappa_s(SiO_2) = a_1 + a_2T$ with $a_1 = 0.9094$ $Wm^{-1}K^{-1}$ and $a_2 = 1.422 \cdot 10^{-3}$ Wm^{-1} K^{-2} (11). The Kirchhoff transform permits to find the approximations

$$\theta_s^* \approx \theta_D^* + 0.46 \; \theta_D^{*2}$$

and

$$T_D^* \approx 1 + \theta_D^* + 0.12 \; \theta_D^{*2}$$

The accuracy of this approximation within the temperature interval $T(\infty) = 443$ K $\leq T_D \leq 2500$ K is 2 - 3 %.

Figure 11 demonstrates the very fast spreading of the deposit in lateral direction. The velocity of lateral growth is strongly influenced by the width of the threshold, δT_{th}. The saturation in spot diameter, which occurs when the temperature near the edge of the spot becomes smaller than the threshold temperature for deposition, is in agreement with experimental observations (7). It becomes evident that the saturation in width takes place much faster than the saturation in height. The change in the (normalized) temperature distribution during the growth process is shown in the lower part of the figure. It shows that within a short time the surface temperature becomes almost uniform over the surface of the spot.

Figure 12 shows the growth of Ni spots deposited from $Ni(CO)_4$. In this case the thermal conductivity of the deposited Ni and of the substrate was kept constant with $\kappa^* (T(\infty)=300$ K$) = 30$. The other parameters employed in the calculations were $T(\infty) = 300$ K, $k_o^* = 1.1 \cdot 10^{13}$, $\Delta E^* / T(\infty) = 45$ and $AI_o^* = 1.33$. Here, no threshold was assumed so that $T_{th} / T(\infty) = 1$. The kinetic parameters correspond to the pyrolytic decomposition of $Ni(CO)_4$ (1). Due to the absence of a threshold, there is no abrupt saturation in the width of the Ni spots. Furthermore, the ratio of spot height to spot width is much larger than in the case of W. This is related to the much higher

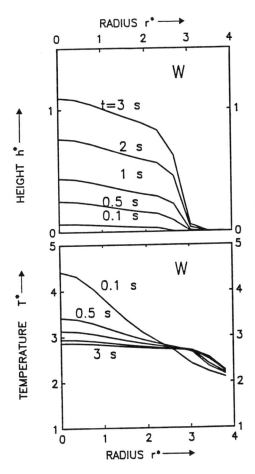

Figure 11. Normalized height of W spots calculated for various stages of growth as a function of the (normalized) distance from the laser-beam center. The parameters employed are typical for laser-CVD of W from WCl_6 + H_2 (see text). The lower part of the picture shows the evolution of the (normalized) surface temperature distribution (Reproduced with permission from reference 9. Copyright 1993 Elsevier Science Publishers B.V.).

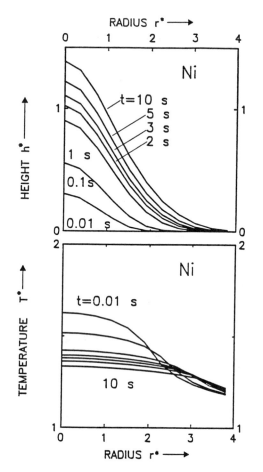

Figure 12. Same as Figure 11 but for Ni spots deposited from
Ni(CO)$_4$ (Reproduced with permission from reference 9. Copyright
1993 Elsevier Science Publishers B.V.).

activation energy, ΔE^* ; it yields a significantly higher growth rate in the center than near the spot edge. This behavior is in qualitative agreement with experimental observations (1,12). The change in temperature distribution related to the growth of the spot is similar to that oberved for W.

4.2.2 Direct writing of lines. Figure 13 shows contour lines (left side) and isotherms (right side) calculated for various stages of direct writing of W lines deposited from WCl_6 + H_2 . The laser beam is switched on at time t = 0. The scanning velocity employed was $v_s^* = 2.0$, the absorbed laser-beam intensity $AI_o^* = 20.6$. The other parameters are the same as in Figure 11. The figure shows that the stationary solution is achieved only after a rather long time. The larger width of the stripe observed in the initial phase of growth (short times) is related to the fact that energy losses due to heat conduction along the stripe are not yet effective. This behavior becomes evident also from the isotherms. It is in agreement with experimental observations.

Within the regime of parameters investigated, the shape of the calculated W stripes remains always uniform, i.e. shows no oscillations in height or width (1).

Calculations of the kind presented in Figure 13 permit to derive the parameter $\xi = r_D /a$ for different scanning velocities [see also (4.6)]. From these calculations we find $1.2 < \xi < 1.5$. Experimentally we find for the WCl_6 + H_2 system $\xi \approx 1.25$.

The calculations performed in (9) show that the discrepancy between the model presented in Sect. 3.2 and the present model is about 30% within the range of reasonable parameters.

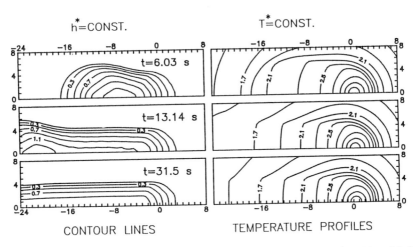

CONTOUR LINES   TEMPERATURE PROFILES

Figure 13. Contour lines (left side) and isotherms (right side) calculated for W stripes produced by laser direct writing. The laser beam is switched on at t = 0 (Reproduced with permission from reference 9. Copyright 1993 Elsevier Science Publishers B.V.).

5. Conclusion

Theoretical models on the gas-phase kinetics and the simulation of growth permit to explain many features observed in pyrolytic laser-CVD of microstructures. Such calculations together with experimental data allow an optimization of the deposition process with respect to specific applications.

Acknowledgement: We wish to thank the "Fonds zur Förderung der wissenschaftlichen Forschung in Österreich" for financial support.

Literature Cited:

(1) Bäuerle, D. Chemical Processing with Lasers; Springer Series in Materials Science; Springer: Berlin, Heidelberg, 1986; Vol. 1.
(2) Toth, Z.; Kargl, P.B.; Grivas, C.; Piglmayer, K.; Bäuerle, D. Appl. Phys. B 1992, 54, 189.
(3) Bäuerle, D.; Luk'yanchuk, B.; Piglmayer, K. Appl. Phys. A 1990, 50, 385.
(4) Kirichenko, N.; Piglmayer, K.; Bäuerle, D. Appl. Phys. A 1990, 51, 498.
(5) Luk'yanchuk, B.; Piglmayer, K.; Kirichenko, N.; Bäuerle, D. Physica A 1992, 189, 285.
(6) Kirichenko, N.; Bäuerle, D. Thin Solid Films 1992, 218, 1.
(7) Kullmer, R.; Kargl, P.; Bäuerle, D. Thin Solid Films 1992, 218, 122.
(8) Arnold, N.; Kullmer, R.; Bäuerle, D. Microelectronic Engineering 1993.
(9) Arnold, N.; Bäuerle, D. Microelectronic Engineering 1993.
(10) Handbook of Chemistry and Physics; Weast, R.C., Ed.; 59th ed. CRC Press: Boca Raton, FL, 1978/79.
(11) Heraeus Inc., Germany: data sheet Q-A 1/112, 1979.
(12) Piglmayer, K.; Bäuerle, D. In Laser Processing and Diagnostics II; Bäuerle, D.; Kompa, K.L; Laude, L.D., Eds.; Les Editions de Physique: Les Ulis 1986), pp 79-84.

RECEIVED April 1, 1993

Chapter 19

Laser-Driven Synthesis of Transition-Metal Carbides, Sulfides, and Oxynitrides

Gary W. Rice[1]

Exxon R&E Company, Annandale, NJ 08801

Laser pyrolysis of transition metal (Cr, Mo, W, Mn, Fe, Co) carbonyls and substituted carbonyls yielded high surface area powders of the title species. A cw CO_2 laser beam intersected a stream of reactant gas, with absorption of energy by ethylene or ammonia as co-reactants, or ethylene as a sensitizer. Powder particle diameters ranged from 10 to 100 nm, decreasing as the melting point of the major product increased. Syntheses of bimetallic products and selectivity toward carbon, nitrogen, oxygen and sulfur incorporation are also discussed.

Lasers, as high precision probes, can give deep insight into the microscopic details of chemical reactions. Lasers are ideal for controlled initiation of reactions, and allow high sensitivity for detection of transient intermediates and stable products that is selective in both space and time. However, lasers as high power industrial tools are mainly found in non-chemical applications such as cutting, drilling, welding and surface heat treating. The potential of lasers to drive or control large scale chemical processing has been reviewed periodically (1-3), with the recurring conclusion that the cost requires high value added products for economic viability. However, in CO_2 laser synthesis of ultrafine ceramic powders from silane (4), a process evaluation suggested that reagent cost was dominant. That work also showed that laser synthesized powders yielded bulk ceramics whose properties surpassed those of materials from conventional precursors.

[1]Current address: Engelhard Corporation, 2655 Route 22 West, Union, NJ 07083

0097–6156/93/0530–0273$06.00/0

The work described here was undertaken to explore pyrolytic laser synthesis of fine powders from transition metal organometallics. Compared to slower thermal syntheses, laser driven reactions give high surface areas without contamination, and some chemical and physical properties that appear to be unique. It is very difficult to obtain refractory materials that combine high surface area and surface purity by conventional techniques. Post synthesis processing, such as grinding, inevitably introduces contaminants that bind strongly to surfaces. Laser synthesis yields high surface area powder in a single step, without contact between the reaction zone and the reactor wall. The reactions employ gaseous organometallics, so transfer into the reaction zone is itself a purifying distillation step. Laser synthesized powders have high purity surfaces with respect to unwanted alkali and alkaline earth metals, for example, although achieving precise stoichiometry and high crystallinity in the bulk materials are difficult.

The laser synthesis technique has been described elsewhere (5). Briefly, a gaseous reactant stream crossed the beam of a CO_2 laser. A luminous flame initiated at the crossing point yielded fine particles which were collected on a membrane filter. Additional gas flows were provided to prevent spreading of the reactant stream and accumulation of powder on cell windows. The need for reactant vapor pressures of about 100 torr was not unduly restrictive. Volatile organometallic compounds exist for all the first row transition metals from titanium to nickel, as well as for many of the second and third row metals.

Reactions were driven with a cw CO_2 laser built for the purpose with optional line-tunability. It produced 200 W as a free-running laser on the 10 P(20) line at 944 cm^{-1}, or up to 120 W when line tuned. Studies of metal carbide synthesis employed C_2H_4 as both a reactant and photon absorber. Equivalent results were obtained with 200 W laser power at 944 cm^{-1}, and with 120 W at the absorption maximum of C_2H_4.

TRANSITION METAL CARBIDE SYNTHESIS

Pyrolysis of $Fe(CO)_5$ and excess C_2H_4 yielded fine powders which were analyzed in detail. TEM showed that the particles had diameters less than one micron, while XRD indicated Fe_3C was the major phase. XRD line broadening suggested crystallites were substantially smaller than the particles. BET surface area analysis indicated that the particles were not porous. XPS showed that surfaces were carbidic iron and excess carbon, free of oxide phases. However, Mössbauer showed that up to 20% of the iron was present as α- and γ-Fe. Though fine iron powder is pyrophoric, these materials were air stable. Further, heating in 10% O_2 in a TGA-mass spec showed that the powders were stable to oxidation to over 200° C. These results suggest a particle formation process consistent with the known reactivities $Fe(CO)_5$, C_2H_4, and probable intermediates, and with rapid heating of the reactant mixture. Since the first order dissociation of $Fe(CO)_5$ is orders of magnitude faster than that of C_2H_4 (6), progressive CO loss is the likely route to α-Fe early in the process. (This phase is the major product when the synthesis is run at lower temperature.) This phase will dissolve only about 0.03 atom% C (7). As the temperature rises, more extensive incorporation of carbon from C_2H_4, produces the γ-Fe phase with up to 8 atom% dissolved C, and then Fe_3C. Continued hydrocarbon pyrolysis after depletion of gaseous iron species yields the carbon overlayer. Some polynuclear aromatics probably condense on the particles during cooling.

Although the iron carbide powders were air stable, they were highly reactive in more interesting ways. As hydrocarbon synthesis catalysts from CO and H_2, they had both better activity and selectivity than conventional unpromoted iron and iron carbide catalysts (8). Synthesis of significant fractions of hydrocarbons with carbon numbers above 9, and particularly of C_{10} and higher olefins, had not previously been demonstrated with unpromoted iron-based catalysts. These laser synthesized powders also showed unprecedented selectivity for synthesis of hydrocarbons other than CH_4 from CO_2 and H_2. These unexpected properties were attributed to the combination of high surface area materials and freedom from surface contamination.

As a final comment on laser pyrolysis of $Fe(CO)_5/C_2H_4$ mixtures, it was shown that the composition of the solid product was related to that of the gaseous byproducts, indicating that feedback control of a continuous synthesis would be possible. The amount of CO in the gas phase after pyrolysis is directly related to the amount of iron deposited in the solid. The source of carbon in the solid was C_2H_4, and its main gaseous byproduct was C_2H_2. A number of experiments showed that the ratio of C_2H_2 to CO in the gas phase correlated well with the ratio of carbon to iron in the solid product. Real-time spectroscopic measurement of the C_2H_2:CO ratio could be used for feedback control of the process.

Laser synthesis experiments using C_2H_4 and other transition metal carbonyls and carbonyl derivatives gave interesting results and some surprises. $Cr(CO)_6/C_2H_4$ gave a chromium carbide whose XRD indicated it was a previously unknown cubic phase. With $Mo(CO)_6$ and $W(CO)_6$, the products were carbides in the composition range $MC_{1.2-2}$, with oxygen contents less than 0.4%. As with the $Fe(CO)_5$ pyrolyses, there was good selectivity in excluding oxygen from the CO ligands from the products. TEM showed that the Mo and W powders had particle diameters less than 2 nm. The 50-fold decrease in diameter, as compared to laser synthesized Fe_3C, was attributed to the much higher melting points of Mo and W carbides.

Pyrolysis of $CH_3Mn(CO)_5/C_2H_4$ gave, surprisingly, a product which was MnO by both XRD and elemental analysis. Mn is between Fe and Cr on the periodic table, and the reactant was reasonably similar to $Fe(CO)_5$ and $Cr(CO)_6$, so a carbidic product was expected for it as well. Moving to the right of iron in the table, two syntheses were attempted with cobalt compounds. $(C_5H_5)Co(CO)_2/C_2H_4$ gave a powder which analyzed as CoC_7, indicating that the carbon from the cyclopentadienyl ligand was retained during pyrolysis. The C_5 ring could participate in polynuclear aromatic and soot formation during pyrolysis. No other experiments were conducted with cyclopentadienyl complexes.

$Co(NO)(CO)_3/C_2H_4$ yielded a metallic cobalt powder. The laser synthesis temperature is above the stability limit for cobalt carbides. The XRD of the product had lines for ß-Co only. This is a metastable high temperature phase. Previous attempts to make it gave products contaminated by a low temperature hexagonal phase (9). Evidently, the quenching rates that occur naturally in laser synthesis are faster than those obtained with considerable effort by conventional thermal methods. Like cobalt, nickel has no stable carbide at laser synthesis temperatures. Pyrolysis of $Ni(CO)_4/C_2H_4$ should give a metallic product, but this experiment was avoided because of the hazards of handling nickel carbonyl.

Several laser syntheses were conducted using C_2H_4, $Fe(CO)_5$ and one other transition metal organometallic to produce bimetallic carbides. The metal content was about 75% iron in each case. These materials were made, in part, because bimetallic systems are catalytically important. The structural results were interesting. While $CH_3Mn(CO)_5/C_2H_4$ gave MnO, and $Co(NO)(CO)_3/C_2H_4$ gave ß-Co, both $Fe(CO)_5/CH_3Mn(CO)_5/C_2H_4$ and $Fe(CO)_5/Co(NO)(CO)_3/C_2H_4$ gave bimetallic M_3C phases. Mn_3C and Co_3C are known, so Mn and Co were apparently incorporated into solid solutions with Fe_3C. By contrast, $Fe(CO)_5/Cr(CO)_6/C_2H_4$ gave M_7C_3. Cr_3C is unknown, but Cr_7C_3 and Fe_7C_3 are, so the structure was defined by the less abundant metal in that case.

TRANSITION METAL SULFIDE AND OXYNITRIDE SYNTHESIS

In addition to laser syntheses of carbides, attempts were made to prepare sulfides and nitrides. Only the former were successful. $Mo(CO)_6/H_2S/C_2H_4$ gave an air stable product that was approximately MoS_2, with less than 2% total carbon and oxygen present. Here, C_2H_4 acted mainly as a sensitizer for the reaction of $Mo(CO)_6$ and H_2S.

These reactions were conducted at lower pyrolysis
temperatures than the carbide syntheses. The greater
reactivity of H_2S compared to C_2H_4 probably favors
sulfides over carbides.
A bimetallic sulfide, with oxygen contamination, was
prepared from $Mo(CO)_6/Co(NO)(CO)_3/H_2S/C_2H_4$. XRD
indicated that $CoMo_2S_4$ was the major crystalline product,
but elemental analyses of samples from two preparations
revealed oxygen contents of 4.2 and 10.6%. The high
oxygen level is attributed to the combination of NO and
molybdenum, since both $Co(NO)(CO)_3/C_2H_4$ and $Mo(CO)_6/C_2H_4$
gave products with little oxygen. It should be noted
that preparation of these particular sulfides was
attempted because of the importance of molybdenum
sulfides in hydrodesulfurization catalysis. Both cobalt
and carbon can act as promoters, so some contribution of
carbon from C_2H_4 was not considered a problem.
Transition metal nitride syntheses were attempted
using NH_3 as reactant and absorber, operating the laser
at 115 W on the 10 P(34) line at 934 cm^{-1}. $Fe(CO)_5/NH_3$
gave a pyrophoric black powder, while $W(CO)_6$ gave an air
stable black powder. XRD indicated an Fe_4N/Fe_3N mixture
and W_2N, respectively, but elemental analyses showed that
both materials were oxynitrides with more oxygen than
nitrogen. The oxygen must be derived from the CO
ligands. Oxygen incorporation was not a problem in
most of the carbide syntheses, where excess C_2H_4 and its
hydrocarbon byproducts maintained highly reducing
conditions.

SUMMARY

The laser synthesis experiments described here were
performed to address two questions. First, what kinds of
materials can be prepared using the technique? Second,
do the materials, once made, have any interesting or even
unique properties?
The first question has a complex answer. Many
materials were discussed here, and those reported
previously include TiO_2 (10), ZrB_x (11), silicon ceramics
(4), and mixed Fe/Si carbides (12). In the iron–carbon
system, for example, the main product can be either α-Fe
or Fe_3C, and the Fe:C ratio can be varied over a wide
range. The carbide syntheses had good selectivity in
that little oxygen from the CO ligands was incorporated
in products. Ligand chemistry presents difficulties,
though. Oxygen derived from NO, carbon from $[C_5H_5]^-$, and
(under less reducing conditions) oxygen from CO were all
found in products. The ligands required to obtain
volatility for reasonable synthesis rates bring their own
chemistry to the pyrolysis process.
The answer to the second question is simple: Yes.
Fe_3C and other iron carbide phases are well known and
have been widely studied, yet the combined catalytic

activity and selectivity of laser synthesized iron
carbide was unprecedented (8). Unusual properties of
known materials can be expected in other cases as well.

REFERENCES

1. Kaldor, A.; Woodin, R. L. Proc. IEEE **1982**, 70, 565.
2. Kleinermanns, K.; Wolfrum, J. Angew. Chem. Int. Ed.
    Engl. **1987**, 26 38.
3. Woodin, R. L.; Bomse, D. S.; Rice, G. W. Chem. Eng.
    News **1990**, 68, 20.
4. Flint, J. H.; Haggerty, J. S. Proc. SPIE **1984**, 458,
    108.
5. Rice, G. W.; Woodin, R. L. Proc. SPIE **1984**, 458,
    98.
6. Lewis, K.; Golden, D. M.; Smith, G. P. J. Am. Chem.
    Soc. **1984**, 106, 3905.
7. Kosolapova, T. Ya. Carbides; Plenum Press, New York,
    NY, 1971.
8. a.   Rice, G. W.; Fiato, R. A.; Soled, S. L.: **U.S.**
         4,659,681, **1987**.
    b.   Rice, G. W.; Fiato, R. A.; Soled, S. L.: **U.S.**
         4,668,647, **1987**.
    c.   Fiato, F. A.; Rice, G. W.; Miseo, S.; Soled, S.
         L.: **U.S.** 4, 687,753, **1987**.
    d.   Rice, G. W.; Fiato, R. A.; Soled, S. L.: **U.S.**
         4,788,222, **1987**.
9. Troiana, A. R.; Tokich, J. L. AIME Trans. **1948**, 175,
    728.
10. Rice, G. W. J. Am. Cer. Soc. **1987**, 70, C117.
11. Rice, G. W.; Woodin, R. L. J. Am. Cer. Soc. **1988**,
    71, C181.
12. Gupta, A.; Yardley, J. T. Proc. SPIE **1984**, 458,
    131.

RECEIVED January 29, 1993

Chapter 20

Laser-Induced Coalescence of Gold Clusters in Gold Fluorocarbon Polymer Composite Films

Paul B. Comita[1], Wolfgang Jacob[2], Eric Kay[1], and Rong Zhang[1]

[1]IBM Research Division, Almaden Research Center, 650 Harry Road, San Jose, CA 95120
[2]Max-Planck-Institut fur Plasmaphysik, Garching, Germany

The formation of conducting films from composite films comprised of gold clusters in plasma polymerized polyfluorocarbon (PPFC) is described. A focussed, visible laser beam is used to coalesce the gold clusters within the PPFC matrix. Heating the composite with the laser causes the film to collapse with a loss of weight due to decomposition and volatilization of the polymer. Under the appropriate laser power and scanning conditions, coalescence of the gold particles results in a conducting metal line, exhibiting close to bulk metal resistivity.

Laser-induced pyrolytic film decomposition of thin solid films has been examined extensively for the patterning of thin metallic films on solid substrates. (1,2) This method consists of coating a substrate with a thin organometallic film, generally composed of a polymer containing a metal, a glass derived from an organometallic molecule, or a composite or mixture of materials. Local heating of the thin solid film with a laser beam leads to dissociation of the film and deposition of a solid material with the evolution of gaseous byproducts. Laser sintering of powders has also been used recently to coat surfaces for fabricating 3-dimensional objects (3). In this process, an object or surface is coated with a powder, and a laser is directed at the region in space where material is to be added to the surface. The laser causes individual powder particles to consolidate or sinter, giving rise to a solid layer. Subsequent layers are then added and further processed. The selective laser sintering approach to solid fabrication has been applied to ceramics, metals, and polymers, with the largest emphasis to date on polymers. This technique is not advantageous for metals, due to the extremely high temperatures needed to sinter metallic powders, due in part to the high conductivity of the metallic surface formed.

We describe here a new technique consisting of laser-induced coalescence of a metal/polymer composite. This technique takes advantage of the physical annealing process of metal particles, and the optical and physical properties of the thin dielectric films as the transformation matrix (see Figure

0097–6156/93/0530–0279$06.00/0
© 1993 American Chemical Society

1). The coalescence of gold clusters embedded in a fluorocarbon polymer has been examined with differential scanning calorimetry, thermogravimetric analysis, and optical spectroscopic methods. The microstructure and DC conductivity of the films are analyzed as a function of gold volume fraction for as deposited as well as laser annealed films. The ability to generate conducting lines with a direct write technique, using a scanning focussed laser beam in a serial process, is demonstrated under ambient air conditions. Scan rates are typically very fast, in the 0.1 mm/s range and the films obtained are a fraction of the thickness of the unreacted solid film. Periodicities in the generation of the continuous gold film are also briefly described.

Film Preparation and Characterization

Gold containing plasma-polymerized fluorocarbon films were deposited in a capacitively coupled, coplanar rf (13.56 MHz) diode reactor system (see Figure 2). This experimental system has been described in more detail in earlier reports (4,5). The system consists of a vacuum chamber pumped by a Turbomolecular pump to a base pressure below 10^{-6} Torr, with the rf power applied via a matching network to the isolated electrode. The samples are fastened to the grounded electrode. It is well known that in such a configuration a sheath with a substantial sheath voltage develops in front of the isolated (small) electrode. Due to the high sheath voltage this electrode (further on called target) is submitted to intense ion bombardment leading to the removal of atoms from the target by physical sputtering. In our case the target consisted either of a gold or a teflon (Polytetrafluoroethylene: PTFE) target to deposit gold containing or pure polymer films, respectively.

A tetrafluoroethylene (C_2F_4) / argon gas mixture was injected into the discharge system with control of the total flow rate. Gold-containing films were deposited at a constant dc bias voltage of 1 kV at the gold target. For the deposition of pure polymer films a teflon target was used. The films were deposited on quartz substrates which are clamped to the temperature controlled, grounded electrode. It has been demonstrated earlier that the polymer deposition rate depends strongly on the sample temperature (6,7). Therefore it is of crucial importance for the reproducibility of the film deposition to carefully control the sample temperature. In these experiments the sample temperature was held constant at 15º C. The average deposition rate with this setup was 0.8 nm/s.

The key processes leading to the deposition of metal-containing fluorocarbon polymers have been reviewed by Kay and Dilks (8) and more recently by Kay (7). We only state here that the volume fraction of the metal in the deposited film can easily be changed by adjusting the ratio of C_2F_4 to argon in the processing gas mixture and keeping all other external parameters constant. The gold volume fraction p of each film has been computed from its density d(p) by using the relation

$$p = \{d(p) - d(polymer)\}/\{d(Au) - d(polymer)\}$$

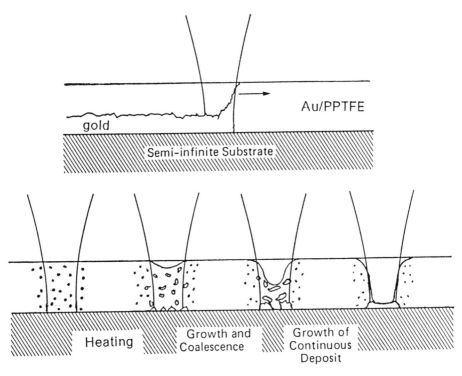

Figure 1. Laser-induced coalescence of a Au/PPFC film using a scanning laser beam (top figure), and a stationary laser beam (lower figure). The coalescence phenomena take place concurrently at different locations within the beam for a scanning beam. With a stationary beam, the heating, collapse, and coalescence to a continuous film takes place sequentially in time, allowing these aspects to be examined in a time resolved experiment.

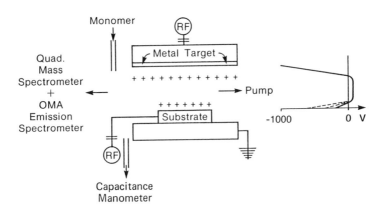

Figure 2. A schematic diagram of the rf diode reactor used to deposit the Au/PPFC films. The diagram at the right depicts the sheath voltage with reference to the location of the electrodes in the reactor.

where d(Au) and d(polymer) are the bulk gold and polymer densities, respectively. d(p) has been determined from the film mass by weighing the sample before and after deposition and from the film volume calculated from the known area and the film thickness which was measured by the stylus method. For the gold density we used the literature value d(Au) = 19.3 g/cm^3 and for the polymer d(polymer) = 2.0 g/cm^3 was used (9). The overall precision of p is estimated to be ± 0.03.

Many physical properties of these gold-containing films have been previously studied (4-10), and will not be described here. However, some of the physical properties are pertinent to the generation of conducting films from the as-deposited dielectric film, and these properties are examined briefly here. The size distribution of gold clusters in the polymeric matrix is in the range of 20-80 Å. For regions which are somewhat smaller than the diameter of the substrate electrode, the dispersion of gold clusters within the polymer matrix is seen to be quite uniform by TEM studies (see Figure 3). The electrical properties have been investigated by Perrin et al. (4) and by Laurent et al. (6). These films are very good insulators at a very low gold volume fraction, similar to pure PPFC. At high volume fraction of Au, the films are very good conductors, approaching the conductivity of bulk gold (see Figure 4). The electrical percolation threshold lies at $p_c = 0.37 \pm 0.03$. In a narrow range of p around this percolation threshold the electrical properties of the films change from an insulator to a conductor which is accompanied by a drop of the electrical resistivity from about 10^5 Ω-cm at p = 0.3 to about 10^{-4} Ω-cm at p = 0.5 (4,5).

Perrin et al. (10) investigated the optical properties of similar films over a wide range of gold volume fractions and correlated the optical properties with the microstructure of the films. We measured the optical transmission of our films in the range from p = 0 (pure polymer) to about p = 0.2. The normal incidence optical transmittance spectra were recorded in a wavelength range from 200 nm to 3000 nm. The films were deposited on quartz samples and the film thickness was chosen to keep the maximal absorbance at 200 nm below 2.

Laser-Induced Coalescence

Laser-induced coalescence of the gold clusters was achieved with the TEM$_{00}$ mode of a Spectra Physics Model 22-05 or American Laser Model 909 argon ion laser operating at 514 nm. The beam was expanded (3X), collimated, and focused through a microscope objective (0.2 NA). The measured beam diameter is approximately 5 ± 1.0 μm at the 1/e intensity points and was approximately Gaussian. Beam scanning was achieved by moving the substrate relative to the fixed final objective on a set of XY translation stages. The coalescence could be viewed in real time via a trinocular nosepiece located above a beam-splitting cube, on which a CCD camera with associated optics was mounted.

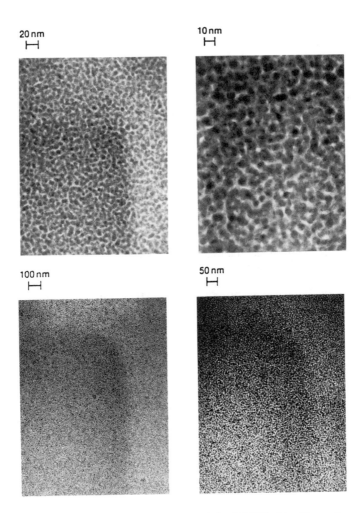

Figure 3. Photographs of TEM images of Au/PPFC thin films at p = 0.2. At lower magnification the uniformity can be easily observed, and at high magnification (top right) the clusters can be seen to be approximately 50 Å in diameter. At this gold volume fraction, which is very close to the percolation threshold, and due to the thickness of the film it is difficult to observe the separation between individual metal clusters.

The extent of cluster coalescence can be controlled by scanning speed (dwell time) as well as the laser power. A number of thin films below percolation threshold have been fabricated, and in all cases the laser was able to induce coalescence of the clusters to form a conducting line (see Table I). In the region of p = 25, where the as deposited resistivity is in 10^{12} $\mu\Omega$cm, the laser-induced coalescence can form a conducting line with resistivity within 10% of the resistivity of bulk gold. The scanning velocity of the beam was not optimized for speed, but the highest scanning velocities achieved were up to .2 mm/sec. The large differences in the resistivities in the laser annealed lines, as shown in Table I, can be accounted for by incomplete coalescence of the as-deposited composite film and/or the morphology of the laser-annealed film (11).

At the higher p (> 4%) clusters coalesce and the film collapses with a quasi-Gaussian profile, similar in shape to the laser intensity. (see Figure 5). The depth of the collapse is dependent on the dwell time and laser intensity, but at very low powers collapse is complete for most films. At p = 4%, we have obtained data on the extent of film collapse at constant dwell time (or scan speed), but with varying incident power (see Figure 5). The profiles show that the film collapse occurs in a continuous fashion, with the extent of collapse dependent on the incident power at constant dwell time. No build up of polymer residue is noted at the side of the groove except at the lowest laser power. This behavior was observed for all p except p = 0 that we have studied. For scan speeds and powers in the region of those specified in Table I, a continuous metallic film can be formed (see Figure 6). The periodic features seen in Figure 6 are simply due to the surface morphology of the film, where there is a change in the reflectance of light. The periodicity of the features arises from an autothermal instability in the laser writing process, and is a feature which we are presently examining. An elemental analysis of the film by energy-dispersive x-ray analysis in the laser-annealed region shows that within the analytical precision of this technique that the deposit is virtually fluorine-free, i.e. the conducting lines are metallic gold (see Figure 7). Bulk annealing of the Au/PPFC films has been described in previous work where it was shown that bulk heating led to film collapse from a metastable polymer state, with concomitant microstructural changes in the size and shape of the gold clusters. However, in the case of the bulk annealing experiments, all of the polymer is not removed (or volatilized).

With no gold microclusters (p = 0), the pure PPFC film has low optical absorbance at 514 nm. High fluence is therefore necessary to achieve any transformation of the polymer in the visible region. At high fluence, we observe a region of collapse with large amounts of solid debris surrounding the site. This finding indicates an ablative type of polymer removal when no metal clusters are present. Additionally, at very low p, the film does not coalesce in a straightforward manner as it does at higher loadings, but forms filaments or threads of gold which are oriented primarily perpendicular to the scan of the laser beam. For p > .04, but for very high power and slow scan speeds, the coalescence becomes an unstable process, leading to gaps in the gold film that is formed. The residue of gold that remains near the gaps

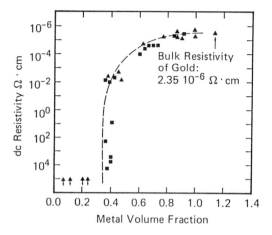

Figure 4. DC resistivity of the as deposited films as a function of gold volume fraction.

Table I. Resistivity of Laser Annealed Au/PPFC Films

Au Volume Fraction	Film Thickness (Å)	Power (mW)	Scan Speed (μ/s)	Line Thickness (nm)	Resisitivity (μohm-cm)
0	14000	-	-	-	10^{16}
4	23000	20	100	500	9.1×10^5
16	10000	50	10	160	3.66
21	1700	148	200	40	4.6×10^4
20	7000	104	10	100	2.70
23	6000	198	10	200	2.66
24	16000	190	200	800	4.83
28	13000	40	20	820	26.4
30	12000	40	20	720	97.6

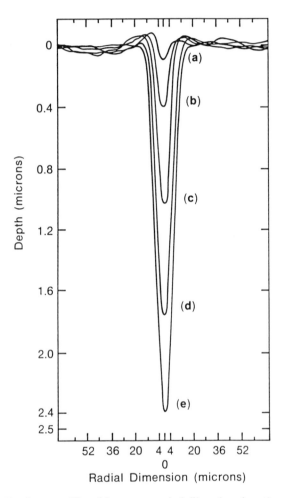

Figure 5. Surface profile of laser-annealed film showing the extent of polymer removal and film collapse with increasing laser power and constant scan velocity. The scan velocity is 100 μ m/s and the incident laser power (mW) is (a) 5.8; (b) 11.6; (c) 29.4; (d) 58.8; and (e) 90.9.

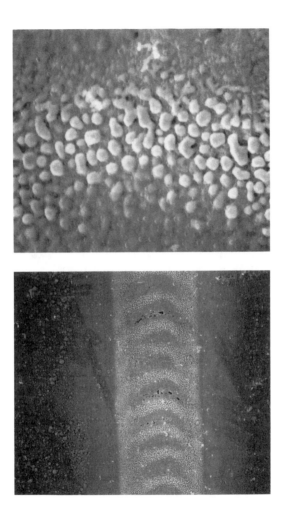

Figure 6. An SEM photograph of the laser annealed Au/PPFC. The periodic features are due to the morphology of the film. In this film, the p = .24 with an as deposited film thickness of 16000 angstroms. The laser power was 76 mW at 514 nm. In the upper photograph, the linewidth is approximately 10 microns. The lower photograph is at 20,000 X magnification, and shows the continuous nature of the film.

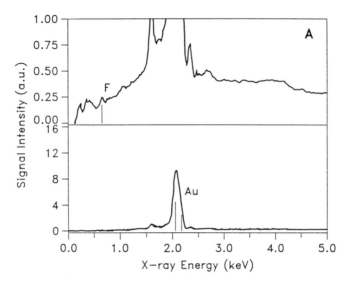

Figure 7. Energy dispersive x-ray analysis of the laser-annealed film. The major line is gold, with trace amounts of other elements, including fluorine.

Table II. Surface Temperature Rise during Laser Coalescence

Power (mW)	Radius (μm)	Depth (μm)	Temp. Rise (C)
5.8	4.71	0.1	70.6
11.6	7.28	0.4	91.3
29.4	8.57	1.1	196.6
58.8	13.5	1.8	249.6
90.9	18.8	2.4	277.0

are generally spherically shaped nodules, indicative of complete melting of the gold film before a continuous film has a chance to form.

We would like to estimate the maximum surface temperature induced by the laser in order to understand the physical and chemical mechanisms which are possible in the estimated temperature range. For the purpose of making an estimate of the temperature rise, we have assumed that the polymer undergoes simple volatilization, and the gold clusters migrate and coalesce to form a continuous film. We can estimate that with gold clusters of approximately 50Å average diameter and p = 4%, that after removal of 125 nm of film we have an effective p = 100% in the top surface layer, approximately 5 nm in depth. To a first approximation, this layer will absorb the visible light and give rise to a heated region which then volatilizes the film beneath it causing further collapse and additional gold coalescence. The beam diameter is much larger than the thickness of the gold layer, so we can neglect thermal losses through the gold layer, and use the thermophysical properties of the film/substrate. The beam is also larger in diameter than the thickness of the Au/PPFC film, and to a first approximation the thermal conductivity of the SiO_2 can be used as the surrounding heat sink.

The temperature rise can be calculated using the steady state expression (12): $T_{max} = P(1 - R)/\pi r_f K_s$. In this expression P is the incident power, R = 0.64, the reflectivity of Au at 514 nm (13) r_f is the radius of the gold film at the coalescence zone, and K_s = 0.02 W/cmC, the thermal conductivity of the quartz substrate. Using the range of laser fluence that we generate conducting lines with (see Table II), we can estimate that the temperature rise extends from 70C to 277C. The temperature present in the film under irradiation could be somewhat higher than that indicated for a number of reasons which we will enumerate here: 1) the reflectivity may be much lower because the films are not continuous and smooth as they are being formed; 2) the teflon film has a lower thermal conductivity than the substrate, and this will lead to less heat loss; 3) the annealing process is highly exothermic due to the polymer and the gold clusters being initially formed in a metastable state in the plasma deposition process. The additional energy input due to the relaxation phenomena occurring in the matrix can be substantial and we have not taken this into account. We may regard the number calculated as being a lower estimate of the surface temperature maximum actually achieved during the laser annnealing process. However, they give an indication of the approximate temperature range.

In order to gain an understanding of the physical and chemical phenomena involved in the laser-induced coalescence, we have examined the optical properties of the system by using a time-resolved reflectance experiment. In this experiment we measure the specular reflectance at a fixed wavelength during a single coalescence event in the thin film. This is accomplished with an experimental system which consists of an Ar+ laser as the light source, the output of which is modulated by a 25 MHz KDP pockels cell. An experimental set-up very similar to this has been described previously (14). The specular reflectance was collected with a high numerical

aperture lens, dispersed in a monochromator, and detected with a
photomultiplier tube. The reflectance signal was captured with a transient
digitizer interfaced to an IBM PS/2.

The change in optical properties is observed to occur under bulk
annealing at approximately 200C, very close to the glass transition (T_g) of
PPFC. The optical change is correlated with the coalescence of the metal
clusters, and can be described as taking place with the collapse of the
polymer phase and migration of small clusters to form larger metal clusters.
It has been determined previously that the size of the clusters (30-50 Å) can
give rise to a strong change in transmittance in the visible region of the
spectrum, and this is expected to induce a change in the reflectance of the
sample as well, through a modification of the optical constants of the film.
This effect is being further examined to quantify the change in the optical
properties.

The time-resolved data was obtained for the change in specularly
reflected light for thin film samples with moderate gold loadings. Again,
these are films which are below the percolation threshold, so that they are
useful as a dielectric film or matrix for the resulting conducting patterns. We
have found that the reflectance change can be fit to a exponential function,
and from this fit we are able to calculate a a half life of 455 μsec (at 10 μm,
and 150 mw). The time scale for complete collapse of the composite film to
a continuous gold film can be seen in Table I, for a 10 μm beam. The beam
velocity is 100 - 200 μm/sec at the fastest scan rates. The complete physical
collapse to a continuous metal line thus takes place on the order of 100 msec.
This indicates that the large reflectance changes occur in the initial stages of
coalescence before the film completely collapses. The approximate time for
spheroidization of two gold spheres can approximately be calculated by a
previously described analytical expression (15). This treatment uses the
surface energies as well as the diffusive flux of metal atoms to roughly
predict the time of sintering and the time of complete spheroidization of two
contacting particles. This can be calculated to be approximately 1 μsec at
150 C for a 40 Å cluster size with a relatively straightforward model (15).
This time is below that which has been measured for the reflectance change,
but is more closely approximated by a particle with a larger radius. We
expect that the rate limitation is therefore either coalescence of large clusters,
which have already coalesced from their initial state, or simply migration of
the clusters together, by the pyrolysis or volatilization of the polymer film.

Summary and Conclusions

In this work, it is shown that bulk heating or annealing of the film leads
to film collapse from a metastable polymer state, with concomitant
microstructural changes in the size and shape of the gold clusters.
Laser-induced coalescence of the gold clusters gives rise to metallic lines with
approximately the linewidth of the 1/e intensity points of the laser beam.
The extent of film collapse can be controlled be both scanning speed or dwell

time as well as the laser power. At high gold loadings (> 4%) the film collapses with a quasi-Gaussian profile, similar in shape to the laser intensity. The depth of the collapse is dependent on the dwell time and laser intensity, but at very low powers collapse is complete for most films. The polymeric matrix undergoes simple volatilization, and the gold clusters coalesce to form a continuous film. To a first approximation, the metal absorbs the visible light and gives rise to a heated region in the top surface which then volatilizes the film beneath it causing further collapse and additional gold coalescence. At the very lowest powers the resistance is far above that for a bulk gold sample, and as the power increases, the resistivity approaches that of bulk gold.

The technological significance of this work lies in the ability to directly generate microcircuit patterns in and on a teflon-like carrier. The method is far simpler than those recently described in that the films are formed in a one step dry process, and the metal line patterns are then directly written in a single subsequent step. The patterning step can take place in an ambient air environment. Gold conductors are equivalent in conductivity to copper, but are more robust to oxidation, and thus do not need passivation. In addition, the pure polymer films can be subtractively patterned with a single step laser exposure. The films can thus be built up into a 3-dimensional structure with metal lines and/or vias, making this an extremely versatile system for microelectronic circuitry or packaging applications.

REFERENCES

1. Price, P. E.; Jensen, K. F. *Chem. Eng. Sci.* **1989 44** 1879-1891.
2. Fisanick, G. J.; Gross, M. E.; Hopkins, J. B.; Fennell, M. D.; Schnoes, K. J.; Katzir, A. *J. Appl. Phys.,* **1985 57**: 1139-1142.
3. Marcus, H. L.; Beaman, J. J.; Barlow, J. W.; Bourell, D. L. *Amer. Ceramic Soc. Bull.,* **1990 69**: 1030-1031.
4. Perrin, J.; Despax, B.; Hanchett, V.; Kay, E. *J. Vac. Sci. Technol. A* **1986, 4**, 46.
5. Laurent, C.; Kay, E.; Souag, N. *J. Appl. Phys.* **1988, 64**, 336.
6. Kay, E.; Hecq, M. *J. Appl. Phys.* **1984, 55**, 370.
7. Kay, E. *Z. Phys. D* **1986, 3**, 251.
8. Kay, E.; Dilks, A. *J. Vac. Sci. Technol.* **1981, 18**, 1.
9. Laurent, C.; Kay, E. *J. Appl. Phys.* **1989, 65**, 1717.
10. Perrin, J.; Despax, B.; Kay, E. *Phys. Rev. B* **1985, 32**, 719.
11. Comita, P. B.; Zhang, R.; Jacob, W.; Kay, E., unpublished work.
12. Carslaw, H. S.; Jaeger, J. C. *Conduction of Heat in Solids* 2nd Ed., Oxford University Press, Oxford, 1959 p 264.
13. Palek, E. D. ed. *Handbook of Optical Constants of Solids,* Academic Press, Orlando, 1985.
14. Comita, P. B.; Price, P. E.; Kodas, T. T. *J. Appl. Phys.* **1992, 71** , 221.
15. Blachere, J. R.; Sedehi, A.; Meiksin, Z. H. *J. Materials Sci.* **1984, 19** , 1202.

RECEIVED February 10, 1993

Chapter 21

Laser-Assisted Chemical Vapor Deposition from the Metal Hexacarbonyls

K. A. Singmaster[1] and F. A. Houle[2]

[1]Department of Chemistry, San Jose State University, San Jose, CA 95192–0101
[2]IBM Research Division, Almaden Research Center, 650 Harry Road, San Jose, CA 95120

Laser induced deposition of thin films from the group VI hexacarbonyls can be accomplished by either photolysis of the gas phase precursor (photochemical deposition) or by laser heating of a substrate in the presence of the organometallic vapor (thermal deposition). Towards elucidating the microscopic processes relevant to photochemical and thermal deposition, a systematic study of compositions of films deposited by cw-257 nm and cw-514 nm irradiation have been performed. UV photolysis is known to lead to partial decarbonylation of the hexacarbonyl; additional surface photodissociation and dissociative chemisorption of CO produce films which have high levels of C and O impurities. On the other hand laser heating of a substrate in the presence of the vapor with visible light produces significantly cleaner metal films. Under these conditions dissociative chemisorption of CO is not observed. Thermal modeling and the relationship between this work and low coverage surface studies will also be discussed.

The technique of laser assisted chemical vapor deposition (LCVD) offers great promise for localized film growth at microscopic scale (1). In LCVD a substrate held in a cell backfilled with metallorganic vapor is exposed to a laser source; metal containing material is deposited onto the surface of the substrate. Film deposition is the result of photochemical and/or thermal processes. Choosing a laser wavelength that corresponds to light absorption by the metallorganic vapor and/or surface absorbed species can result in photochemical reactions which lead to film deposition. If, on the other hand, the substrate absorbs the

NOTE: This chapter was originally presented at the fall 1991 meeting of the Electrochemical Society, Inc., held in Pheonix, AZ.

laser energy, thermal reactions of the precursor on the substrate surface can produce metal containing films. Separation of these two mechanisms of deposition is not always possible and requires careful choice of precursor, substrate and laser conditions.

The group VI metal hexacarbonyls have the potential to be excellent precursors for film deposition by ultraviolet photolysis; they are reasonably volatile, have high cross sections in the UV and dissociate in the gas phase with near unity quantum yield to form partially carbonylated species (2-6). By using low power UV laser sources and choosing an appropriate substrate it is possible to accomplish photochemical deposition in the absence of laser heating. On the other hand, the hexacarbonyls are transparent to visible light thus by using a visible laser and a substrate that absorbs visible light it is possible to observe thermal deposition in the absence of photochemical processes.

Films deposited by photochemical deposition with low power UV lasers are of poor quality, containing large levels of carbon and oxygen impurities (5-7). High levels of impurities produce large resistivities making these films useless to the microelectronics industry. Cleaner films can be obtained with higher power densities (typically obtained with pulsed laser sources) which can produce significant surface heating (7,8).

Few studies have been performed which investigate the composition of films thermally deposited from the metal hexacabonyls. Studies with $W(CO)_6$ indicate that clean metal films can be deposited on 2 micron thick Si substrates (514 nm source), while only 75% metallic deposits were obtained on quartz (9,10). The decrease in impurities observed for thermally deposited films as compared to photochemically deposited films is qualitatively consistent with the known surface chemistry of CO on W single crystals, where surface temperatures greater than 1000 K result in recombination of C and O and desorption of CO from the surface (11).

This paper outlines systematic composition studies we have performed on both photochemical and thermal deposition from the group VI hexacarbonyls. The goal of this work is three fold: to provide a complete composition data set for these films under controlled vacuum and analysis conditions, to provide insight into the methods by which contaminants are incorporated into these films, and to contrast the results obtained to previous surface and gas phase studies of metal hexacarbonyls.

Experimental Conditions

All films were deposited on Si(100) or (111) doped n- or p-type, cleaned in acid oxidant bath and left covered by its native oxide. The light source for the laser heating work was an Ar ion laser focused to a 10 micron spot; for the photochemical work the frequency was doubled and the UV irradiation focused to a 5 micron spot. Under these conditions incident power densities ranging from 40 to 3,700 W/cm^2 were available at 257 nm (UV) while 1 - 4 MW/cm^2 was utilized for the visible irradiation (514 nm). Two deposition cells were used:

a stand alone cell which could be evacuated to the mid 10^{-4} Torr range, and an ultrahigh vacuum cell with a base pressure in the 10^{-9} Torr range. The latter could be coupled to a vacuum suitcase for transportation to the analysis chamber without air exposure. For all depositions the cells were backfilled with the vapor of crystalline $M(CO)_6$ (M = Cr, Mo or W) which had been subjected to several freeze-pump-thaw cycles. No buffer gases were used. Room temperature vapor pressures for the three hexacarbonyls are; 125 mTorr for Cr, 85 mTorr for Mo and 17 mTorr for W. Absorption coefficients at 257 nm were measured for all three materials (12).

Three sets of measurements were carried out for each metal for both photochemical and thermal deposition conditions: growth in low vacuum and exposure to air prior to analysis (LV/air), growth in high vacuum and exposure to air (HV/air) and growth in high vacuum with transfer in vacuum for analysis (HV/vac). All films were characterized with a scanning Auger multiprobe using a 10 keV electron beam. Depth profiles are obtained with a 2 keV Ar ion sputtering beam (6,2).

Calibration of beam damage and preferential sputtering effects was accomplished by using standard oxides and are outlined in Ref. 6. The beam conditions chosen significantly hinder all but instantaneous beam damage effects. Cr_2O_3 was the only oxide that did not undergo preferential sputtering of oxygen under the conditions of analysis, thus O/M ratios for sputtered W and Mo films only provide a lower limit to the oxygen content.

Results

Tables are included for both photochemical and thermal deposition. Only HV/vac data is tabulated. For LV/air and HV/air data see Ref. 6 and 12. Each entry is an average of compositions from 12 - 50 films. Peak-to-peak height ratios of O (KLL, 503 eV) to Cr (LMM, 529 eV), O to Mo (MNN, 186 eV), O to W (MNN, 1736 eV), C (KLL, 272 eV) to Cr, C to Mo and C to W are reported together with the spread in measurements. The percent compositions correspond to the average values and are provided for purpose of comparison to previous studies. Data obtained for authentic oxide and carbide samples and predicted peak height ratios for selected stoichiometries are also included at the end of each table. Comparison of predicted ratios with the experimental data permits assessment of approximate stoichiometries for the films.

Photochemical Deposition. Photochemically deposited films exhibit deep ripples characteristic of deposition with a linearly polarized laser beam under conditions where photo-induced surface reactions are rate limiting (12,13). This permits comparison of surface and gas phase photochemical and photophysical processes. Ripples were observed at all power densities used in the photochemical studies. Deposition rates ranged from 3,000 A/s to 10 A/s depending on the precursor and the power density (12).

Table I, II and III contain the Auger data for Cr, Mo and W HV/vac films deposited from the corresponding hexacarbonyl by photolysis with 257 nm light. Depth profiling showed all films to be homogeneous in composition after removal of the top 1 - 2 nm, so data for sputtered films were averaged over the profile. Data for unsputtered films reflect compositions of the near surface region. No significant variations in composition were found over the laser power density range used, so data for all intensities are combined.

Because of extensive gas phase photolysis during deposition a thin (5 nm) metal containing film covers the substrate. Analysis of this off spot material is

Table I. Auger data for HV/vac Cr films photochemically deposited

	O/Cr	C/Cr	O	C	Cr
FILMS					
unsputtered	2.5 ± 0.6	0.23 ± 0.07	54	19	28
sputtered	1.4 ± 0.3	0.23 ± 0.07	39	25	36
off spot	2.3 ± 0.1	0.43 ± 0.03	43	33	24
Cr_2O_3 Standard					
unsputtered	2.4 ± 0.1		65		35
sputtered	2.0 ± 0.1		61		39
Predicted					
Cr_2O_3	1.9		60		40
CrCO	1.2	0.28	33	33	33
CrC_2O_2	2.4	0.57	40	40	40

Table II. Auger data for HV/vac Mo films photochemically deposited

	O/Mo	C/Mo	O	C	Mo
FILMS					
unsputtered	1.5 ± 0.4	0.63 ± 0.10	27	46	27
sputtered	0.8 ± 0.2	0.59 ± 0.10	16	53	30
off spot	1.5 ± 0.3	0.73 ± 0.15	24	52	24
MoO_3 Standard					
unsputtered	5.7 ± 0.4		79		21
sputtered	3.4 ± 0.4		69		31
Predicted					
MoO_3	4.5		75		25
MoCO	1.5	0.34	33	33	33
MoC_2O_2	3.0	0.68	40	40	40

Table III. Auger data for HV/vac W films photochemically deposited

	O/W	C/W	O	C	W
FILMS					
unsputtered	1.8±0.4	0.7±0.2	20	33	47
sputtered	0.9±0.2	0.5±0.1	12	30	58
off spot	1.8±0.3	0.7±0.2	20	33	47
WO₃ Standard					
unsputtered	6.4±0.5		60		40
sputtered	3.8±0.4		43		57
WC Standard					
unsputtered		1.0±0.1		50	50
sputtered		1.2±0.2		55	45
Predicted					
WO₃	12.8		75		25
WCO	4.3	0.98	33	33	33
WC₂O₂	3.0	0.68	40	40	40

listed in all three tables. Comparison of HV/vac and HV/vac off spot films probes differences in surface reactions of $M(CO)_x$ with and without incident UV photons. No significant differences are found in Mo and W systems, but the Cr off spot films show roughly double the amount of carbon and oxygen.

All carbon line shapes were either graphitic or carbidic, indicating that any CO groups remaining in the films are fully dissociated.

Auger data for LV/air and HV/air (reported elsewhere) show that film compositions depend strongly on deposition conditions (6). The LV/air films have as much oxygen in them as the thermodynamically stable oxide. Comparison of LV/air and HV/air films show that oxygen containing background gases react with the films during deposition decreasing the amount of carbon in them. Air exposure of films deposited under high vacuum conditions (HV/air) results in much higher oxygen content than HV/vac films. In addition, the carbon distribution becomes inhomogeneous, with the bulk concentration being much lower than that of the surface.

Thermal Deposition. Film morphologies and topographies varied widely for thermally deposited films (14). Rippled film structures were not observed. Due to the varied topographies only an estimate of deposition rates for Mo were measured (2,000 - 6,000 A/s).

Table IV incorporates all the HV/vac data for Cr, Mo and W films thermally deposited with 514 nm light. The compositions are spatially inhomogeneous so data for film centers and edges are included. Film compositions remained constant after the first 1 - 2 nm of film were sputtered off, thus sputtered data includes a variety of depth profile data. The data for centers and edges exhibited no variation in composition over laser power density used, so data for all intensities are combined.

Metal-containing material was observed on the substrates in regions far from the films as well as in between films. The thickness of this deposit depends on the distance from the spot and the irradiation time of the spot, and was sometimes thick enough to be sputtered (> 10 nm). Sputtering was only performed for Mo.

Although oxides of W and Mo are well known to undergo preferential sputtering of oxygen, it is highly unlikely that this occurs in the thermally deposited films. Both Mo and W centers exhibit drastic reductions in oxygen content within the first 2 nm. Typical W and Mo oxides lose some but not all oxygen within the first few nm. Clean film centers were observed for sputtered material (HV/vac); this composition was not significantly affected by a five minute air exposure (HV/air).

For Mo and W the presence of background gases during deposition (LV/air) only added relatively small levels of oxygen to the films while in the case of Cr the well documented formation of Cr_2O_3 was observed. The lack of significant oxygen incorporation during air exposure of HV/air films is attributed to the density of the films. Differences between edge and center data arise only during sputtering; the surface compositions are uniform indicating that an additional layer of gas phase material decomposes on the film surface after the irradiation is terminated.

Table IV. Auger data for thermally deposited HV/vac films

Film	Unsputtered					Sputtered				
	C/M	O/M	O	C	M	C/M	O/M	O	C	M
Cr Films										
center	0.23±0.15	1.6±0.8	42	26	32	0.0±0.0	0.3±0.3	19	0*	81
edge	0.23±0.10	1.6±0.4	42	26	32	0.08±0.05	0.8±0.3	33	15	52
Mo Films										
center	0.46±0.15	1.2±0.4	25	43	32	0.0±0.0	0.0±0.0	0*	0*	100
edge	0.58±0.30	1.4±0.4	26	47	28	0.10±0.10	0.2±0.2	8	21	77
off spot	0.97±0.25	2.1±0.4	27	54	19	0.32±0.05	0.4±0.1	12	43	71
W Films										
center	0.44±0.20	1.8±0.8	22	24	53	0.0±0.0	0.0±0.0	0*	0*	100
edge	0.47±0.15	1.8±0.8	22	25	53	0.0±0.0	0.0±0.0	0*	0*	100

* - below detection limit of Auger

It is important to note that oxygen-free film centers for Cr films were observed on some occasions. The data on the table represent averages over many films.

An Hv/vac run using a power density of 0.7 MW/cm^2 at 257 nm was performed to determine the effect simultaneous laser heating had on the composition of photochemically deposited film. This power density represented a 200 fold increase from that used in the photochemical work. Unfortunately, window deposits significantly and rapidly attenuated the beam so the extent of heating was most likely significantly less than one would hope. Significant differences between high power UV and low power UV compositions were not observed. Additional discussion regarding these studies can be found in reference 14.

Discussion

Photochemical Deposition. Although exposure to background gases during and after deposition significantly alter the contamination levels of the films and are of great relevance to possible applications, this discussion will be limited to HV/vac films. It is these films that provide insight into the deposition process and possible clues to methods that might result in lower impurity levels.

It has been suggested that film growth from the metal hexacarbonyls is mediated by gas phase photolysis, that is, surface reactions of $M(CO)_4$ and $M(CO)_3$ are responsible for film growth at 257 nm. Under the present conditions, gas phase reactions are saturated and surface photochemical reactions are rate limiting. Examination of HV/vac film compositions suggests that the film stoichiometries are consistent with CrCO, MoC_2O and $WC_{<1}O_{<1}$. Away from the laser beam (off spot) stoichiometries are consistent with Cr_2O_2, MoC_2O and $WC_{<1}O_{<1}$. Thus it appears that condensation of $M(CO)_4$ and $M(CO)_3$ leads to loss of about 2 to 3 CO groups (or CO and CO_2 for Mo) via a thermal route. The remaining C and O cannot be removed photochemically for the Mo and W system, as judged by film compositions in illuminated areas. In contrast, there exists a well-defined photochemical route for loss of one additional CO from Cr. It should be noted that even though the thermal and photochemical pathways are not chemically distinct in the Mo and W systems, the photochemical reaction rate is much faster.

It is interesting to compare these results to studies of surface photolysis of $M(CO)_6$ condensed on single crystal Si, Mo or Rh surfaces at low temperatures (15). Although details of experimental procedures varied, the data consistently showed that decarbonylation under UV light is incomplete and that the CO groups remaining on the surface were not dissociated. Raising the temperature to 300 K resulted in the loss of additional CO. Dissociated CO still remained on the surface. In the present study, carried out under film growth conditions rather than monolayer adsorption, the initial adsorbate is already decarbonylated, the surface temperature is near ambient, and the surface is an amorphous metal oxycarbide. Despite the marked differences in reaction conditions, it appears that at least qualitatively the increased extent of CO loss and complete dissociation of remaining CO groups may be accounted for by the higher surface temperature. The surface composition may also affect the chemistry since preadsorbed C and/or O is known to inhibit dissociation of Mo and Cr and may thus facilitate CO desorption during deposition (16,17). One of the key differences between the film and surface studies is that the metal hexacarbonyl photoproducts arriving at the surface of the growing film can undergo subsequent photodissociation reactions, whereas primary photoproducts formed directly on single crystal surfaces at reduced temperatures do not. In addition the extent of C and O incorporation into the films is much higher than what would have been expected from coadsorption studies of CO/Mo and CO/W (6).

Contamination of photochemically deposited Cr, Mo and W films by incomplete removal of CO, reaction with background gas during deposition and reaction with air is so efficient that we were unable to find reaction conditions under which pure metal films were formed in the absence of heating.

Thermal Deposition. It is evident that the elemental compositions of HV/vac films are not uniform across the film diameter; centers contain clean metal while the edges have C and/or O impurities. All three hexacarbonyls are transparent to 514 nm light; gas phase photochemistry does not contribute in any way to the film deposition process. Film deposition is the result of laser heating of the Si/SiO_2 substrate and the growing film with subsequent thermal decomposition of the hexacarbonyl. Comparison to chemical vapor deposition work, although useful, must be done carefully since these depositions occur in the presence of carrier gas. These gases typically contain trace impurities and in some cases are chemically active. Kaplan and d'Heurle performed CVD studies on $W(CO)_6$ and $Mo(CO)_6$ in the temperature range between 570 K and 820 K (18). Very low levels of carbon impurities were observed at T > 770 K. The carbon content increased significantly at lower temperatures. Oxygen content was not investigated. Others have also reported appreciable levels of C and O impurities incorporated into CVD films prepared at T < 770 K from Mo and W hexacarbonyls (19). The CVD work suggests that in our Mo and W laser experiments temperatures greater than 770 K might be obtained at the center of the Mo films while the edges of the film and off spot material are below 770 K. Since the films are larger than the focal diameter of the laser it is not surprising to observe increasing degrees of contamination as a function of distance from the film center, representing a radial decrease in temperature.

In order to correlate composition and local temperature of the deposited films and to compare the laser deposition and CVD studies, radial temperature profiles for conditions corresponding to steady state film growth have been calculated. A stochastic method was used which incorporated both thermal and optical parameters for the films. The algorithm was tested with both deterministic approaches and experimental data. Additional details on this approach can be found in reference (14). Temperature profiles for a series of films were calculated. For W films deposited at 2.3 - 2.8 MW/cm^2, the calculated center temperatures are in the order of 900 to 1000 K while film edges reached temperatures in the 500 K regime. Since both edge and center data for W show no detectable impurities, these profiles indicate that temperatures around 500 K can produce clean W films. More interesting results are obtained for Mo. Direct comparison between profiles and film data indicate that impurities for a variety of laser power densities (1.5 - 4.0 MW/cm^2) are not detected until temperatures below 400 K are reached. This is significantly lower than what was expected from the CVD studies. Similar results were obtained for incorporation of C impurities in Cr films. (Oxygen content in Cr does not serve as a reliable marker for impurities since Cr is known to be a very efficient scavenger of oxygen.)

Once local temperatures are available it is possible to compare the film results to low coverage surface studies of CO and hexacarbonyls on single crystal surfaces. Heating a W(110) surface saturated with partially carbonylated species and bound CO produced by the thermal decomposition of $W(CO)_6$ results in both desorption and dissociation of CO (300 - 400 K). Complete removal of C and O from the W(110) surface is accomplished at around 1200 K (11,20). Studies of CO adsorbed on W and Mo provide a range of temperatures over which dissociated CO recombines and is fully removed from the surface: for Mo(100), T > 1200 K (16,21); for polycrystalline W, T > 1600 K (22); for W (100), T > 1300 K (23). Due to experimental difficulties very little is known about recombinative desorption of CO on Cr surfaces. Comparison to the films indicates that loss of all CO from the films is obtained well below any of the temperatures observed for surface studies. It is unlikely that recombinative desorption of CO is kinetically efficient in these films because the C and O concentrations are low. Thus it appears that all CO is lost through direct thermal desorption, that is, the dissociation channel is unimportant when films grow at temperatures higher than 400 K (24). This conclusion is supported by mechanistic simulations described in full elsewhere (14). This result is far from what surface scientists have suggested as the temperature range required to thermally produce clean films (well above 1000 K) and serves to illustrate that direct film studies are essential towards the understanding of the mechanisms involved in deposition.

Recently Houle and Yeh have performed mass spectroscopic studies on the laser thermal decomposition of $Cr(CO)_6$ on Si during film growth (25). Their apparatus allows for the observation of surface generated species in the absence of collisions thus providing insight into surface reactions relevant to the deposition process. They observed two growth regimes: a pre-growth regime characterized by limited decomposition of the precursor and a growth regime in which near total consumption of the hexacarbonyl occurs. No evidence was found for recombinative desorption of C and O, providing independent support for conclusions drawn from the film studies.

Summarizing, laser thermal deposition provides cleaner films than have previously been observed for cw and pulsed UV laser CVD of the hexacarbonyls. The temperatures obtained in the films are well below those predicted by surface scientists to be necessary for complete removal of CO. Under laser heating conditions the dissociation of CO is severely hindered at all but the lowest temperatures (< 400 K).

ACKNOWLEDGMENTS

The authors would like to thank Jorge Goitia for technical assistance.   KAS gratefully acknowledges summer support from IBM East Fishkill, NSF (grant # RII-8911281), the donors of the Petroleum Research Fund (administered by the American Chemical Society) and the Affirmative Action Faculty Development Program at SJSU.   This work was performed under a joint study agreement between IBM and SJSU.

REFERENCES

1)   Herman, I. P. *Chem. Rev.* **1989**, *89*, 1323.
2)   Solanki, R.; Boyer, P. K.; Collins, G. J. *Appl. Phys. Lett.* **1982**, *41*, 1048.
3)   Flynn, D. K.; Steinfeld, J. I.; Sethi, D. S. *J. Appl. Phys.*, **1986**, *59*, 3914.
4)   Gluck, N. S.; Wolga, G. J.; Bartosch, C. E.; Ho, W.; Ying, Z. *J. Appl. Phys.*, **1987**, *61*, 988.
5)   Jackson, R. L.; Tyndall, G. W. *J. Appl. Phys.*, **1988**, *64*, 2092.
6)   Singmaster, K. A.; Houle, F. A.; Wilson, R. J. *J. Phys. Chem.*, **1990**, *89*, 6864.
7)   Gilgen, H. H.; Cacouris, T.; Shaw, P. S.; Krchnavek, R. R.; Osgood, R. M. *Appl. Phys.*, **1987**, *B42*, 55.
8)   Turney, W.; James, S. G.; Cardinahl, P; grassian, V. H.; Singmaster, K. A. submitted *Chem. of Mat.*
9)   Petzold, H.C.; Putzar, R.; Weigmann, U.; Wilke, I. *Mat. Res. Symp. Proc.*, **1988**, *101*, 75.
10)   Oprysko, M. M.; Beranek, M. W. *J. Vac. Sci. Technol.*, **1987**, *B5*, 496.
11)   Flitsch, F. A.; Swanson, J. R.; Friend, C. M. *Surf. Sci.*, **1991**, *245*, 85 and references therein.
12)   Singmaster, K. A.; Houle, F. A. *Mat. Res. Soc. Proc.*, **1991**, *201*, 159.
13)   Osgood, R. M.; Ehrlich, D. J. *Opt. Lett.*, **1982**, *7*, 385.
14)   Singmaster, K. A.; Houle, F. A. submitted to *J. Phys. Chem.*
15)   Creighton, J. R. *Appl. Phys.*, **1986**, *59*, 410.; Gluck, N. S.; Ying, Z.; Bartosch, C. E.; Ho, W. *J. Chem. Phys.*, **1987**, *86*, 4957.; Cho, C. C.; Bernasek, S. L. *J. Vac. Sci. Technol.*, **1987**, *A5*, 1088.; Germer, T. A.; Ho, W. *J. Chem. Phys.*, **1988**, *89*, 562.
16)   Ko, E. I.; Madix, R. J. *Surf. Sci.*, **1981**, *109*, 221.
17)   Shinn, N. D.; Madey, T. E. *J. Vac. Sci. Technol.*, **1985**, *A3*, 1673.
18)   Kaplan, L. H.; d'Heurle, F. M. *J. Electrochem. Soc.*, **1970**, *117*, 693.
19)   Yous, B.; Robin, S.; Robin, J.; Donnadieu, A. *Thin Solid Films*, **1985**, *130*, 181.; Carver, G. E.; Divrechy, A.; Karbal, S.; Robin, J.; Donnadieu, A. *Thin Solid Films*, **1982**, *94*, 269.
20)   Bowker, M.; King, D. A. *J. Chem. Soc. Faraday 1*, **1980**, *76*, 758.
21)   Felder, T. E.; Estrup, P.J. *Surf. Sci.*, **1978**, *76*, 464.
22)   Rigby, L. J. *Can. J. Phys.*, **1964**, *42*, 1256.
23)   Benzinger, J. B.; Ko, E. I.; Madix, R. J. *Catal.*, **1978**, *54*, 414.
24)   Umbach, E.; Menzel, D. *Surf. Sci.*, **1983**, *135*, 199.
25)   Houle, F. A.; Yeh, L. I. *J. Phys. Chem.* **1992**, *96*, 2691.

RECEIVED March 1, 1993

INDEXES

Author Index

Affiliation Index

Subject Index

A

Agglomeration of metal particles, applications, 136

Alkylselenium compounds, multiphoton dissociation dynamics, 51–55

Arene–chromium tricarbonyls, parity-state selectivity in multiphoton dissociation, 26–47

Arene–Cr(CO)$_3$ complexes, background for parity-state selectivity in multiphoton dissociation, 35–38f

Atoms, gas-phase formation, 220–247

B

Binding, ligand, laser photoionization probes of effects in multiphoton dissociation of gas-phase transition metal complexes, 61–72

Bond dissociation energies coordinatively unsaturated metal carbonyls
measurement procedure, 153,155–159
results, 160–161
niobium clusters, determination, 140–141
photodissociation of metal carbonyls, determination from branching ratios, 102,103f

Bond-selective chemistry, relevance to organometallics, 4

Branching ratios for photodissociation dynamics of metal carbonyls
bond dissociation energy determination, 102,103f
fragment yields vs. photon energy, 101f
measurement, 102

Buffer gas, gas-phase formation of atoms, clusters, and ultrafine particle effect, 225,227–230

Buffer gas quenching as dynamic probe of organometallic photodissociation processes
description, 123
experimental procedure, 123,125
photodissociation of dialkyl zincs, 124–130
photodissociation of ferrocene, 129,131–134

C

Carbide synthesis using lasers, transition metal, description, 274–276

Chemical trapping experiments, 62

Chemical vapor deposition, laser-assisted, See Laser-assisted chemical vapor deposition

Chemisorption on niobium clusters
anticorrelation with ionization potential, 139–140
benzene reactions, 140
hydrogen chemisorption, 139

Classical photochemistry with lasers, 5

Cluster(s), gas-phase formation, 220–247

Cluster-assembled materials, descriptions,136

Cluster synthesis using laser chemistry
description, 12,15
history, 15
method for nascent cluster size distribution analysis, 16–21
size vs. properties, 15

CO elimination
kinetic model of dynamics, 99–100
statistical models of dynamics, 98–99

CO vibrational- and rotational-state distributions, study, 62–63

CO$_2$ laser, gas-phase formation of atoms, clusters, and ultrafine particle effect, 230,231f

Coordinatively unsaturated iron carbonyls, reactions, 151,153,154f,156t

Coordinatively unsaturated metal carbonyls
bond dissociation energies, 160–161
bond dissociation energy measurement procedure, 153,155–159
experimental procedure, 149–151
future work, 162
iron carbonyl reactions, 151,153,154f,156t
reaction kinetics for polynuclear iron carbonyls, 159–160
time-resolved spectra, 151,152f

Coordinatively unsaturated molecules, synthesis in laboratory, 189

Coordinatively unsaturated organometallic compounds
importance of kinetic information, 148
kinetic measurement techniques, 148–149

Production: Meg Marshall
Indexing: Deborah H. Steiner
Acquisition: Rhonda Bitterli
Cover design: Amy Hayes

Printed and bound by Maple Press, York, PA

Highlights from ACS Books

Good Laboratory Practice Standards: Applications for Field and Laboratory Studies
Edited by Willa Y. Garner, Maureen S. Barge, and James P. Ussary
ACS Professional Reference Book; 572 pp; clothbound ISBN 0–8412–2192–8

Silent Spring Revisited
Edited by Gino J. Marco, Robert M. Hollingworth, and William Durham
214 pp; clothbound ISBN 0–8412–0980–4; paperback ISBN 0–8412–0981–2

The Microkinetics of Heterogeneous Catalysis
By James A. Dumesic, Dale F. Rudd, Luis M. Aparicio, James E. Rekoske,
and Andrés A. Treviño
ACS Professional Reference Book; 316 pp; clothbound ISBN 0–8412–2214–2

Helping Your Child Learn Science
By Nancy Paulu with Margery Martin; Illustrated by Margaret Scott
58 pp; paperback ISBN 0–8412–2626–1

Handbook of Chemical Property Estimation Methods
By Warren J. Lyman, William F. Reehl, and David H. Rosenblatt
960 pp; clothbound ISBN 0–8412–1761–0

Understanding Chemical Patents: A Guide for the Inventor
By John T. Maynard and Howard M. Peters
184 pp; clothbound ISBN 0–8412–1997–4; paperback ISBN 0–8412–1998–2

Spectroscopy of Polymers
By Jack L. Koenig
ACS Professional Reference Book; 328 pp;
clothbound ISBN 0–8412–1904–4; paperback ISBN 0–8412–1924–9

Harnessing Biotechnology for the 21st Century
Edited by Michael R. Ladisch and Arindam Bose
Conference Proceedings Series; 612 pp;
clothbound ISBN 0–8412–2477–3

From Caveman to Chemist: Circumstances and Achievements
By Hugh W. Salzberg
300 pp; clothbound ISBN 0–8412–1786–6; paperback ISBN 0–8412–1787–4

The Green Flame: Surviving Government Secrecy
By Andrew Dequasie
300 pp; clothbound ISBN 0–8412–1857–9

For further information and a free catalog of ACS books, contact:
American Chemical Society
Distribution Office, Department 225
1155 16th Street, NW, Washington, DC 20036
Telephone 800–227–5558

Bestsellers from ACS Books

The ACS Style Guide: A Manual for Authors and Editors
Edited by Janet S. Dodd
264 pp; clothbound ISBN 0–8412–0917–0; paperback ISBN 0–8412–0943–X

The Basics of Technical Communicating
By B. Edward Cain
ACS Professional Reference Book; 198 pp;
clothbound ISBN 0–8412–1451–4; paperback ISBN 0–8412–1452–2

Chemical Activities (student and teacher editions)
By Christie L. Borgford and Lee R. Summerlin
330 pp; spiralbound ISBN 0–8412–1417–4; teacher ed. ISBN 0–8412–1416–6

Chemical Demonstrations: A Sourcebook for Teachers,
Volumes 1 and 2, Second Edition
Volume 1 by Lee R. Summerlin and James L. Ealy, Jr.;
Vol. 1, 198 pp; spiralbound ISBN 0–8412–1481–6;
Volume 2 by Lee R. Summerlin, Christie L. Borgford, and Julie B. Ealy
Vol. 2, 234 pp; spiralbound ISBN 0–8412–1535–9

Chemistry and Crime: From Sherlock Holmes to Today's Courtroom
Edited by Samuel M. Gerber
135 pp; clothbound ISBN 0–8412–0784–4; paperback ISBN 0–8412–0785–2

Writing the Laboratory Notebook
By Howard M. Kanare
145 pp; clothbound ISBN 0–8412–0906–5; paperback ISBN 0–8412–0933–2

Developing a Chemical Hygiene Plan
By Jay A. Young, Warren K. Kingsley, and George H. Wahl, Jr.
paperback ISBN 0–8412–1876–5

Introduction to Microwave Sample Preparation: Theory and Practice
Edited by H. M. Kingston and Lois B. Jassie
263 pp; clothbound ISBN 0–8412–1450–6

Principles of Environmental Sampling
Edited by Lawrence H. Keith
ACS Professional Reference Book; 458 pp;
clothbound ISBN 0–8412–1173–6; paperback ISBN 0–8412–1437–9

Biotechnology and Materials Science: Chemistry for the Future
Edited by Mary L. Good (Jacqueline K. Barton, Associate Editor)
135 pp; clothbound ISBN 0–8412–1472–7; paperback ISBN 0–8412–1473–5

For further information and a free catalog of ACS books, contact:
American Chemical Society
Distribution Office, Department 225
1155 16th Street, NW, Washington, DC 20036
Telephone 800–227–5558